D0788718

HISTORY OF AMERICAN ECOLOGY

This is a volume in the Arno Press collection

HISTORY OF ECOLOGY

Advisory Editor
Frank N. Egerton III

Editorial Board
John F. Lussenhop
Robert P. McIntosh

See last pages of this volume for a complete list of titles.

HISTORY OF AMERICAN ECOLOGY

With an Introduction by
Frank N. Egerton

ARNO PRESS
A New York Times Company
New York / 1977

LIBRARY
JAN 10 1980
UNIVERSITY OF THE PACIFIC
367730

Editorial Supervision: LUCILLE MAIORCA

Reprint Edition 1977 by Arno Press Inc.

Copyright © 1977 by Arno Press Inc.

HISTORY OF ECOLOGY
ISBN for complete set: 0-405-10369-7
See last pages of this volume for titles.

Manufactured in the United States of America

Library of Congress Cataloging in Publication Data

Main entry under title:

History of American ecology.

(History of ecology)
CONTENTS: Egerton, F. N. Ecological studies and observations before 1900.--McIntosh, R. P. Ecology since 1900.--McHugh, J. L. Trends in fishery research.--Frey, D. G. Wisconsin: the Birge-Juday era. [etc.]

1. Ecology--United States--History--Addresses, essays, lectures. I. Arno Press. II. Series.
QH104.H57 574.5'0973 77-74229
ISBN 0-405-10399-9

ACKNOWLEDGMENTS

''Ecological Studies and Observation Before 1900'' by Frank N. Egerton and ''Ecology Since 1900'' by Robert P. McIntosh are reprinted from *Issues and Ideas in America,* edited by Benjamin J. Taylor and Thurman J. White. Copyright 1976 by the University of Oklahoma Press.

''Trends in Fishery Research'' is reprinted by permission of the American Fisheries Society.

''Wisconsin: The Birge-Juday Era'' by David G. Frey is reprinted from *Limnology in North America,* © 1963 by the Regents of the University of Wisconsin, by permission of The University of Wisconsin Press.

''The Development of Association and Climax Concepts'' by Lucy E. Braun and ''Recent Evolution of Ecological Concepts'' by Robert H. Whittaker are reprinted by permission of *The American Journal of Botany.*

''H.A. Gleason—'Individualistic Ecologist''' is reprinted by permission of the *Bulletin of the Torrey Botanical Club.*

''History of the Ecological Society of America, 1977'' is reprinted by permission of Robert L. Burgess.

''American Grassland Ecology, 1895-1955'' is reprinted by permission of Ronald Tobey.

INTRODUCTION

In the late 1960s the public became aware of an imminent crisis in both the availability of natural resources and in environmental pollution. Suddenly "ecology" was important. The public use of the term, however, was at slight variance to its previous (and continuing) application to a particular science. In the public mind the term is closely associated with conservation and pollution control. Professional ecologists consider these activities, when they are guided by science, as applied ecology.

This anthology contains nine articles that describe the development of the science of ecology in America. The history of conservation and pollution control, applied ecology, is sometimes discussed in some of these articles, but for detailed treatment of these subjects one should turn to the numerous volumes which are already available on them. The history of the "pure" science of ecology has never been told in any detail, nor that of American ecology in any greater detail than exists in any of these articles taken separately. Taken together, they provide the most detailed history of ecology for any country in the world.

There is a widespread impression that ecology is a very recent science. It is a young science, though not quite so recent as many imagine. Ernst Haeckel coined the term "oekologie" in 1866 to name a science which he felt should exist. However, he knew that various ecological questions had been investigated for some time, and he was not so much trying to invent new lines of investigation as he was trying to organize coherently various types of study that fell under the broad heading of "natural history." Some ecological topics, including the balance of nature and the possible influence of climate upon disease, had been discussed since antiquity. However, in spite of Haeckel's christening of the new science in 1866, it took some decades for it to reach a stage of theoretical cohesion, and more decades for it to begin to mature. In the late 1890s a few biologists began to think of themselves as ecologists, but not until the early 20th century did a recognizable science of ecology take its place alongside other biological disciplines.

This anthology begins with articles by Frank N. Egerton and Robert P. McIntosh which survey the entire history of ecology in America. The other seven articles consider the history of particular aspects of the subject. Only Egerton's article is specifically concerned with the early history, before 1900. That early history is as much tied to cultural history as it is to history of science. Among the notable early contributors to ecological knowledge were Benjamin Franklin, William Bartram, Thomas Jefferson, Noah Webster, Henry Schoolcraft, Daniel Drake, Henry David Thoreau, and George Perkins Marsh, all of whom belong as much or more to America's general culture as to its science. Furthermore, the history of ecological studies before 1900 must be told within the context of the history of exploration, public health, nature appreciation, agriculture, and so forth. In most respects Americans followed along paths laid out by Europeans. Before 1900, only in the development of the biological control of insect pests did Americans lead the way.

When ecology began to be organized, from the 1890s through the 1920s, what emerged were four ecological sciences—plant ecology, animal ecology, oceanography and limnology. As these disciplines have developed through the years they have come to have more and more in common. By 1914 American ecologists had achieved a self-conscious identity which led to the formation of the Ecological Society of America. In his survey of ecology in America since 1900, McIntosh describes both the intellectual and organizational advances.

Early intellectual advances consisted of the identification of important problems for research, establishment of research techniques, the development and the defense of theories, and the publication of textbooks for training ecologists. American ecologists quickly shed their dependence upon European leaders and developed their own leaders, their own research problems, their own publications, and even, to some extent, an American orientation toward ecology. The ecological sciences in America from 1920 to 1950 were quite largely descriptive sciences which attracted only moderate numbers of students. Ecologists by then were acutely aware of America's abuse of its environment and waste of its natural resources, but their knowledge was not yet advanced enough for them to be able to offer comprehensive and sophisticated recommendations to remedy the abuses. And neither the government nor the public offered much financial or psychological encouragement to ecologists to develop this capacity.

World War II stymied the growth of sciences that were not obviously related to the war effort, but by 1950 the American universities had begun to expand their programs to compensate for the paucity of biologists they had trained during the war. With the establishment of the National Science Foundation the federal government began to provide the financial support of research which ecologists now wished to conduct. These two favorable factors to the growth of ecology were soon joined by a third, the public awareness of America's environmental and resource crisis. By the late sixties there was a general appreciation of the importance of ecology which led to a broadening of support in the universities and by the national and state governments.

Particular aspects of this story are subjects of the other seven papers in this anthology. Since there are few published studies on the history of either animal ecology or oceanography in America, J. L. McHugh's accounts of "Trends in Fishery Research" must suffice for both subjects. He begins with the apprehension that existed among fishermen around 1870 that the fish in the eastern part of the country were being depleted. This fear led mainly to uninformed stocking programs, but the resulting failures and difficulties did lead to interest in scientific research. There followed a steady increase in the knowledge of scientific management of sport and commercial fisheries, with an expansion in the amount of research around 1950, as was true for ecology generally.

David G. Frey, in his study of the contributions of Edward A. Birge (1851-1950) and Chancey Juday (1871-1944) to limnology, has, without exaggeration, spoken of "The Birge-Juday Era." Birge published limnological papers for over five decades and Juday for over four. They were a team who published on both the biological and environmental aspects of lakes, and they were responsible for training at the University of Wisconsin many of the subsequent leaders in limnology. Thus Frey's account provides a good indication of progress in limnology from the 1890s to the 1940s.

Four of the articles are on the history of plant ecology. This is a disproportionate representation, but it is a reflection of the disproportionate interest in the history of their science by plant ecologists rather than any bias in selection by the editor. All four articles are concerned with the classification of vegetation into ecologically meaningful units and with theories to explain the replacement through time of one ecological unit by another within the same area. Ronald Tobey has focused upon these issues in the development of prairie ecology, while Lucy Braun and Robert

Whittaker discuss the same issues with reference to the eastern forests. One ecologist stands out in all four articles, Frederic E. Clements (1874-1945), who developed a detailed system of vegetation classification and a theory of climate being the overwhelmingly important cause of vegetation change through time.

E. Lucy Braun (1889-1971) was elected vice president of the Ecological Society of America for 1935, president for 1950, and given a certificate of merit at the Fiftieth Anniversary Meeting of the Botanical Society of America in 1956. She was therefore one of America's leading plant ecologists of her time. She used a modified Clementsian classification in her research and publications for some five decades. Her article explains and defends her perspective and explains its differences from Clements' system. Robert H. Whittaker received the Ecological Society of America's Mercer award in 1967 and is today one of the leading plant ecologists in America. His article, from the same volume as Braun's, explains the difficulties with Clements' system and describes alternative systems from Europe. He also discusses Braun's classification without condemnation, but with less enthusiasm than for European alternatives.

The fourth of these botanical articles is another by McIntosh, on the ecological contributions of Henry A. Gleason (1882-1975). Gleason's ecological work came early in a career later oriented toward systematic botany. As early as 1908, McIntosh points out, Gleason "became convinced that succession could be retrogressive and that the Clementsian concept of succession, as an irreversible trend leading to the climax, was untenable." Gleason offered as an alternative his "individualistic concept" of the plant association (1917) which he continued to defend with decreasing frequency through the following years, but it was not until the 1950s that his early work became widely appreciated by American plant ecologists. Clements dominated the intervening years.

Seven of these nine articles are primarily concerned with the progress of ecological theory and methods. The other two, published here for the first time, are concerned with the social aspects of American ecology. Tobey's study on "American Grassland Ecology, 1895-1955: The Life Cycle of a Professional Research Community" is more about the ecologists who elaborated and refined the Clementsian system than about that system itself. Methodologically, this is the most provocative article in the anthology; it is the most analytical and the least descriptive. It ably defends a thesis concerning the waxing and waning of a domain of research, with conclusions based upon statistical data.

Robert L. Burgess has written the history of the Ecological Society of America. This is an important aspect of the history of American ecology. Historians of science have known for at least several decades that institutional growth is a vital aspect of the growth of a science. A professional society helps a science achieve maturity. Burgess's history provides information on the growth of the Society, indicates who its leaders were, and discusses the various reactions of the Society to the government, to relevant public issues, and to other scientific organizations.

As a group, then, these articles provide a broad and valuable range of information and insights into the intellectual and social history of the ecological sciences in America. Although this is not the story of pollution control and resource management in the past, it undoubtedly is a story which provides a background for understanding progress in resource management in the future.

Racine, Wisconsin
April, 1976

Frank N. Egerton

CONTENTS

ECOLOGICAL STUDIES AND OBSERVATIONS BEFORE 1900

Frank N. Egerton

Reprinted from *Issues and Ideas in America*, edited by Benjamin J. Taylor and Thurman J. White, published and copyrighted by the University of Oklahoma Press, Norman, Oklahoma.

16

ECOLOGICAL STUDIES AND OBSERVATIONS BEFORE 1900

By Frank N. Egerton

The science of ecology intruded upon public awareness in America for the first time in the late 1960's as a result of acute environmental problems. Because of the context in which this awareness developed, the term *ecology* became synonymous with environment and pollution. This was a distortion of the modern historical development of the science in two respects: it placed more emphasis upon the inanimate surroundings than upon the organisms and more emphasis upon the applied than upon the pure science.

INTRODUCTION

The public perception is not much of a distortion of the earlier history of ecological interests, observations, and discussions. When ecology became a formal science at the turn of the twentieth century, however, it necessarily became less diffuse than the body of knowledge on which it was built. Thomas Kuhn has explained why this happens as a science approaches maturity. If the science is to progress beyond the level of common sense perceptions, it must identify a select group of problems that are soluble and ignore others that may be even more interesting but are less readily soluble.[1] That narrower but more profound science will be the subject of the next chapter, by Robert P. McIntosh. The purpose of this chapter is to describe the various kinds of ecological investigations that were carried out in America before 1900 and provided the foundation for the formal science.

Ecology is a science concerned with the interactions between organisms and their environment. The environment is composed of both animate and inanimate components. A rabbit's world consists of grass and foxes and also climate and soil. The history of ecology in its preformal period is similar to but not identical with the history of natural-history studies. In works on natural history written before 1900, one finds ecological discussions embedded within a context of classification and description of species. There were also many concomitant developments that were part of science but not included in most natural histories that are relevant for the history of ecology—discussions of climate and disease, for example.

The New World from the time of Columbus onward presented a challenge to European explorers to describe the differences between what they found here and in their homeland. Howard Mumford Johes found, however, that new observations and experiences did not endow the Europeans with new powers of description.[2] The New World environment could stimulate interest, but for almost the entire period from 1492 to 1900 the observations made in America did not transcend in any important ways the perspectives and understanding that had developed in Europe. As Raymond Phineas Stearns has observed: "Neither scientific knowledge nor the scientific method was *sui generis* in America. Both were tender plants introduced to the new world scene by European proponents, seeded, watered, and fertilized by European patrons."[3] Only after 1850 did American naturalists take the lead in one area of ecological research, the investigation of biological control of undesirable species.

BEFORE 1800

Exploration and Inventory

The sixteenth, seventeenth, and eighteenth centuries were notable for worldwide exploration and colonization. As the maritime nations of Europe sought new products and wealth around the world, there was also a strong demand for accounts of the places explored. Ecological description and analysis would have been very useful to prospective settlers, but unfortunately the explorers had little capacity for providing them. It is true that Aristotle's and Theophrastus' writings in zoology and botany contained ecological observations, and influences of those observations can be found in natural histories by such naturalists as Conrad Gesner in the sixteenth century, John Ray in the seventeenth, and Carl Linnaeus in the eighteenth. However, few explorers were well-trained naturalists, and most of those who published accounts of their discoveries merely described new species that might be of either economic or ornamental interest to their countrymen in Europe.

The literature of exploration from the three centuries before 1800 is too extensive to survey fully its ecologically relevant accounts here. That does not seem desirable anyway, because no clear-cut ecological tradition emerged from this literature. On the other hand, some of the literature does provide an indication of the level of ecological awareness at various times, and it apparently stimulated some interest in ecological understanding. Therefore, a few of the ecologically more significant examples of this literature and their authors are discussed below.

Thomas Harriot (1560–1621) came to the first English coastal settlement on Roanoke Island with Grenville's expedition of 1585–86, and afterward he published *A Briefe and True Report of the New Found Land of Virginia . . . .* It contained descriptions of useful plants and animals that were either native to America or were European and could grow in the New World. The significance of his work lies in the fact that it has so little in it of ecological significance. It seems representative in that respect of the exploration literature of the sixteenth century. Explorers of that century seem to have had less ecological awareness than later explorers, and also they were likely to remain in the New World a shorter time and to write shorter books. Harriot's only observation of ecological interest is his brief comparison of the lands farther inland from the coast in terms of production. He found

the soyle to bee fatter; the trees greater and to growe thinner; the grounde more firme and deeper mould; more and larger champions; finer grasse and as good as ever we saw any in England; in some places rockie and farre more high and hillie ground; more plentie of their fruites; more abondance of beastes; the more inhabited with people.[4]

Because his purpose was to promote settlement, he did not attempt to ascertain why the interior was more productive. It was enough for him merely to discover that it was.

Captain John Smith (1580–1631) shared Harriot's promotional interest in America, but he also lingered longer and had more to say. His characterizations of Virginia and Massachusetts coastal regions were not detailed, but did convey an idea of the climate, the fertility of the soil, the terrain, the vegetation, and the animals. It was not yet ecology, but it was certainly an environmental picture. Sometimes he also attempted ecological interpretations of what he saw, as illustrated in this description:

Virginia doth afford many excellent vegetables, and living Creatures, yet grasse there is little or none, but what groweth in low Marshes: for all the Countrey is overgrowne with trees, whose droppings continually turneth their grasse to weeds, by reason of the rancknes of the ground, which would soone be amended by good husbandry. The wood that is most common is Oke and Walnut.[5]

It is clear that the exploration literature was stronger on ecological observations about plants than about animals. That is commonly true for the literature down to 1800 and even beyond. The dependence of plants upon soil, water, terrain, and climate was more apparent than the dependence of moving animals upon features of their environment, except for fish, whose dependence was so obvious that it provoked little comment.

John Josselyn (ca. 1608–75) was an adventurer who went to Scarborough, Maine, in 1663 to live with his brother. Eight years later he returned to London to publish his observations. His *New England Rarities . . .* (1672) is notable for being devoted primarily to the "birds, beasts, fishes, serpents, and plants." His strongest interest seems to have been in their practical uses, and many were mentioned without ecological observations. Sometimes, however, he did make such observations, particularly about the plants, which he knew more extensively than the animals. Typical is his account of White hellebore: "The first Plant that springs up in this Country, and the first that withers; it grows in deep black Mould and Wet, in such abundance, that you may in a small compass gather whole Cart-loads of it."[6] This part of his account is followed by an explanation of how the Indians used it to heal wounds.

Later in the seventeenth century, as the French moved deeper and deeper into the Great Lakes region, a young explorer, Louis Armand de Lom d'Arce, Baron de Lahontan (1666–1715) recorded in some detail his experiences in *New Voyages to North America . . .*, which became popular as a source of information and adventure. Although he was accused during his lifetime of having embellished his adventures, his interest in nature seems to have been genuine. He gave in some detail accounts of the plants and animals he observed, and his description of the Lake Erie region is at least a superficial ecological survey:

The Lake Erriè *is justly dignified with the illustrious name of Conti; for assuredly 'tis the finest Lake upon Earth. You may judge of the goodness of the Climate, from the Latitudes of the Countries that surround it. Its Circumference extends to two hundred and thirty Leagues; but it affords every where such a charming Prospect, that its Banks are deck'd with Oke-Trees, Elms, Chesnut-Trees, Walnut-Trees, Apple-Trees, Plum-Trees, and Vines which bear their fine clusters up to the very top of the Trees, upon a sort of ground that lies as smooth as one's Hand. Such Ornaments as these, are sufficient to give rise to the most agreeable Idea of a Landskip in the World. I cannot express what vast quantities of Deer and Turkeys are to be found in these Woods, and in the vast Meads that lye upon the South side of the Lake. At the bottom of the Lake, we find wild Beeves upon the Banks of two pleasant Rivers that disembogue into it, without Cataracts or rapid Currents. It abounds with Sturgeon and white Fish; but Trouts are very scarce in it, as well as the other Fish that we take in the Lakes of* Hurons *and* Ilinese. *'Tis clear of Shelves, Rocks, and Banks of Sand; and has fourteen or fifteen fathom Water. The Savages assure us, that 'tis never disturb'd with high Winds, but in the Months of December, January, and February, and even then but seldom.*[7]

As the English settlements became more extensive, the English explorers remained longer in one area and gradually wrote more and more detailed books. John Lawson came to America in 1700 (deciding at the last moment that it would be more exciting than a trip to Rome) and remained for eight years, during which time he acquired the title surveyor-general of North Carolina. He returned to London and published his *A New Voyage to Carolina* (1709), and then came back. While continuing his explorations in 1711, he was murdered by Tuscarora Indians. As a promoter of the colony, Lawson was interested in all of its production, its climate, and its agriculture. He provided a more detailed account of the plants and animals than had any previous Englishman. His accounts of plants have much more ecological interest than do those of animals. For example, he found large oaks of four or five sorts

very common in the upper Parts of both Rivers; also a very tall large Tree of great Bigness, which some call Cyprus, *the right Name we know not, growing in Swamps. Likewise Walnut, Birch, Beech, Maple, Ash, Bay, Willow, Alder, and Holly; in the lowermost Parts innumerable Pines, tall and good for Boards or Masts, growing, for the most part, in barren and sandy, but in some Places up the River, in good Ground, being mixt amongst Oaks and other Timbers. We saw Mulberry-trees, Multitudes of Grape-Vines, and some Grapes which we eat of.*[8]

He also provided a favorable evaluation of the climate.[9]

Mark Catesby (1683–1749) traveled more broadly in the same region as Lawson, but Catesby came to America especially to study and collect plants and animals, and he provided more information than ever before about them in his two-volume *The Natural History of Carolina, Florida and the Bahama Islands* . . . (1729–47). Although a major portion of the work is devoted to descriptions and illustrations of animals, as in previous works, practically all he wrote of ecological interest concerned plants. He did, however, convey his ecological perceptions of animals by illustrating them against a background of an associated plant.[10]

After placing his region geographically, Catesby discussed the climate and pointed out some disparities between American and European climates:

The Northern Continent of America is much colder than those Parts of Europe which are parallel to it in Latitude; this is evident from the mortal Effects the Frosts have on many Plants in Virginia, that grow and stand the Winters in England, tho' 15 Degrees more North; and what more confirms this is the violent and sudden freezing of large Rivers.[11]

Although he may not have fully understood it, he also pointed out the moderating effect of the ocean upon the temperature of the adjacent land.

Catesby distinguished three main types of soil in eastern Carolina, "distinguished by the Names of *Rice Land, Oak* and *Hiccory Land*, and *Pine barren Land*." They were distinguished partly by soil texture and partly by water content, but they were identified by the plants growing on them. "*Oak* and *Hiccory Land*" was characterized by

those Trees, particularly the latter, being observed to grow mostly on good Land. This Land is of most Use, in general producing the best Grain, Pulse, Roots, and Herbage, and is not liable to Inundations; on it are also found the best Kinds of Oak for Timber, and Hiccory, *an excellent Wood for Burning. This Land is generally light and sandy, with a mixture of Loam.*[12]

Catesby traveled inland into the Piedmont, and although he probably did not reach the Appalachian Mountains, he did collect some information about the topography, soil, and trees in both these regions.[13] In his section on agriculture Catesby discussed the species that succeeded in America and those that did not. The latter, such as the fig and the orange, were often found to be killed by the winter frosts.[14]

Although European naturalists would continue to make important explorations in America for more than a century after Catesby, native American naturalists soon joined in the explorations and gradually supplanted the Europeans in importance. The general pattern of this replacement has been well described by George Basalla.[15]

John Bartram (1699–1777) was a successful Pennsylvania farmer who became more interested in natural history and horticulture than in farming. He was America's first prominent native naturalist, and he became known and respected by European naturalists. He recorded his observations in letters and in journals that he kept during two trips, the first through Pennsylvania and New York to Lake Ontario in July and August, 1743, and the second through the Carolinas, Georgia, and Florida from July, 1765, to April, 1766. Only his first travel journal and extracts from the second were published during his lifetime. The ecological observations in his first journal are all of the same sort: indications of the species of plants that he encountered in particular kinds of places, often specified according to soil quality, moisture, and relative elevation.[16] These observations were more extensive than Catesby's but not more informative.

In his second journal Bartram occasionally offered some interpretation along with his descriptive observations, such as this one concerning soil formation:

landed on a high bluff, on the east-side of the river, at Johnson's Spring, a run of clear and sweet water, then travelled on foot along thick woody but loamy ground, looking rich on the surface by reason of the continual falling leaves, and by the constant evergreen shade rotting to soil, as the sun never shines on the ground strong enough to exhale their virtue before their dissolution, as under deciduous trees.[17]

This observation takes on ecological implications when tied to his frequent correlations between particular soil traits and plant species, but Bartram did not have a sufficiently analytical mind to pursue the subject further.

Because Bartram's farm was just outside Philadelphia and because he was respected by European naturalists, it was to be expected that when they came to America they would visit him. One who did and who drew upon Bartram's knowledge in his own writings was Pehr Kalm (ca. 1715–79). Kalm had studied natural history under the great Swedish professor Carl Linnaeus, one of the important founders of ecology. This training and his own scientific capabilities enabled Kalm to make the most extensive and perceptive ecological observations of America during the eighteenth century. His ecological perspective was further enhanced by his mission. He was sent to America by the Royal Swedish Academy of Sciences to discover useful plants that could grow in Scandinavia. He had to pay attention to soil, climate, and other environmental factors, as well as to utility.

The ecological observations in Kalm's *Travels into North America . . .*, published in Swedish in three volumes between 1753 and 1761, are so numerous that only a selection can be discussed here. His treatise was published in a German translation in 1754–64, but it did not appear in English until 1770–71. This delay may have been due to his unfavorable comments about English and Dutch settlers in America. His sympathies lay with the Swedish and French settlers, and he complained of the neglect of agriculture and inappreciation of science among the English settlers to the point that his English translator, the German naturalist John Reinhold Forster, accused him of bias.[18] Kalm's expressions of disapproval must have lessened considerably the appreciation of his work among the British and Dutch Americans.

The most extensive of Kalm's ecological observations concerned the correlation of distribution of plants with their habitat, about which he was more thorough than Bartram (he thought Bartram's book too superficial to have merited publication).[19] Kalm's book had the merit of using the Linnaean scientific names for species, enabling others, both then and now, to identify them readily. Kalm stayed in the vicinity of Philadelphia from September, 1748, until early June, 1749, when he traveled to New York City, then up the Hudson to Niagara, and on to Quebec and Montreal. About three-quarters of his observations were made in southeastern Pennsylvania and in New Jersey, but his northern trip enabled him to compare the plants and animals of the milder and colder regions.

For the Philadelphia region he made a survey of the trees and shrubs and listed fifty-eight of them according to abundance. He did not state how he determined abundance, but since this was long before the development of statistical methods of sampling, he must have depended upon his impressions. For many, but not all of them, he also stated where they grew. For example:

1. Quercus alba, *the white oak in good ground.*

6. Acer rubrum, *the maple tree with red flowers, in swamps.*

7. Rhus glabra, *the smooth leaved Sumach, in the woods, on high glades, and old corn-fields.*

9. Sambucus canadensis, *American Elder tree, along the hedges and on glades.*[20]

Often Kalm's observations on a species were detailed enough to be characterized as ecological life histories. *Rhus glabra* he found to be the commonest sumac; it seldom grew above three yards high; its berries remained on the shrub all winter; and the berries were eaten by boys with impunity. "This tree is like a weed in this country, for if a corn-field is left uncultivated for some years altogether, it grows on it in plenty, since the berries are spread everywhere by the birds."[21]

This last statement is among the earliest observations on plant succession. Brambles, he also noted, are among the first species to invade abandoned fields.[22] The Swedish settlers, as well as Kalm, realized that it was useful to know not only which types of soil a plant came from when transplanting plants but also that certain plants can indicate by their presence the existence of good soil for crops.[23]

One of the interesting questions that Kalm asked Bartram was whether Catesby had claimed correctly that members of plant species are smaller in the northern than in the southern parts of their range. Bartram replied that there are some species suited to southern regions of which this is true, but that other species are most suited to northern regions, and among these the smaller specimens grow in the south. Kalm's subsequent experience supported Bartram's answer. Kalm found the sugar maple to be a common and tall plant in Canada, but in New Jersey and Pennsylvania it grew less than a third as high and was found only "on the northern side of the blue mountains, and on the steep hills which are on the banks of the river, and which are turned to the north." The sassafras tree, on the other hand, grew tall and thick south of forty degrees latitude, but between forty-three and forty-four degrees "it hardly reached the height of two or four feet, and was seldom so thick as the little finger of a grown person."[24]

Kalm also wrote good accounts of a number of mammals that included perceptive ecological information. In discussing the gray squirrel, he reported where they nest, the kinds of nuts they eat, the fact that hogs, Indians, and white men raid their winter stores, that after heavy snows they sometimes starve because they cannot dig down to their stores, that they are a serious pest in both maize fields and barns, that they eat not only the acorns but sometimes also the flowers of oaks, and that rattlesnakes can reputedly charm them into jumping into the snakes' mouths. It was also reported that gray squirrels migrate out of the high country before severe winters, but Kalm discounted this weather-predicting capacity and suggested instead that the occasional migrations were actually caused by a scarcity of nuts in the high country.[25]

Although America did not produce in the eighteenth century any native son who could describe ecologically the plants and animals as well as Kalm, in the second half of that century there did emerge a few capable observers who wrote notable works on the production of certain regions, and these works contained some ecological observations. William Bartram (1739–1823) was America's most capable naturalist of the eighteenth century. He followed in his father's footsteps both in his interests and in the sense that he accompanied his father on trips to Florida and elsewhere. Yet William's *Travels Through North & South Carolina, Georgia, East & West Florida* (1791) shows that he was far from a replica of his father. William had quite a different psyche, and also much more to say. William was a romantic.

His *Travels* is a famous book because it is a spellbinding adventure story about a sensitive person wandering through a tropical Garden of Eden. The book is a dazzling accomplishment because the events themselves are only occasionally exciting. Bartram's style was so nearly perfect that a high level of interest is sustained throughout. Samuel Taylor Coleridge and William Wordsworth found it a treasure chest of poetic imagery.[26] Apparently in his schooling Bartram had been as fascinated by Homer, Thucydides, Livy, Cicero, and perhaps Shakespeare, as he had been by Catesby and Linnaeus. His gift of description enabled him to carry descriptive ecological surveys to a new level of detail. Here is only a portion of his description of an island off the coast of Savannah, Georgia:

The intermediate spaces, surrounding and lying between the ridges and savannas, are intersected with plains of the dwarf prickly fan-leaved Palmetto, and lawns of grass variegated with stately trees of the great Broom-Pine, and the spreading ever-green Water-Oak, either disposed in clumps, or scatteringly planted by nature. The upper surface, or vegetative soil of the island, lies on a foundation, or stratum, of tenacious cinereous coloured clay, which perhaps is the principal support of the vast growth of timber that arises from the surface, which is little more than a mixture of fine white sand and dissolved vegetables, serving as a nursery bed to hatch, or bring into existence, the infant plant, and to supply it with ailment and food, suitable to its delicacy and tender frame, until the roots, acquiring sufficient extent and solidity to lay hold of the clay, soon attain a magnitude and stability sufficient to maintain its station. Probably if this clay were dug out, and cast upon the surface, after being meliorated by the saline or nitrous qualities of the air, it would kindly incorporate with the loose sand, and become a productive and lasting manure.

The roebuck, or deer, are numerous on this island; the tyger, wolf, and bear, hold yet some possession; as also raccoons, foxes, hares, squirrels, rats and mice, but I think no moles; there is a large ground-rat, more than twice the size of the common Norway rat. In the night time, it throws out the earth, forming little mounds, or hillocks. Opposoms are here in abundance, as also pole-cats, wild-cats, rattlesnakes, glass-snake, coach-whip snake, and a variety of other serpents. Here are also a great variety of birds, throughout the seasons, inhabiting both sea and land.[27]

This wealth of detail can sometimes be overpowering rather than enlightening.

Bartram's powers of dramatic evocation are illustrated in his account of a "forest" of century plants visited by swarms of butterflies and bees. Looking closely, he observed the drama of a spider catching and overpowering a bumblebee.[28] Whatever ecological conclusions might possibly have emerged are lost because the focus is upon drama rather than upon trying to understand how nature functions. He did not even shrink from describing smoke coming from the nostrils of a fighting alligator.[29] This focus upon drama rather than upon the meaning of interactions was enhanced by his tendency to anthropomorphize: "The bald eagle is a large, strong, and very active bird, but an execrable tyrant: he supports his assumed dignity and grandeur by rapine and violence, extorting unreasonable tribute and subsidy from all the feathered nations."[30]

Fortunately, Bartram's integrity and love of nature prevented him from embellishing his discussions with invented details, but it seems clear that his success as an enchanter diminished his significance as a scientist. His was undoubtedly a more appealing personality than Kalm's, but Kalm's more prosaic orientation enabled him to penetrate deeper into the significance of what he saw.

Climate and Plants, Animals, People

Climate and Disease. Most of the early ecological observations were descriptive. Descriptions have remained an important part of ecology throughout its history, but a science needs a program to guide the collection of data if it is to penetrate beneath superficialities. The first ecological program that served to guide investigation was established within the context of medical research.

Since antiquity one of the most influential of medical writings was the Hippocratic treatise *Airs, Waters, and Places.* It opens with an explicitly ecological program of correlations between environmental conditions and sickness:

Whoever wishes to pursue properly the science of medicine must proceed thus. First he ought to consider what effects each season of the year can produce; for the seasons are not at all alike, but differ widely both in themselves and at their changes. The next point is the hot winds and the cold, especially those that are universal, but also those that are peculiar to each particular region. He must also consider the properties of the waters; for as these differ in taste and in weight, so the property of each is far different from that of any other. Therefore, on arrival at a town with which he is unfamiliar, a physician should examine its position with respect to the winds and to the risings of the sun. For a northern, a southern, an eastern, and a western aspect has each its own individual property. He must consider with the greatest care both these things and how the natives are off for water, whether they use marshy, soft waters, or such as are hard and come from rocky heights, or brackish and harsh. The soil too, whether bare and dry or wooded and watered, hollow and hot or high and cold. The mode of life also of the inhabitants that is pleasing to them, whether they are heavy drinkers, taking lunch, and inactive, or athletic, industrious, eating much and drinking little.[31]

This was written when diseases were generally believed caused by an imbalance of four humors—blood, phlegm, yellow bile, and black bile. Changes in climate were thought to influence the balance. For example, winter is predominantly cold and wet, and therefore there is a strong tendency for a superfluidity of phlegm. Summer is predominantly hot and dry, with a tendency for a superfluidity of yellow bile.

Beginning in the middle of the seventeenth century the humoral theory was slowly replaced by the perception of diseases as specific entities, though what these entities were was not clear. The shift away from a purely Hippocratic concept of disease did not, however, cause a loss of confidence or interest in the Hippocratic program of correlating environmental conditions with diseases. Interest in this program actually grew stronger, because, coincidentally, the barometer, thermometer, and other meteorological instruments were being developed.[32] These instruments provided the opportunity to increase the precision of the correlations and thereby increase the likelihood of obtaining a precise understanding of the environmental conditions that caused specific diseases. This program of research was encouraged in the writings of the English physicians Thomas Sydenham (1624–89) and Janes Jurin (1684–1750).[33]

The English clergyman and naturalist John Clayton (1657–1725), who came to Virginia in 1684, packed barometers and thermometers to make such studies, only to lose them at sea. Nevertheless, he reported from tidewater Virginia that in July and August the breezes ceased and the heat became "violent and troublesome." In September "The Weather usually breaks suddenly, and there falls generally very considerable Rains." When this happened, "many fall sick, this being the time of an Endemical Sickness, for Seasonings, Cachexes, Fluxes, Scorbutical Dropsies, Gripes or the like, which I have attributed to this Reason."[34]

The Scottish immigrant physician–naturalist–civil servant Cadwallader Colden (1688–1776) made much more extensive observations on weather and disease in America. In his "Account of the Climate and Diseases of New York" (written in 1723 but not published until 1811) he pointed out that spring comes to New York much later than it comes to England. The people in New York "are subject to pleurisies and inflammatory fevers, as in all other countries, upon the breaking up of hard winters; but not so much as in Pennsylvania and in the countries to the southward." He found July, August, and early September the sickliest period of the year, with intermitting fevers, cholera morbus, and fluxes being prevalent. The intermitting fevers were less prevalent than they were farther south, but the fluxes were more so, apparently because the poor of New York ate more watermelons than they did in Philadelphia. He also believed this difference in the prevalence of fluxes to be due to differences between the waters of New York and Philadelphia, New York's being more brackish and harder. Consumptions and diseases of the lungs were uncommon in New York because its air was clear and its elasticity strong. He found autumn to be America's healthiest season, the weather being mild and dry. Winter was long and often extremely cold, and the people were subject to rheumatic pains and pleurisies. In the essay he associated certain diseases with adverse qualities of the air, others with adverse qualities of water, and still others with adverse temperature.[35]

Colden stated in his "Observations on the Fever Which Prevailed in the City of New York in 1741 and 2" that the fever recurs annually in the hot months because of noxious vapors arising from "the stagnating filthy water. . . . not only different kinds of vapours are raised from a different fermentation in the stagnating fluids, but they raise likewise different fermentations in the animal fluids; hence, different kinds of fevers are produced in different constitutions of the air." He observed that the part of New York City mainly afflicted by the summer fevers was built on a swamp and also along the docks. He suggested that the city undertake the construction of a drainage system, which, in fact, was done and did lead to an improvement in the health of the area.[36]

In a letter written in 1745, Colden concluded that, while the swamps and dock areas enhanced the spread of yellow fever, the disease must initially have been imported on the ships that used the docks. He also suspected that the various malignant fevers reported from Europe were probably the West Indian yellow fever modified by climate.[37] He did not, however, go to the extreme that Benjamin Rush was to go a half century later of collapsing all fevers into different stages of one kind.[38] For Colden diseases were of different species, just as plants and animals were.

Following the lead of Dr. William Douglass, of Boston, Colden subjected an epidemic of "throat distemper" to an analysis of relevant environmental factors. He investigated its presumed origin, the geography of its dissemination, and the situations of its greatest frequency. He found it most common among children, the poor, those living in rural areas, those eating pork, those who were scorbutic, and those living on wet grounds. Sensible though this approach may have been, it did not lead to any greater understanding of the disease (now believed to have been a mixture of diphtheria and scarlet fever).[39]

On another occasion Colden drew the interesting analogy between insect galls and animal parasites, but the apparent perceptiveness of this speculation must be balanced by another of his beliefs, that scurvy is contagious.[40] Colden could ask many perceptive questions, but he could not establish causal relationships.

Another medical practitioner who emigrated from Scotland, John Lining (1708–60), settled in Charleston, South Carolina, in 1730. He made a much more thorough attempt to record weather data for correlation with disease than had been previously attempted in America, or perhaps anywhere else:

> *What first induced me to enter upon this Course, was, that I might experimentally discover the Influences of our different Seasons upon the Human Body; by which I might arrive at some more certain Knowledge of the Causes of our epidemic Diseases, which as regularly return at their stated Seasons, as a good Clock strikes Twelve when the Sun is in the Meridian; and therefore must proceed from some general Cause operating uniformly in the returning different Seasons.*[41]

Following the examples set by Santorio Santorio in 1614 and James Keil in 1723, Lining concluded that the correlations would have to include not only incidence of disease and climatic changes but also physiological responses to climate.[42] Therefore, he recorded not only temperature, barometric pressure, and rainfall but also his weight at morning and night, his food and drink, urine and stools, perspiration and pulse rate, and daily exercise.

He began recording all these data in 1737 and continued for fifteen years. His first paper on the subject (1743) stressed the sudden changes in temperature over wide ranges at Charleston as a possibly significant factor. In following papers he refined his techniques of measurement, but without ever getting around to demonstrating the assumed correlations between weather changes and diseases. In his paper of 1753 he shifted the justification for his studies from a focus on disease to one on the environment. He included a monthly and seasonal summary of rainfall for every year from 1738 to 1752 and commented that if this series of data was extended for a half century it "might be of use in discovering to us the changes made in a climate, by clearing the land of its woods."[43]

In 1748, Charleston was struck by an epidemic of yellow fever. Lining's decade of observations enabled him to rule out the possibility that this epidemic had its origin in particular weather conditions. He concluded that it was an infectious disease because it spread throughout the city but not into the surrounding countryside.[44]

Pehr Kalm was not a physician, but the question of the relation of climate to disease interested him, and, whatever his deficiencies in medical knowledge, they were probably compensated for by his ability to reason about environmental factors. His reservations about America and Americans did not cause him to blame precipitously either climate or habits for diseases. He judged both factors as important, but only among a series of others. For example, he had heard that the air made teeth decay in America, but he pointed out that the teeth of Indians lasted longer than those of white settlers and therefore that other factors must operate. He thought that both tea drinking and eating hot food were relevant adverse factors.

There was a general belief that the climate was ameliorating, but there was also some suspicion that fevers were becoming more common. These presumed trends could be reconciled by emphasizing the increase of sharp weather changes as a cause, as in Lining's train of thought. But Kalm also found some old people who believed that fevers were not proportionately more numerous but were simply encountered more often because the population had increased. That it was unhealthy to live near stagnant bodies of water was a widespread assumption to which Kalm adhered. Since, however, there were some healthy regions near stagnant waters, he felt that the water could not be a sole cause. To him the aggravating factor seemed to be the "intemperate consumption of fruit."[45] Later he found that the French Canadians ate as many watermelons as the Pennsylvanians did, and he saved his hypothesis with the additional assumption that the greater heat of Pennsylvania made watermelons more dangerous there than in Canada.

Kalm also suspected that the consumption of tea, coffee, chocolate, sugar, and strong liquors contributed to the unhealthiness and shortness of life in America. Still another factor he considered as possibly relevant was the decline of odoriferous plants from the forests because of grazing cattle. The plant odors might have rendered harmless "the noxiousness of the effluvia from putrifying substances."[46] When he traveled to Albany, he also worried that the drinking water, containing "Monoculi," might cause diseases.[47]

From his description of the "Monoculi," they were apparently mosquito larvae, and one is inclined to imagine that at last Kalm was on the right track. Since, however, he was not prepared to verify experimentally any of his speculations, there was no way for him or his readers to know that his suspicion of mosquitoes would prove more justifiable than his fear of Pennsylvania watermelons. His observational ecology was no more effective in solving the mystery of disease than were the ecological medical correlations of Colden or Lining.

Though Lining never demonstrated a relationship between particular diseases and changes in weather, he impressed a fellow Scottish immigrant and colleague, Lionel Chalmers (1715–77), with the value of his researches. For a time they were partners. Chalmers was to make some measurements of his own that enabled him to provide a chart of monthly maximum and minimum temperatures and rainfall data for the years 1750 to 1759. He also supplemented his own data in his *Account of the Weather and Diseases of South Carolina* (1776) by republishing one of Lining's charts.[48]

Chalmers was a less ambitious weather recorder than Lining, but he did attempt to establish a closer connection between climate and disease than Lining had. He began his book with a chapter on the climate, water, and soil of South Carolina. Having come from a cooler climate with less luxuriant vegetation, he was impressed by South Carolina's heat, stagnant waters, and fertile soils. He believed that the land had been unhealthier before the settlers cleared the land than it was in his day, because there would have been more exhalations from decaying organic matter and more moisture in the air. On the other hand, much of the cleared land had been dammed up for rice fields, indigo extraction, and mills, causing "such multitudes of fish and reptiles of various kinds [to] perish, that, for a long time after the air is tainted, with the putrid *effluvia.*"[49]

Chalmers described the topography of the Charleston region and then explained the relationship between its topography and its climate:

As a south wind blows from the warmer latitudes and sweeps over a great extent of sea, it must be always hot and moist. That which comes from the south-west and west must be sultry and moist in the summer, as it passes over large spaces of heated, marshy, overflowed or wood-lands; and in the winter it will bring damps or rain, being fraught with the exhalations that are made from the above soils as well as with those vapours which are collected and condensed by the high bleak mountains that lie behind us.[50]

After describing the climate, Chalmers devoted much of his treatise to the discussion of correlations between the weather at different seasons and the diseases that were most prevalent during particular seasons. One example will show what he was able to achieve:

If the weather be either sultry and showery, or cloudy and close, and sometimes calm, intermixt with a gentle southerly wind, as often happens toward the end of summer, low and what are called nervous *and* putrid *fevers will appear, more especially among corpulent people and others who are of a weak or lax habit.*[51]

Such correlations could not lead to any causal understanding because the disease entities were too vague.

Although such studies as those by Colden, Lining, and Chalmers could not be readily published during the Revolution, a number of other physicians had similar interests, and in 1792, Dr. William Currie (1754–1828), of Philadelphia, was able to draw upon his own and their records to compile *An Historical Account of the Climates and Diseases of the United States of America.* There was no uniformity in the data, since they had been collected by different physicians with no common plan. Nevertheless, the book gave some indication of the weather and diseases state by state from Maine to Georgia. Although the conclusions to be drawn from this juxtaposition of geographical, climatic, and disease data were neither startling nor altogether clear, the book as a whole certainly reinforced the conviction that climate and general environment are relevant to the incidence of disease.

In 1793 there occurred in Philadelphia "one of the most devastating outbreaks of pestilence ever recorded on this side of the Atlantic."[52] Yellow fever carried off a tenth of the population, and a heated controversy arose over whether the disease had been imported from the West Indies or was endemic. The eminent Dr. Benjamin Rush was convinced that it was endemic. He agreed with the proponents of the West Indian origin that the epidemic started in the region of wharfs, but he attributed it to exhalations from putrified coffee on a wharf near Arch Street. The coffee was removed, but the epidemic continued. The Philadelphia College of Physicians examined the available evidence and concluded that yellow fever had never been endemic in America and appeared always to have been imported on ships.[53]

This controversy was to continue even more strongly as an epidemic of yellow fever again struck Philadelphia in 1797. Controversy, as Karl Popper has pointed out, is important to the advancement of scientific understanding, and this controversy provided the opportunity for re-examining assumptions and increasing both ecological and medical understanding.[54] In the twentieth century a research program might be designed to demonstrate the adverse effects of some environmental factor by testing the effects upon laboratory animals of variations in one environmental factor after another. In the 1790's knowledge of physiology and chemistry was not sophisticated enough for this approach to seem realistic.

In the closing years of the century America's most famous lexicographer, Noah Webster (1758–1843) joined the controversy on the side of endemic origin, and he produced a better documentation of evidence than had previously existed. His approach was systematically to collect and evaluate past evidences and to collect statistical data from American physicians. He sent a questionnaire to physicians throughout the states requesting the following information:

The time of the appearance and disappearance of any epidemic disease, with its general history.

The places where it first occurs to be described, in regard to land and water, height of land, construction of the city or streets, position as to points of compass, woods, morasses, &c. The classes of people most generally affected.

The general state of the seasons, as to heat, and cold, drought and moisture.

The time of earthquakes, meteors, lumen boreale, and all singular celestial appearances—with unusual tempests, especially when accompanied with hail—all compared with the lunar phenomena.

The appearance of unusual insects of all kinds, and any circumstance attending them.

Diseases among cattle, sheep and other animals.

Sickness and death of fish of all kinds.

Volcanic eruptions, with the phenomena preceding, attending and following them.[55]

This was a commendable effort to increase the objective data on which a decision could be made. From his returns Webster published *A Collection of Papers on the Subject of Bilious Fever, Prevalent in the United States for a Few Years Past* (1796). What he received, however, was not so much objective data as information supporting the disease theories of the responding physicians. This approach was democratic but not decisive. The available information would have to be arranged and evaluated from one point of view, and this Webster achieved in his two-volume *Brief History of Epidemic and Pestilential Diseases*... (1799). He stated that he had undertaken his inquiries with an impartial mind and that only after examining the evidence had he concluded that yellow fever was endemic. That may be true, but if so, he was pleased that his conclusions coincided with what he perceived to be best for American commerce. If yellow fever was endemic, it would be unnecessary to require the long periods of quarantine demanded by those, like Dr. Currie, who thought the disease was imported on ships from the West Indies.[56]

Webster's theory of disease was not original. He shared the general consensus that certain diseases can have an environmental origin and then become infectious. He believed that adverse weather changes could cause epidemics and that such changes might in turn be responses to comets, earthquakes, volcanic eruptions, and atmospheric electrical disturbances. He, like Dr. Rush, was skeptical of the trend of the time to name many different kinds of diseases. Webster believed that certain presumed disease entities were actually only mild or severe forms of other diseases or that certain diseases manifested themselves differently in different localities. He suspected that Old World plague and New World yellow fever were the same disease, which affected people differently in the different climates.[57]

Neither Webster nor his opponents were close to solving the problem of the origin of disease. Yet the debates they waged helped maintain an interest in environmental factors and influences.

Phenology. Phenology is a branch of ecology dealing with the correlation between climate and biological activity. The term itself is of comparatively recent origin, being first used in German in 1853 and in English in 1875. The term is easily confused with phrenology, the nineteenth-century term for the science of correlations between the shape of a person's head and his mental attributes. In the eighteenth and nineteenth centuries the term generally used until replaced by phenology was *calendar of flora.* This term described the form of such studies, but it was often too restricted because animals as well as plants were observed and reported.

Phenology does not occupy an important place in the

modern science of ecology because such studies today would usually be undertaken in relation to a specific question, which would necessitate providing the data in a different context. However, in the eighteenth and nineteenth centuries, when naturalists began to grope toward the organization of an ecological science, phenology held an important place in their thinking.

Two kinds of prior interests provide some reinforcement for phenological studies. Farmers' almanacs have been written for five thousand years, and throughout history one or another of them has been in demand.[58] Botanists have never been strongly motivated to contribute to this form of literature, though a few of them did so in the eighteenth century. The other interest was in drawing correlations between climate and the prevalence of diseases. This kind of study was recommended in antiquity by the Hippocratic treatise *Airs, Waters, and Places*. Nevertheless, the matter was not much investigated before the eighteenth century.

No doubt Linnaeus, who was both physician and professor of natural history, was aware of these two areas of previous study, and his own interest in phenology must have depended in part upon them. Linnaeus likely conveyed to Pehr Kalm the idea that the times of blooming, leafing, and fruiting are significant data for the life histories of plants. In addition, Kalm's mission to discover American species that might thrive in Scandinavia alerted him to the practical importance of such data. He was first to pay much heed to these phenomena in America, and his *Travels* emphasized the significance of such information.

Beginning in February, 1749, Kalm carefully noted the seasonal unfolding of biological phenomena. Purple grackles began to reappear on February 23. The flowers of the hazel began to open on March 12 and on March 13, those of the alder, *Dracontium foetidum*, and *Draba verna*.[59] This information was enhanced by meteorological tables for the Philadelphia area, which Kalm compiled for August, 1748, to September, 1749 (while he was in Canada in the summer of 1749, Bartram collected the data for him).[60]

Kalm wrote his *Travels* in chronological sequence from his diary, and sometimes he discussed a subject more than once and from different perspectives. At one point he discussed the difficulties of raising American plants in Europe in relation to climate, and later he discussed the reverse problem of raising European plants in America. The main difficulty with raising American plants in Europe was that most of them bloomed late and their fruits did not ripen before autumn frosts. Kalm evidently was speaking, or assumed he was, of plants that were being transplanted into conditions rather similar to their native situation. There were several factors that he felt were actually different and could account for the difficulty. The first was that, although the winters in Pennsylvania and northward were as severe as those in Sweden, the American winters were much shorter. Next, the summers in Pennsylvania were much longer than they were in Sweden, it being hot in the former from April to October but in the latter only in June and July. America's late-blooming flowers are correlated with its longer hot season. However, he noted that even in America some of these would not always bring their seeds to maturity before cold weather set in. He found the wisdom of God, nevertheless, in the fact that these were perennial species. Another contributing factor in the late blooming of American species was that most of them would have grown in forests before they were cleared by the settlers, and there the dead leaves and the shade from the trees would have created seasonally late and slow-growing conditions.[61]

In a later discussion he pondered why cultivated European fruit trees imported into America bloomed earlier than similar native American species. He suggested that the key to this puzzle might lie in the more fluctuating American climate. In Europe, the plants could safely bloom as soon as it turned warm, but in America this could lead to death of the flowers, since the early warmth could be followed by a cold spell. Therefore, the late-blooming American species were better suited for the American climate than were the earlier-blooming European species.[62] Kalm did not attempt to verify his hypothesis experimentally. The removal of a problem from the field to the greenhouse or laboratory for solution was not often attempted in the eighteenth century, or even very often in the nineteenth century. Nevertheless, his discussion of his hypothesis is one of the subtlest examples of ecological reasoning in the eighteenth century.

Another interesting and innovative example of his use of phenological information occurs in a scientific paper that he published in Swedish in 1759 but that did not appear in English translation until 1911. It was on the passenger pigeon, then one of the most abundant of American birds. He discussed nine important food plants for these birds, indicating when their fruits became available in Pennsylvania or in Canada. It was possible that the movements of these birds might be correlated to some extent with the maturing of their food supply rather than only with climate, but Kalm did not have sufficient information to discuss this possibility.[63]

Meanwhile, in Sweden, Kalm's former teacher, Linnaeus, had organized the collection of information useful for determining the time to plant crops in Sweden. This information was published in 1753 under the title *Vernatio Arborum*, and it was followed in 1756 by the better-known *Calandarium Florae*, which provides phenological information for Upsala, Sweden, for the whole year. The English naturalist Benjamin Stillingfleet was impressed enough by the latter work not only to translate it into English, but also to make his own calandar of flora for Stratton, in Norfolk, and to extract from Theophrastus' *Historia Plantarum* a calandar of flora for ancient Athens. Stillingfleet included all three in a collection of Linnaean dissertations that he published in English, and it was apparently this collection and Kalm's *Travels* that introduced the idea of phenology to Americans.[64]

Thomas Jefferson (1743–1826) at times seemed more interested in natural history than in politics. He was well read in the literatures of natural history and agriculture, and he was actively interested in his crops and gardens at Monticello. He also studied the weather. In his *Notes on the State of Virginia* he based his chapter on climate upon meteorological data that his friend the Reverend James Madison, a professor of mathematics, collected at Williamsburg during

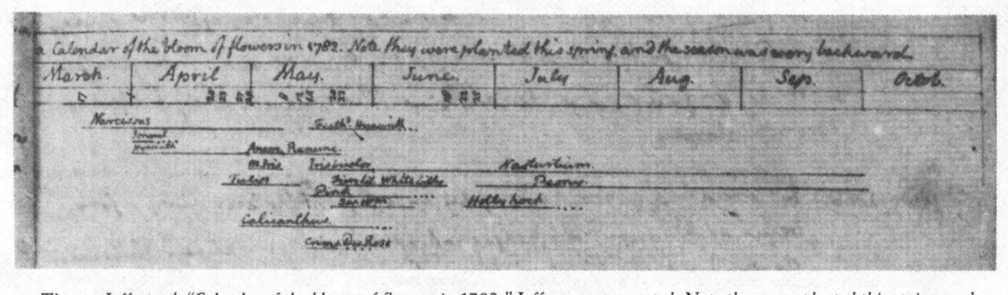

Calendar of the bloom of flowers in 1782. Note they were planted this spring. and the season was very backward.

March	April	May	June	July	Aug	Sep.	Octob.

Thomas Jefferson's "Calendar of the bloom of flowers in 1782." Jefferson commented: Note they were planted this spring, and the season was very backward." From Thomas Jefferson, Garden Book, 1776–1824, *page 25. Reproduced courtesy of the Massachusetts Historical Society.*

the years 1772 to 1777 and upon some data that they collected simultaneously at Williamsburg and Monticello. Even America's Independence Day did not distract Jefferson from his observations, for he recorded the temperature on July 4, 1776, at 6:00 and 9:00 A.M. and at 1:00 and 9:00 P.M. He even bought a new thermometer that day, for 3 English pounds 15 pence. He also bought a barometer on July 8 for 4 pounds 10 pence. The War of Independence was ultimately to interrupt the coordinated weather investigation, however, because the British soldiers robbed Madison of his thermometer and barometer.[65]

Jefferson seems to have begun collecting weather data systematically in 1776, but he began entries in his *Garden Book* a decade earlier. His *Garden Book*, not published until 1944, was a combination of phenological data and planting diary. For example, the entry for March 23, 1767 was:

Purple Hyacinth & Narcissus bloom.
sowed 2. rows of Celery 9.I. apart.
sowed 2 rows of Spanish onions & 2.d° of Lettuce.[66]

In 1782 he constructed an interesting chart to indicate the lengths of time various species of garden flowers remained in bloom.

In his *Notes on the State of Virginia*, which he completed early in 1785, he showed an awareness of the close relationship between climate and the distribution of plants. This subject became of great practical interest to him while he was serving as an American diplomat in France. He made there some investigations similar to Kalm's in America. Jefferson decided that the climate of southern France was similar enough to that of South Carolina and Georgia for it to be worthwhile to send to America seeds from French crops, particularly rice and olives. Although his attempts to establish these crops in America were not successful, his reconaissance was almost as sophisticated as Kalm's. Jefferson was elected an honorary member of the South Carolina Society for Promoting and Improving Agriculture and Other Rural Concerns, and the long letter that he sent to William Drayton of that society on July 30, 1787, is of particular interest. One passage illustrates well his awareness of the relationship of climate and topography and distribution of plants:

My journey through the southern parts of France, and the territory of Genoa, but still more the crossing of the Alps, enabled me to form a scale of the tenderer plants, and to arrange them according to their different powers of resisting cold. In passing the Alps at the Col de Tende, we cross three very high mountains successively. In ascending, we lose these plants, one after another, as we rise, and find them again in the contrary order as we descent on the other side; and this is repeated three times. Their order, proceeding from the tenderest to the hardiest, is as follows: caper, orange, palm, aloe, olive, pomegranate, walnut, fig, almond. But this must be understood of the plant only; for as to the fruit, the order is somewhat different. The caper, for example, is the tenderest plant, yet, being so easily protected, it is among the most certain in its fruit. The almond, the hardiest, loses its fruit the oftenest, on account of its forwardness. The palm, hardier than the caper and orange, never produces perfect fruit here.[67]

Jefferson's brief experience at collecting simultaneous weather data with Madison may have given him the idea to do the same with biological phenomena. In June, 1790, while serving the federal government in New York, he sent his daughter Maria information about the reappearance of whippoorwills, swallows, and martins and the growth of peas and strawberries. He requested similar information from her about these at Monticello. She replied that

we had peas the 10th of May, and strawberries the 17th of the same month, though not in that abundance we are accustomed to, in consequence of a frost this spring. As for the martins, swallows, and whip-poor-wills, I was so taken up with my chickens that I never attented to them.[68]

He tried again the following spring this time from Philadelphia, and he finally received some information on the weather and fruit trees from his son-in-law Thomas Mann Randolph, Jr., but nothing about the robins, bluebirds, or frogs.[69]

Having many interests and activities, Jefferson never got around to publishing his phenological studies. However, he was fond of discussing his scientific interests and encouraging others to undertake the detailed investigations for which he never found the time. In a letter on phenology to the botanist Jacob Bigelow in 1818, Jefferson summarized the phenological data he had collected during the previous seven years at Monticello.[70] Those years he had spent in the White House, 1801 to 1809, had not erased his enthusiasm for the subject.

Samuel Williams (1743–1817), who for a while taught physical sciences at Harvard University, was especially interested in meteorology. He seems to have planned a comprehensive investigation of the climate and phenology of the United States, for in 1786 he sent to prospective observers a plan for collecting these data. The plan was, no doubt, interrupted by his dismissal from his professorship on grounds of forgery.[71] He moved to Vermont, where he wrote his useful *Natural and Civil History of Vermont*, first published in 1794. In the chapter on climate he gave not only a chart showing the seasonal progress of various cultivated plants in Vermont but also a capable account of the differences in climate from Charleston, South Carolina, to Rutland, Vermont. In his chapter on native animals he also provided a list of birds that are permanent residents and those that are migratory. For the latter he gave arrival and departure dates. Comparable information on birds was also published by William Bartram in 1791 and Benjamin Smith Barton in 1799.[72]

Climate and Inferiority. Captain John Smith, John Lawson, and Mark Catesby were among the early authors who praised America's climate and fertility. Pehr Kalm was unenthusiastic, but the widely read naturalist Comte Buffon initiated a controversy that lasted throughout the second half of the eighteenth century and beyond. It concerned the quality of the American climate and the possible negative influences of that climate upon its inhabitants.

Kalm's negative feelings about the English Americans have already been mentioned. Among other things he felt that their agriculture was wasteful in comparison with the intensive practices of European farmers. Yet he was not inclined to judge negligence as the sole cause of agricultural difficulties in America. Climate was obviously a potent environmental factor, though it was not always easily separated from the other factors. He reported that the Americans had failed to establish a textile industry not only because cloth could be obtained easily from England but also because "the breed of sheep which is brought over degenerates in process of time, and affords but a coarse wool." Cattle, horses, and hogs also degenerated: "The climate, the soil, and the food, altogether contribute their share towards producing this change."[73]

Even the settlers seemed affected. Kalm thought that they matured earlier than their European relatives but also were weaker and died earlier. However, Indians of old age had been known in the past—before the introduction of Old World diseases and rum—and so this degeneracy could not be blamed entirely on the weather. He also found that the Negroes had not grown lighter in color after 130 years of American slavery. But, even though Kalm could not separate the effects of weather from other factors, he did believe that the sudden changeableness of the weather and seasonal shifts from hot summers to cold winters were important health hazards.[74]

Comte Georges Louis Leclerc de Buffon (1707–88), writing at about the same time as Kalm, published the most influential pronouncements on the degenerating effects of the American environment. His first discussion relating to the subject appeared in "The Natural History of Man" (1749). In its final section, "Of the Varieties of the Human Species," he expressed the general European sense of superiority over the other peoples of the world. He supposed the cause of the differences to be environmental, believing that if white men lived long enough in the tropics and adopted the habits of black men, they would eventually become like them. Among the Indians of America it had been reported that occasionally a white child is produced, as also occurred among dark peoples elsewhere. From this fact Buffon deduced that white was the primitive color, but that skin color "may be varied by climate, by food, and by manners, to yellow, brown, and black, and which, in certain circumstances, returns, but so greatly altered, that it has no resemblance to the original whiteness . . ."[75]

Buffon later developed the idea that European animals had migrated to America and degenerated into the American species. This idea seems to have been an extension of his above assumption from people to animals. He published his speculations about human changes before Kalm published his *Travels*. However, Kalm's observations were brought to the attention of the French intelligensia by Abbé Arnaud in 1761, about the time that Buffon extended his theory to animals, and it may be that Buffon depended to some extent upon Kalm in his assertion that

> *All the animals which have been transported from Europe to America, as the horse, the ass, the ox, the sheep, the goat, the hog, the dog, &c. have become smaller; and those which were not transported but went thither spontaneously, those, in a word, which are common to both Continents, as the wolf, the fox, the stag, the roebuck, the elk, &c. are also considerably less than those of Europe.*[76]

The same conditions that produced small, weak animals and men, he claimed produced large reptiles and insects. Those conditions were supposedly lesser heat and greater humidity, with high mountains, extensive forests, and many bodies of water. He developed this line of reasoning to the point that he could claim inferiority for all American productions, regardless of the changes in climate from one region to another.

Other European authors picked up Buffon's claim and developed their own variations of it.[77] The Americans were not convinced. Jefferson published a strong and effective rebuttal in his *Notes on the State of Virginia.* Jefferson realized that Buffon was vulnerable both in theory and in command of relevant facts. He collected his own data to show that American mammals were not usually smaller than similar European species but, indeed, were often larger. Jefferson also challenged Buffon's assumption that American species represented degenerate European species by claiming that members of any species can be modified only within limits by changes in soil, climate, and food. Jefferson could also challenge Buffon's assumption that cold and moisture caused American species to be of inferior size by citing another of Buffon's discussions, which stated that colder and more humid countries seemed better for oxen than hot dry ones.[78] Not content with the arguments in his book, Jefferson in 1786 sent back to America to obtain antlers of American deer, elk, and caribou, which he presented to Buffon in 1787. Buffon found them convincing but died before he could publish a further discussion on the subject.

Jefferson was not the only American to reply to the degeneracy arguments. Somewhat similar to his were those of Samuel Williams, who in his *Natural and Civil History of Vermont* (1794) provided additional evidence from the weight of American mammals that they were more often superior than inferior to comparable European species. He also argued that the European species would never have emigrated to America if the conditions here had not been more favorable than the ones they left.[79] Benjamin Franklin also entered the fray in a modest way by pointing out at a party in Paris that the American guests were taller and more robust than the Frenchmen present, and in 1780 he collected weather observations that seemed to indicate that England and France were more humid than America.[80]

This controversy in retrospect may seem to represent mere chauvinism argued within the realm of environmental speculation. That was certainly one aspect of the controversy, but, on the other hand, it must have served the useful function of encouraging Americans to increase their understanding of their own climate and its influences upon all forms of life in America.

Changing Environment and Changing Climate. America's climate in the seventeenth and eighteenth centuries was not simply a matter of idle conversation. It was of vital concern to settlers and would-be settlers. The early settlers in New England found the winters colder than expected, and the first settlers in the southern colonies found the summers hotter than expected. Although they soon learned to cope with the unfavorable aspects of the climate, weather remained a serious concern, for the country was strongly dependent upon agriculture, and there was a widespread belief that climate played an important role in health and sickness.

There was an understandable alertness to any changes in climate—perhaps an overalertness, for sometimes the changes appear to have been exaggerated. As early as 1688, John Clayton in tidewater Virginia was reporting, "I have been told by very serious Planters, that 30 or 40 years since, when the Country was not so open, the Thunder was more fierce."[81] He believed the climate could be favorably modified by wise agricultural planning. The existing plantations, he complained, were too large. If each planter had smaller holdings, he could clear the woods more effectively and drain the swamps, thereby imrpoving the circulation of fresh air and decreasing the unhealthy ferments in the air.[82]

Clayton's general idea that the amount of vegetation in an area would affect the climate soon received support from Dr. John Woodward, of London, a professor of medicine and cosmology. He conducted experiments on plant growth and concluded that the water evaporated from plants could cause high hunidity in heavily forested regions. That, he stated, had formerly been the condition in America, but as the forests were burned or destroyed to make way for homes and agriculture, the air became drier and more serene.[83] This conclusion met with little if any skepticism in America until the very end of the eighteenth century.

Pehr Kalm was the best historian of American land use of the mid-eighteenth century. He thought that the clearing of forests and draining of swamps must have modified the climate, and he asked elderly people whether the climate had changed since their youth. He found one old Swede who thought not, but the others felt that there had been changes. They believed that both the winter and the spring had formerly come earlier. The Delaware River had formerly froze about mid-November, but in the 1740's it did not freeze until mid-December. They generally agreed that formerly the winters had been colder but that the climate had changed less rapidly.[84]

Dr. Hugh Williamson (1735–1819), a physician who did research in the physical sciences and taught mathematics at the College of Philadelphia, had an interest in the question from the standpoints of medicine, meteorology, and patriotism. His arguments were mainly meteorological and very ingenious. He argued that as the land was cleared for cultivation it would reflect back into the immediately adjacent air more heat than would the forest, thereby increasing the warmth in the winter. On the other hand, because heat rises, he thought that the same situation would not necessarily make the summers hotter than before. He also suggested that the increasing warmth in the wintertime would decrease the difference between the land and ocean temperatures, which should also decrease the ferocity of the winds during winter.[85]

Jefferson briefly discussed the amelioration of the climate in his *Notes on the State of Virginia*, citing the fact that the rivers that were once known to freeze over in winter seldom did so any longer and that the snows that once accumulated in the mountains and caused spring floods no longer accumulated, and the rivers no longer flooded.[86]

The estimates of the changes wrought in the climate because of cultivating the land were made, up to this point, in terms of temperature and humidity, and the conclusions were favorable. During the second half of the eighteenth century chemists developed the capability of identifying different kinds of gases, and in 1772, Joseph Priestley discovered that plants restore the respirability of vitiated air.[87] Trees were not merely useful providers of fuel, lumber, and shade. The Marquis François Jean de Chastellux appreciated this point and cast doubt upon the Americans' wisdom in proceeding as rapidly as they were in eliminating their forests. He thought it beneficial to cut down forests where the swamps could be drained, but in eastern Virginia that could never be fully accomplished, and therefore the trees were needed to absorb the "mehpitic exhalations."[88] This concern for the value of trees to absorb noxious gases from swamps was also expressed in scientific papers read by Dr. Benjamin Rush in 1785, Dr. Thomas Wright in 1794, Dr. William Currie in 1795, and Dr. Adam Seybert in 1798.[89]

Samuel Williams decided to add precision to this discussion by quantifying the climatic influence of forests. He noted that the rainfall figures for North America were about twice those for Europe at the same latitude (he did not cite his sources of information, nor did he mention that his figures contradicted Benjamin Franklin's). This difference he assumed was caused by the great amounts of water evaporating from America's forests. He measured the evaporation rate during a six-hour period from a maple twig having two leaves and one or two buds—16 grains troy weight—and then cut down the thirty-foot tree and counted its leaves—21,192. He then estimated the density of trees in an acre of forest—640—and calculated that an acre of this forest would emit 3,875 gallons of water to the atmosphere in twelve hours. In a similar manner he measured the amount of "air" emitted by the same twig with leaves and buds (before the cutting) and calculated that an acre of forest emits 14,774 gallons of "air" in twelve hours. He pointed out, however, that not all forests would have the same density of trees as the one on which his calculations were based.

The existence of forests affects not merely the water and gases of the atmosphere but also the temperature. To ascertain the magnitude of this effect, Williams measured the temperature of the soil ten inches below the surface in an open field and in a Vermont forest for thirteen days, from May 23 to November 16:

Time		*Heat in the Pasture*	*Heat in the Woods*	*Difference*
May	*23*	*50°*	*46°*	6° [sic]
	28	*57*	*48*	9
June	*15*	*64*	*51*	*13*
	27	*62*	*51*	*11*
July	*16*	*62*	*51*	*11*
	30	*65½*	*55½*	*10*
August	*15*	*68*	*58*	*10*
	31	*59½*	*55*	*4½*
September	*15*	*59½*	*55*	*4½*
October	*1*	*59½*	*55*	*4½*
	15	*49*	*49*	*0*
November	*1*	*43*	*43*	*0*
	16	*43½*	*43½*	*0*

He was impressed by the ten or eleven degrees difference in June and July, because he suspected that the winters had increased in warmth by that average amount since the settlement of America. He discerned a direct correlation between the temperature of the soil and of the atmosphere because, he reasoned, the heat of the soil would heat the atmosphere to about the same temperature. As the land was cleared, the pasture temperature would come to dominate rather than the woods temperature. Unfortunately for this argument, he failed to notice that there was no difference in soil temperatures between pasture and woods for October and November. His data might support an argument for increasingly warmer summers but could not support an argument for increasingly warmer winters. Furthermore, Williams also compared pasture and woods soil temperatures on January 14, 1791, and found the pasture frozen to a depth of three feet five inches, whereas the soil in the woods was thirty-nine degrees.[90]

Down to the end of the eighteenth century there seems to have been virtually unanimous agreement that the American climate was becoming milder as the land was brought into cultivation from forests and swamps. In his study of the relationship between climate and disease, Noah Webster acquired a familiarity with the historical evidence on which that assumption was built. In 1799 he called into question the evidence for climatic change that Jefferson, Williams, and several others had offered. Webster chose to argue the issue not just for America but also for Europe, since some had also argued that the European climate ameliorated as its forests and swamps were cleared from the Middle Ages onward. Webster brought no new data to the controversy, and there were not sufficient data to show that the climate had not ameliorated. However, Webster was able to show that there were difficulties with the reasoning of Williams and other authors on the subject and that their case could not be presumed proved.[91]

The Balance of Nature

The balance of nature is the oldest ecological theory. It is a theory that has never been worked out in detail, but the basic assumptions were expressed in antiquity, beginning with Herodotus and Plato, and the theory developed by accretion throughout subsequent history. There is not much to say about it before the time of Kalm's visit to America, because the writings on American natural history had been long on observation and short on theory. In 1749, Kalm's teacher, Linnaeus, outlined an ecological science based on the balance of nature concept, though his name for it was "*oeconomia naturae.*"[92]

Although Kalm was in America when Linnaeus published his outline for a new science, Kalm would surely have learned something about it while studying under Linnaeus. In America, Kalm found that one species after another, including bear, beaver, deer, crane, duck, goose, passenger pigeon, mosquito, and grass, was known to have declined in population within the memory of the settlers. In some places the water table had also been lowered. He heard that "about sixty or seventy years ago, a single person could kill eighty ducks in a morning; but at present you frequently wait in vain for a single one." The reasons for the decline seemed clear. Before the white men arrived, there had been only a few Indians, and they lacked guns. The white men with guns had a greater ability to kill, and they indiscriminately took eggs, mothers, and young, "because no regulations are made to the contrary. And if any had been made, the spirit of freedom which prevails in the country would not suffer them to be obeyed." Other important factors he pointed out were the increasing population and the clearing of the land.[93]

Not only were certain species becoming rarer in the settled regions but some, such as the beaver, had actually disappeared. As time went on, trappers in Canada had to go farther and farther westward and northward to find them. Kalm did not predict what the ultimate fate of the beaver might be, but the situation also became known to Buffon, who in 1756 commented that "if the human species, as is reasonable to suppose, shall, in the progress of time, people equally the whole surface of the earth, the history of the beaver, in a few ages, will be regarded as a ridiculous fable."[94]

Because of his practical orientation, Kalm's ecological observations were analytical as well as descriptive. For example, when he learned that Pennsylvania farmers burned off the ground cover of leaves in the forest in March to allow the grass to grow more rapidly for their livestock, he admitted the advantage but nevertheless concluded that it was outweighed by the disadvantages:

All the young shoots of several trees were burnt with the dead leaves, which diminishes the woods considerably; and in such places where the dead leaves had been burnt for several years together, the old trees only were left, without any wood. At the same time all sorts of trees and plants are consumed by the fire, or at least deprived of their power of budding; a great number of plants, and most of the grasses here, are annual; their seeds fall between the leaves, and by that means are burnt: This is another cause of universal complaint, that grass is much scarcer at present in the woods than it was formerly; a great number of dry and hollow trees are burnt at the same time, though they could serve as fewel in the houses, and by that means spare part of the forests. The upper mould likewise burns away in part by that means, not to mention several other inconveniencies with which this burning of the dead leaves is attended. To this purpose the government of Pennsylvania have lately published an edict, which prohibits this burning; nevertheless every one did as he pleased.[95]

Among the other inconveniences that Kalm may have had in mind may have been the depletion of odoriferous plants that would counteract the noxious effluvia from putrifying substances, which elsewhere he attributed to the grazing of cattle in the forests.[96]

Implicit in the balance-of-nature concept was the assumption that different species were in some numerical proportion to each other, but it must have been clear to some that these relationships were not inflexible. It was obvious from the start that the Indians had not fully populated the New World. Free land was one of the most important factors in the immigration of Europeans to America. Kalm noted that America's white population grew rapidly, because with easily available land there was no inhibition to early marriage.[97]

Benjamin Franklin expressed the situation and its implications very well in his "Observations Concerning the Increase of Mankind, Peopling of Countries."[98] Because of the early marriages and large families he guessed that America's population might double every twenty years, and he predicted that within a century there would be more Englishmen in America than in England. One of the main objectives of his essay was to point out that the English settlers alone could easily and quickly populate the country by themselves and that the country would be more harmonious if Negro slaves and Europeans of "swarthy complexion" were excluded. Since he compared the capacity of human beings to increase to that of plants, by implication he seemed to suggest that the human population in America would eventually stabilize in the same way that plant populations had.

Franklin's ideas about stocking the country with people might in some way have been influenced by a childhood experience that he related to Kalm. His father, Josiah Franklin, had observed in the Boston area that in one river there were many herrings that swam upriver to spawn in the spring time, while in a neighboring river there were never any herrings. Kalm continued:

This circumstance led Mr. Franklin's father who was settled between the two rivers, to try whether it was not possible to make the herrings likewise live in the other river. For that purpose he put out his nets, as they were coming up for spawning, and he caught some. He took the spawn out of them, and carefully carried it across the land into the other river. It was hatched, and the consequence was, that every year afterwards they caught more herrings in that river; and this is still the case. This leads one to believe that the fish always like to spawn in the same place where they were hatched, and from whence they first put out to sea; being as it were accustomed to it.[99]

This is surely one of the earliest records of wildlife stocking in America, and it shows a belief that the rivers, like the land, might not always contain as much life as possible. This insight appears to run counter to the general assumptions of the balance-of-nature concept. The concept, however, was so loosely formulated that it was difficult to be skeptical of it on the basis of a few superficial experiences.

On a later occasion, in 1774, Franklin was pleased to learn from Joseph Priestley that vegetables restore the air that is spoiled by animals. Franklin replied that this was one more feature in a rational system of complementary relations, which included animals eating vegetables and then fertilizing them with their wastes.[100] The rational system to which Franklin referred was clearly the balance of nature, here envisioned in terms of natural cycles of matter rather than in terms of a balance in the populations of species.

While writing *Natural and Civil History of Vermont*, the mathematically minded Samuel Williams was to develop an interest in the abundance of different kinds of animals of Vermont, and when he could obtain the information, he introduced into his accounts of the mammals their periods of gestation, numbers of litters a year, and numbers of offspring a litter as means of accounting for their relative abundance. While rebutting Buffon's arguments that American mammals had degenerated from European species, Williams grew used to the assumption that American species had migrated to America, and he tied this idea explicitly to the balance of nature in his argument against Buffon's idea that America's environment was less conducive to fertility than was Europe's:

Whatever be their multiplying power, it would require a long period of time, before they would arrive at that increase of numbers, in which their progress would be checked, by the want of food. They would naturally spread over the whole continent, before they arrived to such a state. This they had done in every part of America, when it was first discovered by the Europeans. . . . what is the greatest number of quadrupeds, that the uncultivated state of any country will support, we have no observations to determine. But it seems probable, that the maximum *had already taken place. . . . if we may judge of the energy with which* [*nature*] *acts, from the effects of her multiplying power, the conclusion will be, that in no country has she displayed greater powers of fecundity than in America.*[101]

With Jefferson's and Williams' defense of the fertility of America and its forms of life against the slander of foreigners, it is not surprising that they did not share Kalm's and Buffon's doubts about the survival of America's wildlife. Although Jefferson and Williams must have been aware of the slaughter of America's animals and their decreasing numbers, they held tenaciously to the platonic tenent that no species can ever become extinct.[102]

THE NINETEENTH CENTURY

Exploration and Inventory

From 1590 to the time of Kalm and the two Bartrams there was a definite increase in the detail and perceptiveness of ecological observations in the literature of naturalists' travels and surveys of natural productions. By 1800 only a small portion of the North American continent had been explored in much detail by naturalists. The literature of exploration and inventory of North America continued to appear throughout the nineteenth and twentieth centuries with persisting ecological significance. This literature was useful for ecological observations but usually not, however, for ecological theory, and therefore there are diminishing returns on its significance for progress in ecological understanding.

This is evident in the *History of the Expedition under the Command of Lewis and Clark* (published in two volumes in 1814). A survey of the natural productions was one of the important objectives of the trip, and no doubt the natural-history reports would have been more extensive and of higher quality if Meriwether Lewis had not died tragically in 1809 and if Benjamin Smith Barton had been able to fulfill his subsequent agreement to prepare the scientific report of the expedition. Nevertheless, these misfortunes aside, it is not clear that the scientific reports could have added much to the progress of ecology. The expedition did make daily weather records,[103] occasional phenological notes, and observations on the kinds of trees encountered in river valleys. These data and the specimens of plants and animals they brought back were valuable additions to knowledge. Nevertheless, an exploratory expedition that traversed the continent could not be expected to improve upon the quality of observations published from the more knowledgeable and leisurely explorations of trained naturalists like Kalm and William Bartram.

Shortly after the Lewis and Clark expedition returned to civilization, however, the possibilities for explorers to contribute to ecology was enhanced by the example set by one of the greatest scientific explorers of all times, Alexander von Humboldt. The *Personal Narrative* of his travels through Latin America (published in three volumes in Paris between 1814 and 1825) contained the same sort of ecological observations Kalm had made, and Humboldt was probably the first naturalist whose observations exceeded Kalm's both in ecological quality and quantity. His *Personal Narrative* inspired the explorations and accounts of many naturalists, including Charles Darwin, Alfred Russel Wallace, and Henry Walter Bates.[104]

Humboldt's most useful work for ecology was probably his *Essai sur la Géographie des Plantes* . . ., written with his colleague Aimé Bonpland and published in French and German editions in 1807. He urged that plants should be investigated from the standpoint of the different zones and elevations they inhabit; the degrees of atmospheric pressure, temperature, humidity, and electrical charge of the atmosphere in which they live; and their growth patterns—whether they are solitary or social species. He explained how the terrain and climate enabled the same plants to exist continuously from Canada at low elevations to Mexico at high elevations, and how the Mediterranean Sea and the Pyrennees prevented the spread of African and Spanish plants northward.[105] Accompanying the essay was a large chart of two volcanic mountains in Equador whose sides were divided into vegetation districts. On each side of the chart were columns of figures indicating elevation, gravitational attraction, barometric pressure, maximum and minimum temperature, and other physical variables. Using these columns, one could locate the magnitude of any of the environmental factors in relation to the plants listed on the sides of the mountains. One could thus presumably determine the physical requirements for any species listed.

Humboldt's correlational science of biogeography was too ambitious to be fully accepted by other naturalists of the nineteenth century. In the first place, he had not demonstrated the importance of some factors, such as atmospheric pressure and electrical charge for the distribution of species. In the second place, field studies could not yet be organized for collecting elaborate data over large areas. As the American people moved westward in the second half of the nineteenth century, however, some naturalists developed an interest in the correlation between species distribution and temperature and rainfall patterns. Some indebtedness to Humboldt appears to exist in the resulting studies.

On his return from Latin America to Europe, Humboldt stopped off for six weeks in the United States in the summer of 1804. He met Barton and included in an appendix to the *Essai* some data Barton had obtained indicating that species of plants grow farther northward on the eastern side of the Alleghany Mountains than they do on the western side, which Barton attributed to the warmer climate of the eastern side. Barton had a few pages set into galleys in 1809, but his study was never completed.[106]

After a generation went by, another American, Charles Pickering (1805–78), wrote "On the Geographical Distribution of Plants" (1830), which summarized and applied to America some of the understanding of the subject that was developing in Europe.[107] In 1838 he accompanied the United States Exploring Expedition under Lieutenant Charles Wilkes to the far Pacific, and in 1854 he published some observations on the geographical distribution of the plants and animals he had observed. He also summarized his ideas on the subject in 1859 and 1860, though some of the ideas were immediately eclipsed by arguments in Darwin's *Origin of Species*.[108]

Henry R. Schoolcraft (1793–1864) accompanied the 1820 expedition to the Great Lakes led by Governor Lewis Cass of Michigan, and in 1831 and 1832 he led expeditions of his own to the same general region. His *Narrative Journal of Travels*... of the Cass expedition can serve as a representative example of the exploration literature during the first half of the nineteenth century. He was particularly interested in stratigraphy and mineralogy, and his observations were strongest for those subjects. His observations on plants and animals appear to be reliable and useful, but they were mainly descriptive and not very detailed. For example, he characterized a sandy plain west of the Huron River as "covered principally with a growth of yellow pine. Among the shrubs and plants, the pyrola rotundifolia, or common winter green, is very abundant, and we here first noticed a creeping plant called *kinni-kinick* by the Indians, which is used as a substitute for tobacco." He was also alert for phenological changes:

> *On leaving Buffalo, on the 6th of May, the blossoms of the peach tree were not yet fully expanded, and the petals of the apple were just beginning to swell. On reaching Detroit, two days afterwards, the leaves of the peach blossom had fallen, and those of the apple had passed the heighth of their bloom. Gardening also, which had not commenced at Buffalo, we found finished at Detroit, and the half grown leaves of the beach, the maple, the common hickory,* (juglans vulgaris), *and the profusion of wild flowers on the commons, gave to the forests and to the fields the delightful appearance of spring. These facts will go farther in determining upon the differences of climate, than meteorological registers, which only indicate the state of the atmosphere, without noticing whether a corresponding effect is produced upon vegetation.*[109]

While these observations were perceptive and useful, they can be considered at best as on a par with Kalm's, made seventy years earlier. Schoolcraft's later books on Indians undoubtedly did more for American literature by serving as a source of materials and inspiration for Longfellow's "Hiawatha" than his early exploration reports could do for ecology.[110]

A more ambitious effort was undertaken by Louis Agassiz (1807–73), the Swiss naturalist who came to America in 1846 and rose to a position of prominence in American science. Agassiz had come to know Humboldt in Paris in 1831 and 1832, and he was eager to follow Humboldt's example of undertaking a natural history exploration. He organized an expedition to Lake Superior for the summer of 1848, the members consisting of a few naturalists and other young "gentlemen." Although the expedition was afield for less than three months, and although Agassiz was handicapped by dogmatic and simple-minded ideas on biogeography that he owed to Georges Cuvier's influence rather than to Humboldt's, the resulting book is of ecological interest.

As a natural-history survey Agassiz'a book, *Lake Superior*... (1850), differs from previous efforts, such as Catesby's of the Carolinas or William Bartram's of Florida, in that it is more thorough because it is a joint effort of several authors of varying expertise. It also differs from the splendid survey of the natural history of New York State that James Elsworth Dekay supervised in the 1840's in that it is confined to a restricted, somewhat homogeneous region. The main emphasis in *Lake Superior* is placed upon an inventory of productions, with a catalogue of shells by Alpheus A. Gould, of birds by J. Elliot Cabot, of Coleoptera by John L. Leconte, and of Lepidoptera by Thaddeus William Harris and chapters by Agassiz on vegetation, fishes, reptiles, fossils, the lake basin, and geology. Cabot also wrote the introductory "Narrative of the Tour" but drew heavily upon materials from Agassiz in doing so. The narrative is more oriented toward plants and animals than Schoolcraft's, but it is not otherwise more ecologically notable. The book falls short of being an ecological survey of the Lake Superior region, but is a suggestive step toward such a survey.

The most interesting chapters for the history of ecology are not Agassiz's on fishes and reptiles—where his expertise lay—but rather his two chapters on vegetation. These chapters, more than anything else he wrote, show the benefit of Humboldt's influence. The first vegetation chapter discussed the environmental factors that seemed to influence the distribution of species. Unlike Humboldt, however, Agassiz did not believe that the known factors were sufficient to account for distributions, and therefore he also invoked a "Supreme Intelligence." The second vegetation chapter compared the vegetation of the northern shores of Lake Superior with that of his native Swiss Jura: "Making full allowance for the influence of the lake, and leaving out of consideration a small number of species peculiar to North America, there remains about Lake Superior a subalpine flora which is almost identical with that of Europe, with which it is here compared."[111]

The explanation for ecologically similar floras in different regions was lost upon Agassiz because he rejected Humboldt's suggestion of the possibility of historical migrations. In identifying the plants from the Lake Superior region, Agassiz obtained the assistance of Asa Gray (1810–88), the Harvard professor of botany who later became both his professional rival and his theoretical opponent concerning Darwin's theory of evolution. Gray had already become interested in the striking similarities between mountain plants in America and Japan. He was to explain these relationships successfully in the 1850's to 1870's by using the concepts of migration, competition, extinction, and evolution that had been developed in England by Charles Lyell, Charles Darwin, and

Joseph Dalton Hooker.[112] Gray thus took the first steps in America toward understanding ecological adaptation as dependent upon evolution by natural selection.

Climate and Plants, Animals, People

Climate and Germs. The theory of contagion arose in antiquity and was transmitted to Europe through a biblical discussion of leprosy and through discussions by several Roman authors. Among the Romans two possible categories of contagion were suggested. The atomic theory, preserved in Lucretius' *De Natura Rerum*, suggested infection by invisible seeds, while Varro, Vitruvius, Columella, and Palladius expressed fears of the poisons and stings of minute swamp animals.[113] During the Renaissance a number of physicians amplified the ancient speculations, most notably Girolamo Fracastoro. In *De Contagionibus* (1546) he presented cogent arguments supporting a theory of contagion by invisible seeds. He did not claim that all diseases are contagious or that contagious diseases arise only by contagion or that the agents of contagion are animate. In 1557, Jerome Cardan suggested that the agents of contagion are alive and reproduce as animals do.

In succeeding years the ideas of Fracastoro won more support than did those of Cardan. The theories of contagion and of the environmental origin of disease were generally viewed as complementary, not contradictory because the agents of contagion were thought to be inanimate. The small minority who supported the idea of *contagium vivum* were prone to wild speculations because they attempted to fit the theory to particular kinds of animals. Of the European authors defending *contagium vivum* before 1800, it will suffice to mention here only the Italian Giovanni Maria Lancisi, whose *De Noxiis Paludum Effluviis Eorumque Remediis* (1717) revived the ancient Roman idea that mosquitoes from the swamps might cause malaria. He did not pretend that he could prove this possibility, and his arguments won few adherents during the eighteenth century. Nevertheless, at the beginning of the nineteenth century some Americans developed an interest in this idea, and Samuel Latham Mitchill published a translation of Lancisi's speculations.[114]

The development of interest in *contagium vivum* in the first half of the nineteenth century can be related to an increasing awareness of parasitism in the plant and animal kingdoms. The first American to publish a defense of *contagium vivum* was Dr. John Crawford (1746–1813), an Irishman who emigrated in 1796 and settled in Baltimore. In 1800 he distinguished himself as one of the first two Americans to use the new Jenner cowpox vaccine. In 1807 and in 1809 he published in local periodicals his germ theory, based upon a presumed parallel between plant parasitism and disease. He gained no noticeable following.[115]

Among the other American physicians who either defended or seriously discussed the possibilities of *contagium vivum* during the next few decades, the most notable were Daniel Drake (1832 and 1850), Josiah Clark Nott (1848), and John Kearsley Mitchell (1849). Daniel Drake (1785–1854) was one of the leading middle western physicians of his time. He taught medicine in Ohio and Kentucky and published a constant stream of medical and other writings, many of them in a medical journal of which he was cofounder and coeditor. He began developing his ideas about pathogenic animalcules in 1832, when he observed an epidemic of cholera in Cincinnati. He thought that the animalcules might be mosquitoes or gnats whose eggs one either inhaled from the air or drank in water. This idea, not greatly different from the speculations of Kalm and Crawford, had the advantage of accounting for the fact that epidemics like cholera, malaria, and yellow fever were associated with wet places where those insects were found. This hypothesis could also account for the sudden spread of epidemics, since insects and other small animals reproduce rapidly.[116] Drake defended his hypothesis further in his *A Systematic Treatise, Historical, Etiological, and Practical, on the Principal Diseases of the Interior Valley of North America* (published in two volumes, 1850–54) which will be discussed in the next section.

Josiah Clark Nott (1804–73), from Columbia, South Carolina, received his medical degree from the University of Pennsylvania and settled in 1836 in Mobile. He soon became the leader of Alabama medicine and wrote extensively on both medicine and race. He argued the case for the animate propagation of yellow fever in the *New Orleans Medical and Surgical Journal* (March, 1848). He reasoned that some insect like the mosquito could best explain the propagation pattern of the disease and that the existence of plant parasites provides a model of possible similarity. A killing frost stops the disease and kills insects, but it should not stop miasmas. He felt sure that yellow fever was not transmitted from one person to another. Nott practiced medicine in New York City from 1868 to 1872. In 1870 a yellow fever epidemic broke out on Governor's Island. On that occasion he presented a report to the city board of health, in which he repeated his insect hypothesis. Since he did not believe that it was transmitted from one person to another, he saw no need to quarantine arriving ships.[117]

John Kearsley Mitchell (1793–1858) was a widely known and respected professor in the Jefferson Medical College in Philadelphia. He rejected the idea of infection by animalcules on the grounds of implausibility, for no poisonous animalcules were known. He found very plausible, however, the possibility that malaria and other epidemic fevers could be caused by fungi, and he developed at some length arguments supporting this hypothesis. He had got the idea in 1829, while contemplating the rapidity with which fungi can rot tree stumps, and, by the time he published his hypothesis in 1849, he could cite the known fungal diseases of corn rust, cattle hoof and horn rot, and muscardine of silk worms.[118]

Although the defense of *contagium vivum* in America can be correlated with an increasing awareness of parasitism, all the arguments were indirect, and until late in the century its defenders were a minority. Some of them defended the quarantine, and they were opposed by a free-trade anticontagion movement that followed in the footsteps of Rush and Webster. The anticontagionists did not deny that any diseases were contagious, but they placed primary emphasis upon the traditionally accepted causes of unfavorable weather, miasmas from swamps, and other sources of filth and decay.[119]

Medical Geography. The interest in correlations between weather and disease that had been so strong in the eighteenth century continued into the nineteenth century and only waned late in the century, when the European science of bacteriology began to exert an influence in America.

Constantin François Chasseboeuf, Comte de Volney (1757–1820) came to the United States in 1795 and traveled for three years through the country, making extensive notes on the rocks, minerals, fossils, physiography, geologic regions, climate, and diseases. He returned to France when a hostility toward Frenchmen arose, and he published his *Tableau du Climate et Sol des Etats Unis* in Paris in 1803. Charles Brockden Brown's English translation appeared in Philadelphia in 1804. The book is of interest for the history of ecology from various points of view, but here the focus will be upon Chapter 10, "Prevailing Diseases in the United States." In discussing the decay of teeth, he was influenced by the views of Kalm. In discussing consumption, intermittent fever, and yellow fever, he emphasized the environmental factors of sudden and violent changes of temperature and "pernicious exhalations" from swamps and from squalid low and wet areas, coupled with the hot, humid summer climate.[120] None of this was new, and it was a bit superficial, but it was related by the author of the leading study of the day on the American environment.

Early in the nineteenth century the correlations between environment and disease became the subject of a government investigation. Although it was rather modest in the amount of research that was involved, this investigation appears to be the first long-term scientific research carried out in America. On April 2, 1814, while the War of 1812 was in progress, Dr. James Tilton (1745–1822), physician and surgeon general of the United States Army, ordered all hospital surgeons to begin recording the weather at their posts. The first to comply was Dr. Benjamin Waterhouse. He had made history in 1800 by being first in America to use the Jenner cowpox in vaccinating against smallpox. He joined the army during the war, and in 1816 he made history less dramatically by submitting a meteorological journal to the surgeon general's office. In 1818, Dr. Joseph Lovell (1788–1836) was appointed surgeon general, and he had Tilton's order fully implemented, stating in his own directive that "the influence of weather and climate upon diseases, especially epidemics, is perfectly well known." His instructions were that "every surgeon should be furnished with a good thermometer, and, in addition to a diary of the weather, should note everything relative to the topography of his station, the climate, complaints prevalent in the vicinity, etc., that may tend to discover the causes of diseases, to the promotion of health, and the improvement of medical science."[121]

In 1826 the surgeon general's office published the first compilation of the evidence, for the years 1822 to 1825, and although it was not extensive enough to solve the mysteries of disease, the report did point out that the meteorological tables could serve, as time went by, to ascertain to what extent the climate changed as more and more land was being converted from forest to cultivation. Data from eighteen military posts were published.[122]

A similar compilation for the years 1826 to 1830 was published in 1840. Assistant Surgeon Samuel Forry (1811–44) compiled the official data and published his own summary accounts in 1841 and 1842. The army medical department had succeeded in obtaining the first statistical data on both climate and incidence of disease in America. Using this data, Forry could dramatically describe the wide ranges of temperature found in the country, with the moderating effects of the Great Lakes and the extremes that occurred from summer to winter in the inland regions away from large bodies of water. He also felt that he could demonstrate some correlations between climate and rates of disease, though admitting the subject was complex. Part of his caution was due to European demonstration that phthisis (tuberculosis) was not, as had been commonly supposed, caused by a changeable climate. Nevertheless, it seemed clear that a warm, dry atmosphere was therapeutic for those having the disease. It was also clear that malaria was much more prevalent in the southern than in the northern regions of the country. His statistics indicated that in many respects the Southwest was the unhealthiest region of the country. In 1840, however, the Southwest was not the presently understood region but the eastern river valleys of Texas.[123]

In retrospect it is clear that even if this was the greatest statistical study ever undertaken in America, the populations of army posts were not large enough to allow clear correlations to be drawn between diseases and climate. This was not understood by Surgeon General Thomas Lawson, however, who obtained a number of Daniell's hygrometers from Europe and also provided each military post with De Witt's conical rain gages. Forry resigned his commission in 1840, and in 1842 was replaced as meteorologist on the surgeon general's staff by James Pollard Espy (1785–1860), who was already a distinguished meteorologist. An interest continued in the correlation between climate and disease, but Espy's successes in meteorology were superior to the problematic results that were obtained from correlating these meteorological data with diseases at the army posts of America. Gradually meteorology became of more importance to the navy and other branches of the Department of War than to the surgeon general, and, although the original correlations were not quickly abandoned, they did decline in significance.[124]

Although some polarization occurred in the first half of the nineteenth century between the majority, the anticontagionist physicians, and the minority, the contagionists, both groups continued to regard as important the correlation of disease with environmental factors. Daniel Drake, an avowed contagionist from 1832, carried out the greatest investigation of this subject that had ever been made in either Europe or America.

Drake's family had migrated from New Jersey when he was a child, and he grew up in a log cabin at Mays-Lick, Kentucky. In December, 1800, at age fifteen, he went to Cincinnati to study medicine by apprenticeship. He fell in

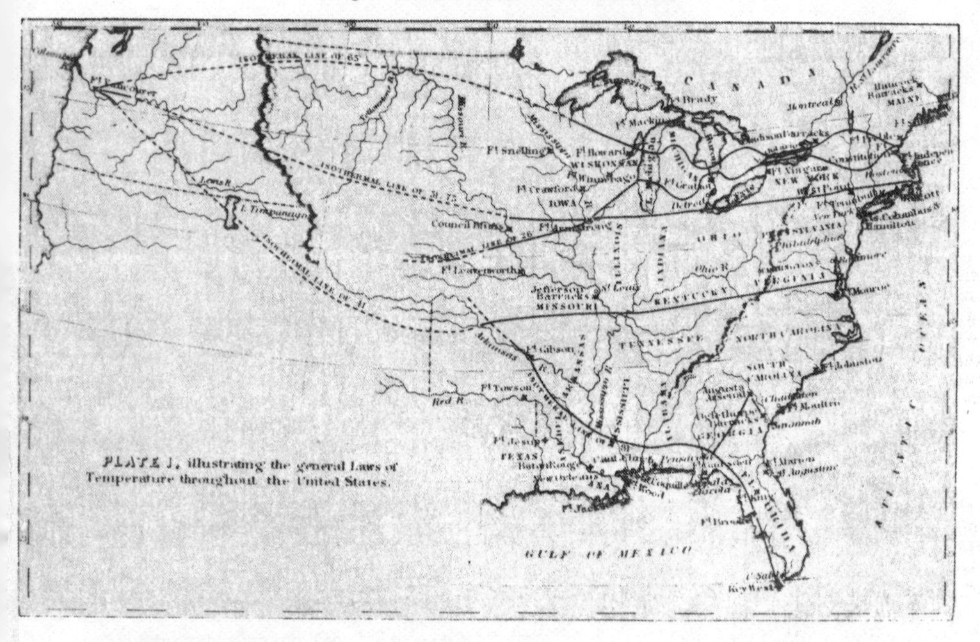

Samuel Forry's temperature map of the United States. From The Climate of the United States and Its Endemic Influences: Based Chiefly on the Records of the Medical Department and Adjutant General's Office, United States Army *(New York, J. & H. G. Langley, 1842.*

love with the town. In 1805 he attended the University of Pennsylvania, hearing the lectures of Rush and Barton. When he returned to Cincinnati, he sent Barton "Some Account of the Epidemic Diseases Which Prevail at Mays-Lick, in Kentucky."[125] This brief paper shows a faithful following of the advice given to physician observers in the Hippocratic *Airs, Waters, and Places.*

Three years later, in 1810, Drake merged his interests in medical geography and in Cincinnati in a sixty-page pamphlet *Notices Concerning Cincinnati.* This pamphlet, recently described as "an ecological survey of the Cincinnati area,"[126] began with accounts of the topography and geology. These were followed by an extensive account of the climate, including month-by-month comparisons of temperature in Cincinnati and Philadelphia, average diurnal variations in temperature, wind movement, precipitation, and frequency of different weather conditions, and a discussion of possible changes in climate between 1785 and 1810. The evidence was slight but seemed to indicate that the summers had remained constant while the winters had become colder. Drake included a "Calendarium Florae" for 1809, though he had also compiled similar observations for 1807 and 1808. Next he gave a brief account of the living conditions and habits of the 2,320 Cincinnatians. The second half of the pamphlet was devoted to the diseases of the area, with the usual attention to the relation between climate and disease. His investigations were pursued with diligence but thus far with little originality.

Five years later, in 1815, he expanded the pamphlet into a 250-page book, *Natural and Statistical View, or Picture of Cincinnati and the Miami Country.* It was a good geographical monograph for the period and was well received. In discussing diseases, Drake continued to speak of "miasmata" and "marsh effluvia."[127] In 1832, however, his experience with a cholera epidemic convinced him that the cause of that disease was animate. He assembled evidence supporting his hypothesis and compared it with the miasmic theory in his book on cholera in 1832. He was to return to the subject with a stronger case in his greatest work, *A Systematic Treatise, Historical, Etiological, and Practical, on the Principal Diseases of the Interior Valley of North America, as They Appear in the Caucasian, African, Indian, and Esquimaux Varieties of Its Population.*[128]

In some respects this work is a vast extension of the earlier book, now encompassing the region of the Great Lakes and occasionally northward, south to the Gulf of Mexico, and from east Texas eastward along the major rivers to Quebec and Pittsburgh. He traveled this region for ten summers before publishing the first volume, and the second volume was published posthumously from his incomplete notes. *A Systematic Treatise* was more limited to medical geography than had been *Picture of Cincinnati.* He could not obtain detailed weather data for such a vast region, and so the climate was described in general terms and supplemented with data on mean, maximum, and minimum temperatures from the localities where those data had been collected. Nor could he provide phenological data for the area. He did, however, devote part of a chapter to the distributions of trees, mammals, and birds in relation to climate.[129]

Volume I was devoted mainly to detailed regional accounts of "medical topography." Drake's regions were mostly defined by river basins, and his attention was largely concentrated upon the likelihood that the terrain was conducive to fevers, mainly malaria. For example:

From Milwaukie to Racine, twenty-five miles, a belt of compact and lofty forest, nourished by the influences of the lake [Michigan], spreads to the distance of two or three miles into the country, beyond which, there are rolling prairies. The site of Racine, in N. Lat. 42°50′, is a part of this wooded plain, elevated from thirty to fifty feet above the lake. In rainy weather, small pools of water form on many parts of its surface. In digging wells, as Doctor Cary informed me, they pass through a bed of sandy loam, and then through a deposit of gravel, into another of blue clay, with pebbles, when pure but hard water is obtained. At the same level, springs burst out from the banks of Root River, which enters the lake, adjacent to the northern side of the town. The valley of this river, for two or three miles up, is about sixty rods in width, and not subject to inundation. Doctor Cary, who had resided in the place ten years, that is, from the beginning of its settlement, informed me, that for the first two years, there was scarcely a case of autumnal fever; in the next two, a number of cases occurred; and in the following year, 1839, it assumed a mild epidemic character, putting on an intermittent type, and proving fatal in a single instance only. The following year it was again epidemic. In both those years, the mouth of Root River was choked up with sand, and its waters rendered stagnant.[130]

Volume II Drake organized into five parts, reflecting his nosological understanding of fevers: autumnal (malaria), yellow, typhous, eruptive (smallpox, chicken pox, measles, scarlet fever, and so on), and phlogistic (meningitis, bronchitis, pneumonia, tuberculosis, and so on). He was interested (following in the Hippocratic tradition) not only in those diseases that were presumably caused by habits, food, and occupation. The demarkations of different diseases according to cause was not then, of course, clearly established. In investigating these diseases, he undertook ecological investigations as they seemed relevant. This was exemplified in his arguments for the animate origin of cholera and malaria.

Medical geography has continued to be an important subject, particularly in the tropics, but Drake's *A Systematic Treatise* represented a high point for the subject in the United States. It remained definitive until rendered obsolete by the development of a definite germ theory of disease. When particular pathogens could be identified, then microbiologists studied the life cycles of the pathogens rather than the general environmental conditions. But, even so, an interest in "healthy climates" has persisted in America.[131]

Phenology. We have seen that the idea of recording the seasonal unfolding of biological phenomena was introduced into natural history by Linnaeus and Kalm in the mid-eighteenth century and that such records had been compiled in America by Jefferson, William Bartram, Williams, and Barton. The presumption was that these *calendaria florae*, as they were called even though records of animals were usually included, were useful to both farmers and naturalists. However, little by way of rationale for them appears to have been published before 1803. In that year Benjamin Smith Barton (1766–1815) discussed them in his *Elements of Botany . . .*, America's first textbook of botany. Barton, who taught botany and natural history at the University of Pennsylvania, briefly reviewed some of the *calendaria florae* that had been published, explained the kinds of data they should include, and described their various uses. While Linnaeus had sought to organize an ecological science around the balance-of-nature concept, Barton apparently believed that it could be done within the structure of *calendaria florae*:

Calendaria Florae, if they be properly kept, form some of the most interesting notices in the natural history of a country. They form, next to the living, the best, picture of the country. They show us, in the most beautiful and impressive manner, the relations of the vegetable and the animal kingdoms to each other, and to the various agents by which they are surrounded, and by which they are affected. They enable us to compare together the climates of different countries or places, which are included within nearly the same latitudes, such as Florida and Palestine, Philadelphia and Pekin, New-York and Rome.[132]

He went on to point out that if the spread of cultivation was changing the climate of America then the evidence should appear in changes in the timing of seasonal phenomena.

The urgings of Barton, along with the examples already published, led other naturalists to compile and publish *calendaria florae*. The one Daniel Drake published for Cincinnati in 1809 has already been mentioned. Barton's student, Jacob Bigelow (1787–1879), who became a prominent Boston physician and professor in Harvard Medical School, also took up the study. He published his data in a different format, however, placing notations on locale, habitat, and month of flowering in the account of each species listed in his *Florula Bostoniensis . . .* (first published in 1814 and twice revised). This was a useful reference work, but the information was not in a form conducive to theoretical conclusions. The desirability of coordinating phenological studies in different places throughout the country was mentioned to Bigelow by the Pennsylvania botanist Gotthilf Henry Ernst Muhlenberg (1753–1815), who may have been aware of

Williams' and Jefferson's interest in the project. Constantin François Chasseboeuf, Comte de Volney had also provided some data on times of blooming and harvests as he traveled in May from Annapolis across the Appalachians to Cincinnati.[133] Bigelow acted on Muhlenberg's suggestion, and in 1818 he published "Facts Serving to Show the Comparative Forwardness of the Spring Season in Different Parts of the United States."[134] This article was based upon replies received in 1817 from correspondents in Charleston, South Carolina; Richmond, Virginia; Louisville and Lexington, Kentucky; Baltimore, Maryland; Philadelphia, Pennsylvania; New York City; Albany, New York; Brunswick, Maine; Montreal, Quebec; and Geneva, Switzerland. All the data were published, and since the peach was reported from all but one location, the flowering dates of that tree were listed along with the latitude and longitude of the different locations. Temperature data were not included, but from the data given Bigelow concluded that "the difference of season between the northern and southern extremities of the country is not less than two months and a half. Difference of longitude does not seem very materially to affect the Floral Calendar within the United States."

Quite likely the floral calendars that Jacob Porter and Stephen W. Williams compiled for Plainfield and Deerfield, Massachusetts, were also prompted by Bigelow. Bigelow sent a copy of his paper to Thomas Jefferson, who responded appreciatively and sent a summary of his own phenological observations for a seven-year period.[135] It is also possible that Bigelow's paper may have alerted Schoolcraft to make his phenological observations, quoted above.

Shortly after Bigelow began to collect his data from different parts of the country, Josiah Meigs (1757–1822), commissioner of the General Land Office, petitioned Congress to authorize the keeping of meteorological records at the twenty land offices throughout the country and to provide the necessary instruments. Congress did not fulfill his request, but even without the legislative authority or instruments he had the forms printed for a meteorological register and on April 29, 1817, sent them to the land-office agents with a request that they record the temperature, wind, and weather three times a day and that they also record the following information:

1. *The time of the unfolding of the leaves of plants.*
2. *The time of flowering.*
3. *The migration of birds, whether from the north or south, particularly swallows.*
4. *The migration of fishes, whether from or to the ocean.*
5. *The hybernation of other animals, the time of their going in winter quarters, etc.*
6. *The phenomenon of unusual rains and inundations.*
7. *The phenomenon of unusually severe droughts. The history of locusts and other insects in unusual numbers.*
8. *Remarkable effects of lightning.*
9. *Snowstorms, hailstorms, hurricanes, and tornadoes; their course, extent, and duration.*
10. *All facts concerning earthquakes and other subterranean changes.*
11. *Concerning epidemic and epizootic distempers.*
12. *The fall of stones or other bodies from the atmosphere; meteors, their apparent velocity, etc.*
13. *Discoveries relative to the antiquity of the country.*
14. *Memorable facts relative to the topography of the country.*

Some of Meigs's returns were published, and no doubt more of them would have been had he not died in August, 1822.[136]

Although Congress was not yet ready to support the study of phenology beyond whatever was done in the surgeon general's office, additional institutional support appeared in 1825 from the Board of Regents of the State University of New York. It organized a state meteorological system by directing the faculty members to keep records of temperature, precipitation, winds, and other related phenomena, such as "the first appearance of flowers and leaves, the beginning of haying and harvesting, first autumnal frosts and snows, appearance and departure of birds of passage, first notice of fireflies, reptiles."[137] Annual abstracts of the returns were compiled for the Report of the Board of Regents by T. Romeyn Beck, principal of the Albany Academy. In 1855, Franklin B. Hough (1822–85) published the returns for 1826 through 1850.[138] One of the teachers under Beck at the Albany Academy was Joseph Henry, who became America's leading electrical scientist and in 1846 the first secretary of the Smithsonian Institution. Henry published in 1851 an appeal for the observation of periodical phenomena, and in 1864, Hough drew upon both New York State and Smithsonian data to publish "Observations Upon Periodical Phenomena in Plants and Animals from 1851 to 1859 . . . ."[139] This was an impressive body of data, but Hough did not attempt to draw conclusions from it—that was left to the reader.

Private individuals also continued to publish observations. Charles Peirce compiled monthly weather summaries for Philadelphia for fifty-seven years and published them, with a brief comparison with European winters, in 1847.[140]

Henry David Thoreau (1817–62) has been called the father of American phenology.[141] Although that claim has been refuted by Leo Stoller (and by the above discussion), it is true that Thoreau's phenological observations have attracted more attention in the twentieth century than those of anyone else. This is so mainly because his writings have a wide appeal and yet are somewhat enigmatic. He expressed hostility toward science at various times, but he made increasingly more and more of the kinds of observations that he disparaged. A number of capable authors have shed light upon this paradox. The interpretation followed here is by Nina Baym. She found the key to understanding Thoreau's attitude toward and interest in science as having been formed by a hope that he could unite science and Transcendentalism. He expressed hostility toward professional scientists because they were not also attempting to create this synthesis, and his hostility was reinforced by his frustration at not being able to achieve this synthesis himself. His attempted synthesis was quite largely within the realm of phenology, in which Baym sees the key to understanding his attitudes and beliefs concerning science.

It seems that Thoreau wanted to go further than Barton, creating not just an ecological science around phenology but a transcendental phenological ecology. Baym's thesis is too complex to be explained here in detail, but her explanation of the significance of this transcendental phenology for Thoreau should be expressed here in her words:

To study nature becomes the program of an action which achieves the reconciliation of man and God. The naturalist lives a saintly life because he must obey the laws he learns to know. . . .

The laws of the universe are a great rhythm, to which man at his most fulfilled moments marches. If a sense for this universal rhythm is implanted in the creature, he always moves with the music. Such a sense is instinct, possessed by animals and possibly by Indians. In most men, rhythm is not implanted, and there is substituted a capacity to learn the music. This substitution is man's curse and challenge; it is the sign that he has been expelled from the garden of Eden, but in true Miltonic fashion it is the sign that there is a peculiarly human, and therefore superior, way to regain it. The way back is through learning, which ultimately approaches instinct. When man has learned the music so thoroughly that he can anticipate it, when he knows what comes next, he will be able to keep time.

Thoreau devoted the rest of his life to learning nature well enough to "anticipate" her. In her cyclical organization he saw assurance that anticipation was possible, for in repetition lay the hope of learning, correcting, refining, and profiting from past error. Certain patterns obtruded immediately: the alternation of day and night, the succession of the seasons. These broad patterns could be quickly learned so that man might keep his life in some rough harmony, such that his ignorance did not destroy him. The next step was to sharpen perception and make out finer patterns and more precise regularities. Spring certainly follows winter, but when? This question subdivides—where, on what day, does the first crocus spring? When do the geese fly over Concord?[142]

To explain how Thoreau came to develop such a goal for his transcendental science would take us out of the history of American ecology and into the history of his psyche. However, both Stoller and Baym have uncovered some of the influences upon his phenological thoughts and the stages that these thoughts went through, and they can be mentioned. Thoreau attended Harvard while Bigelow was on the medical faculty, and Bigelow's *Florula Bostoniensis*, with its phenological data, was Thoreau's guide to the plants of the area. Also influential upon Thoreau's interest in seasonal observations was the *Book of the Seasons, or the Calendar of Nature* by the English druggist-chemist-poet-naturalist William Howitt (1792–1879). The book was published in 1831 and went through seven editions. Thoreau's manuscripts reveal an early acquaintance with this work, and he referred back to it throughout his life. Baym wrote, "Howitt is thinking of strictly practical uses for his tables, but his language is virtually identical to Thoreau's."[143]

The beginnings of Thoreau's scientific program can be found in his early essay "The Natural History of Massachusetts" (1842), in which phenological observations are pervasive:

In the autumn days, the creaking of crickets is heard at noon over all the land, and as in summer they are heard chiefly at night-fall, so then by their incessant chirp they usher in the evening of the year. Nor can all the vanities that vex the world alter one whit the measure that night has chosen. Every pulse-beat is in exact time with the cricket's chant.

In the same essay he also expressed some dissatisfaction with conventional science:

It is with science as with ethics, we cannot know truth by contrivance and method; the Baconian is as false as any other, and with all the helps of machinery and the arts, the most scientific will still be the healthiest and friendliest man, and possess a more perfect Indian wisdom.[144]

Thoreau seems to have thought that he could achieve the understanding he sought by patiently observing nature. He made many seasonal observations while living in his cabin at Walden Pond from July 4, 1845, to September 6, 1847. They were not systematic, however. As time went by, he seems to have felt that the deep insights he sought were eluding him, and he began to make systematic phenological notes in late 1851 or early 1852. In the spring of 1852 he wrote himself memoranda to "observe all kinds of coincidences, as what kinds of birds come [back from the south] with what flowers," and not to overlook the reptiles and frogs. He began a phenological chart to which he continued to add during succeeding years through the spring of 1858.[145] Wrote Baym:

But, reading the late journals and seeing what Thoreau finally did with the material he had collected for the calendar, one senses that Thoreau approached, if he did not accept, the realization that the task, as he had defined it, was impossible. Partly, he needed instruments and manuals to see what could be seen and know what he saw; partly, too, nature was simply not as regular as he had assumed.[146]

Nevertheless, his literary powers did not fail with his scientific program. Two of the last essays, "Autumn Tints" and "Wild Apples," published posthumously in 1863 by his sister, drew upon his store of phenological knowledge, and still later in the century H. G. O. Blake published extracts from Thoreau's journals arranged under the titles *Early Spring in Massachusetts* (1881), *Summer* (1884), *Winter* (1888), and *Autumn* (1892).

Thoreau's phenological observations thus continued to be published down to the time that the formal organization of ecology began to emerge. Where did phenology fit into the new science? Hardly at all. Phenology was one of the possibilities that naturalists seized as a basis for an ecological science, but it did not offer a rich enough program of investigation. Instead, therefore, ecology was founded upon phytogeography, limnology, and entomology. The secrets that eluded Thoreau were never discovered by anyone else.

A page from Thoreau's phenological chart, headed "The Flowering of Plants, accidentally observed in '51, with considerable care in '52; the Spring of '51 being 10 days and more earlier than that of '52. The names those used by Gray.
X Observed in good season)
) in '52
XX Very early)
The XX before the names refers to '52"
From the Berg Collection, courtesy of the New York Public Library.

Nature Appreciation and Conservation

An appreciation of the beauties of nature has been expressed in the literature of all peoples through the ages. One finds it in the exploration and natural-history literature on America from the beginning. Furthermore, Mark Catesby and any number of others who had illustrated America's plants, animals, and scenery had surely expressed through their art their appreciation for the beauty of nature. However, as a literary genre, one finds little nature writing in America before William Bartram's *Travels* (1791). Bartram and the romantic movement in Europe (inspired in part by his book) helped Americans develop a literature of nature appreciation in the nineteenth century. Bartram was also a talented artist and published some of his drawings of plants and animals in his *Travels*, others were included in Barton's *Elements of Botany* (1803), while still others have been published only in recent times.[147]

Prudence, in the early days of the Republic, was an insufficient motive for the preservation of resources because the resources seemed almost limitless. Game laws were passed in the seventeenth and eighteenth centuries to protect the deer, but there was little concern about the passing of the wolf, beaver, and buffalo from the Atlantic states because other forms of wildlife were still common. The conservation movement in America has always owed more to nature lovers than to economists. Both nature appreciation and conservation in America have their histories, and these movements will be discussed here primarily in relation to their encouragement of interest in ecological subjects.[148] Ecology is the obvious companion science to both movements.

Besides William Bartram, other eighteenth-century authors in America who were significant for influential writings expressing an appreciation of nature were St. Jean de Crèvecoeur (1735–1813), whose *Letters from an American Farmer* first appeared at London in 1782, and Philip Freneau (1752–1832), who began writing his nature poems in the 1780's.

Closer to ecological developments, however, was Alexander Wilson (1766–1813), a Scottish weaver-peddler-poet and defender of workers against oppression who found it expedient to immigrate to the United States in 1794. He developed an acquaintance with William Bartram in 1802, and under Bartram's encouragement and with access to his library, Wilson quickly developed an interest in American natural history, particularly the birds. His poetry was not stifled by this new interest. Wilson, in fact, attempted to synthesize his interests in poetry, natural history, and America in a grand epic poem, *The Foresters* (1809–10), which he also illustrated with his own drawings. His poetry was well appreciated by his contemporaries. Van Wyck Brooks observed that "Alexander Wilson's poems were continuously in print throughout the nineteenth century in Scotland, and there was no better poet in America during the years in which he lived and died here (1794–1813)."[149]

Wilson began to study American birds in earnest in 1803, and he soon began to draw them, assisted by Bartram and his niece, Nancy Bartram. Wilson found that his critical faculties, which previously had been used to write satiric verse, could also be used to collect natural histories. He also began writing poems about birds. There are in his nine-volume *American Ornithology* poems on the baltimore oriole, the blue bird, the hummingbird, the kingbird, the wood pewee, the bald eagle, and the osprey. He traveled through the states from Maine to Florida and west to the Mississippi collecting information on birds and simultaneously taking orders for his great work. *American Ornithology* contains Wilson's colored illustrations of all the birds he discussed, and it was the finest publication that had ever been produced in America. Much more than a catalogue of birds, its appeal lay not only in the high quality of Wilson's work and the attractiveness of the volumes but also in a national pride in the work and in the birds described. In several respects, such as his census in 1811 of nesting birds in Bartram's eight-acre garden, Wilson made pioneering contributions to ornithology.

Wilson consciously sought to instill a love of nature in his fellow Americans, but for those who did not respond to that appeal, he also argued for the preservation of birds on grounds of self-interest. In building these arguments, he had to investigate the ecological role of the species, as illustrated in his poem on the kingbird:

> *Yet, should the tear of pity nought avail,*
> *Let* interest *speak, let* gratitude *prevail;*
> *Kill not thy friend, who thy whole harvest shields,*
> *And sweeps ten thousand vermin from thy fields;*
> *Think how this dauntless bird, thy poultry's guard,*
> *Drove ev'ry Hawk and Eagle from thy yard;*
> *Watch'd round thy cattle as they fed, and slew*
> *The hungry black'ning swarms that round them flew;*
> *Some small return, some little right resign,*
> *And spare* his *life whose services are thine!*[150]

That Wilson succeeded in winning the hearts of many to the causes of bird study and preservation is evidenced by the fact that his *American Ornithology* went through nine editions in the nineteenth century.

Among those inspired by Wilson was John James Audubon (1785–1851), whose own interest in birds led to an ambition to outdo Wilson after they met at Louisville, Kentucky, in March, 1810. Although Audubon's drawings and life have become a useful symbol of conservation in the twentieth century, he himself hunted for sport as well as for specimens, and he did not develop the concern for preservation that had distinguished Wilson.

Wilson's message was clear and influential, but it was embedded within poems and natural history accounts. To become still more influential, the concern for the preservation of nature and wildlife needed philosophical elaboration. The elaboration first came from Ralph Waldo Emerson and Thoreau.

Ralph Waldo Emerson (1803–82), who in 1832 resigned from the ministry because of his doubts about Christianity, felt the need for a more comprehensive philosophy of man in nature than either Christianity or society could provide. In his lecture "The Uses of Natural History" (1833) he commented, "We feel that there is an occult relation between the very worm, the crawling scorpions, and man." No doubt this is more metaphysics than ecology, but it was evidently related to both. He was disturbed by Americans' unplanned exploitation of nature, and in 1839 he warned that "this invasion of Nature by Trade, with its Money, its Credit, its Steam, its Railroad, threatens to upset the balance of man, and establish a new, universal Monarchy more tyrannical than Babylon or Rome."[151]

Emerson's expression of a need for a new philosophy of man's relationship to nature was an important step, though couched in metaphysical language. His contact with nature was not close enough, however, for him to be able to do more than give general expression to the need. Thoreau had amused himself as a child by hunting and fishing, and although he gave up these activities as an adult, he exchanged them for that of a lifelong observer. He became friends with Emerson and was influenced by his thought, but without losing his own identity in the process.

Thoreau graduated from Harvard in 1837, with a moderate exposure to science. His interest in nature led him to read works of exploration, including those by Kalm, William Bartram, Humboldt, Schoolcraft, and Darwin. In his natural-history studies he consulted the standard references, including Bigelow's *Florula Bostoniensis* and Wilson's *American Ornithology.*[152] He was, however, discontent with the science he read. As a Transcendentalist he believed that feeling for nature and the interrelatedness between man and nature was mistakenly barred from science (if not barred by some of the above authors). His thoughts on this have been extensively discussed.[153] The discontent can be attributed largely to his insistence upon finding personal meaning in nature, but this does not seem to have been the only factor. He also demanded a science of interrelationships. Some of his ideas about that science can be surmised from the two books he published. In *A Week on the Concord and the Merrimack Rivers* (1849) he presented an essay on fishes, which included both man's impressions of the fishes and interactions with them. Among the more interesting of these accounts is this one:

> *The pickerel* (Esox reticulatus), *the swiftest, wariest, and most ravenous of fishes, which Josselyn calls the fresh-water or river wolf, is very common in the shallow and weedy lagoons along the sides of the stream. It is a solemn, stately, reminant fish, lurking under the shadow of a pad at noon, with still, circumspect, voracious eye, motionless as a jewel set in water, or moving slowly along to take up its position, darting from time to time at such unlucky fish or frog or insect as comes within its range, and swallowing it at a gulp. I have caught one which had swallowed a brother pickerel half as large as itself, with the tail still visible in its mouth, while the head was already digested in its stomach. Sometimes a striped snake, bound to greener meadows across the stream, ends its undulatory progress in the same receptacle. They are so greedy and impetuous that they are frequently caught by being entangled in the line the moment it is cast.*[154]

The objective of the book is to convince us that nature has a valuable meaning for us—one much more of guidance than of material gain. But nature's lessons can be learned only by attentive students.

The same message Thoreau presented, perhaps more forcibly, in *Walden, or Life in the Woods* (1854). The message he conveys effectively by a combination of exhortation and detailed observation. His technique can be illustrated by his two chapters on Walden Pond, which contain a blend of subjective description, narration of his activities, details on natural history, and philosophical comment—all conveyed in a very compelling style. As Thoreau leads us from a consideration of gardening to fishing, to scenery, to color of the pond's water, to a discussion of its bottom, its plants, animals, and man's use of the lake and its surrounding land, a sense of interrelationship is built up that is transcendental but also ecological. The ecological dimension is not merely a point of view; it also includes some precise details. Edward S. Deevey, Jr., has argued rightly that these two chapters constitute the first notable contribution to limnology from America.[155]

Besides tramping through the woods for the fun of it, Thoreau sometimes worked as a surveyor. His store of nature lore seems to have been respected locally, and sometimes while discussing the woodlots he surveyed he was asked "how it happened, that when a pine wood was cut down an oak one commonly sprang up, and *vice versa.*" This question was well suited to his methods of observing nature, and after some time of patient observation and reading he wrote "The Succession of Forest Trees" (1860), a fine contribution to ecology. Kalm had discussed old-field succession and also the role of squirrels in planting acorns, and Thoreau mentioned William Bartram's observation, quoted in Wilson's *American Ornithology*, that the blue jay "is one of the most useful agents in the economy of nature, for disseminating forest trees, and other ruciferous and hard-seeded vegetables on which they feed." Thoreau felt, however, that he was the first to make a comprehensive study of the means by which forests were propagated. He rejected the spontaneous origin of trees and discussed the means by which the seeds of different species were disseminated—pines by wind, cherries by birds, and oaks and birches by squirrels. He then explained why pines are succeeded by oaks: young oaks can grow in the shade of pine trees, but young pines cannot; and why pines spring up when an oak forest is cut: young pines can grow in an open field, but young oaks grow best in shade.[156] The essay ended with some philosophical speculations, but it was not the end of Thoreau's thoughts on the matter. His journal for the same year shows that he had an interest in the practical application of his knowledge to forest management.[157] Were it not for his untimely death at age forty-five, he might have carried these studies further.

Thoreau's following has grown slowly but steadily, and although it appears to be stronger now than ever before, his influences upon nature appreciation, conservation, and ecological perspective in the nineteenth century were not negligible. More conventional approaches to the same issues were needed, however, if a science of ecology was to be developed to help Americans understand and manage their natural environment and resources.

George Perkins Marsh (1801–82) shared Thoreau's love of nature, but, as a lawyer-politician-diplomat, he undertook a more scholarly approach to the task of convincing Americans to conserve their natural resources. His approach demanded an ecological understanding of nature. He was a versatile and learned man who was successful at law but disliked it. He served Vermont in the United States House of Representatives from December, 1843, to June, 1849, and while there he began to express concern about the way Americans were mismanaging their resources.

In 1847 he described before the Agricultural Society of Rutland County the extent of the changes that Vermonters had wrought upon their surroundings. He was concerned about the erosion of soils and the increase of both droughts and floods that followed upon the careless destruction of forests, and he urged a concern for wise management of both forest and agricultural lands. In a letter to Asa Gray dated May 9, 1849, he outlined a proposal for a program of investigation and management of American lands. While serving as American ambassador to Turkey from 1849 to 1853, he became impressed by the long-range consequences of man's careless use of the land. Back in Vermont as railroad commissioner, he wrote a "devastating exposé of corporate irresponsibility and financial skulduggery." This was followed, in 1857, by his report, as fish commissioner, *On the Artificial Propagation of Fish*, in which he surveyed the impact of industry, agriculture, and forestry upon fisheries.[158]

In 1861, President Lincoln appointed Marsh ambassador to Italy, and he remained in that position until his death. He now had time for research and also opportunity to witness the effects of man's modifications of nature since antiquity. In 1860 he had decided to write a book about man's modification of nature, and in 1864 he published *Man and Nature; or, Physical Geography as Modified by Human Action*, which has been called "the fountainhead of the conservation movement" and "the most important and original American geographical work of the nineteenth century."[159] It also contains an elaborate account of the balance of nature and a discussion of man's upsetting of this balance. The first title Marsh proposed for the book was "Man the Disturber of Nature's Harmonies."[160]

Although Marsh wanted to make ecological studies a part of geography, like Linnaeus he saw the balance of nature as the organizing principle:

> *It was a narrow view of geography which confined that science to delineation of terrestrial surface and outline, and to description of the relative position and magnitude of land and water. In its improved form, it embraces not only the globe itself, but the living things which vegetate or move upon it, the varied influences they exert upon each other, the reciprocal action and reaction between them and the earth they inhabit. Even if the end of geographical studies were only to obtain a knowledge of the external forms of the mineral and fluid masses which constitute the globe, it would still be necessary to take into account the element of life; for every plant, every animal, is a geographical agency, man a destructive, vegetables, and even wild beasts, restorative powers.*[161]

Marsh at one point cited Darwin's *Origin of Species*, but it is evident that he failed to appreciate Darwin's discussion of the extinction of species. This is not surprising, considering that Darwin himself failed to realize fully the implications of his own theory of evolution for the balance-of-nature concept. Darwin had, nevertheless, built a strong argument for the extinction of species through competition.[162] The reality of extinction had been accepted by most naturalists since around 1800. Marsh believed that man was the prime cause of extinctions, perhaps the sole cause.[163]

Marsh emphasized that man's impact upon nature was not confined to the direct elimination or diminution of the numbers of species. In the second chapter he discussed the effects of introduced species of plants and animals upon the economy of nature. In the third chapter he discussed the importance of forests for both man and nature and the effects of removal of forests upon the soil, climate, and rivers. In the fourth chapter he discussed the geography of marshes, lakes, and rivers, and in the fifth chapter, the sandy regions of America and Europe. The final chapter is concerned with the environmental impact of large projects, such as the Suez Canal, the draining of the Zuiderzee, and mining.

Man and Nature presented a compelling case for the development of an ecological science in relation to resource use. Marsh's message was widely read, for the book was an instant success, and was reprinted eight times, in 1865, 1867, 1869, 1871, 1874, 1885, 1898, and 1908. A number of changes were made for the 1874 edition, and the title was changed to *The Earth as Modified by Human Action*. There were also English and Italian editions.[164] Marsh also urged the support of ecological research when he was consulted in 1870 about the use of funds for the development of the University of California.[165] Marsh's book and a speech by Franklin B. Hough are credited with leading Congress to establish in 1873 the United States Forestry Commission and government forest reserves. The first forestry commissioner was Hough, who looked upon Marsh as the leader of the movement for forest conservation.[166]

The nature-appreciation and conservation movements continued to gain support in the remaining quarter of the nineteenth century, but these movements did not regularly produce ecological investigations as by-products of their concern. Stephen A. Forbes began studying the food of and interactions between species in the 1870's, and when the Division of Economic Ornithology and Mammalogy was added to the United States Department of Agriculture in 1886, it was charged with collecting such information. While doing so, however, it also carried out systematic eradication programs against rodents, coyotes, wolves, and hawks without having much quantitative information on their feeding and other habits.[167] The concern of mainstream conservationists for preserving certain species of fish and wildlife too quickly reached the desperate stage for reliance upon ecological studies. The buffalo, passenger pigeon, and Carolina parakeet had been descimated, and if other species were to be saved, it was necessary to concentrate upon obtaining legal restrictions to hunting, game preserves, and restocking programs. In the twentieth century all these programs would provide important motivation for ecological research, but when they began, there was a general conviction that good intentions, adequate funds, and common sense would make the programs work.[168]

One man who understood where good intentions without adequate knowledge could lead was John Wesley Powell (1834–1902). He spent his early years in Ohio, Wisconsin, and Illinois and then distinguished himself as a member of General Ulysses S. Grant's staff in the Civil War. He led expeditions sponsored by the Illinois State Natural History Society to the Rocky Mountains in 1867 and 1868. In 1869 he won national recognition by leading a party of ten men in four boats down the canyons of the Green and Colorado rivers, and his account of this adventure has been frequently reprinted ever since. In 1870, Congress established the Geographical and Geological Survey of the Rocky Mountain Region under Powell's supervision. He then directed the surveying and mapping of the designated region. In 1878 he published *Report on the Lands of the Arid Region of the United States*, the purpose of which was to explain the importance of matching land use with environmental conditions. He saw the settlers farming lands unfit for cultivation, and he hoped that disaster could be avoided by increasing the awareness of the conditions of the various arid lands in western America.

James S. Lippincott in 1864 had urged the importance of correlating crops with the temperature of the growing season in different regions of America. In the West, Powell urged a similar consideration for rainfall.[169] He also pointed out the interrelationship of these factors:

Primarily the growth of timber depends on climatic conditions—humidity and temperature. Where the temperature is higher, humidity must be greater, and where the temperature is lower, humidity may be less. These two conditions restrict the forests to the highlands. . . . of the two factors involved in the growth of timber, that of the degree of humidity is of the first importance; the degree of temperature affects the problem comparatively little.

He found that all western lands that might support forests did not, and this he attributed to fire:

The conditions under which these fires rage are climatic. Where the rainfall is great and extreme droughts are infrequent, forests grow without much interruption from fires; but between that degree of humidity necessary for their protection, and that smaller degree necessary to growth, all lands are swept bare by fire to an extent which steadily increases from the more humid to the more aird districts, until at last all forests are destroyed, though the humidity is still sufficient for their growth if immunity from fire were secured. The amount of mean annual rainfall necessary to the growth of forests if protected from fire is probably about the same as the amount necessary for agriculture without irrigation; at any rate it is somewhere from 20 to 24 inches.[170]

Powell was not the first to point to fire as an important environmental factor on the prairies. Caleb Atwater had argued against the idea in 1818, suggesting instead that prairies are the remains of former lake bottoms. In response to his paper, R. W. Wells had countered with observations he had made in areas where fire had indeed destroyed forests and were replaced by grasslands.[171] Powell went further and attempted to determine with surveys the lands that could realistically be maintained under forests and those that could not. He also discussed the environmental requirements for pastureland.[172]

The Utah region was the heartland of the survey he had supervised, and he provided a summary of its lands and vegetative resources. Ecological notes are included, but only insofar as they were of practical value. For example: "*Pinus aristata* is of no commercial value, as it is much branched and spreading with limbs near the base; it grows on the crags at an altitude of from nine to eleven thousand feet."[173]

As Wallace Stegner, one of Powell's biographers, has observed, the importance of the report lies not so much in its contents or in its immediate impact, important though both of these were, but rather in the fact that the report was the beginning of Powell's sustained drive for the intelligent use of western lands based upon an understanding of the environments and the possibilities of plant and animal life within those environments.[174] Marsh had urged this procedure in a general discussion; Powell urged it from the vantage point of intimate knowledge.

Biological Control

The most important problem in applied ecology is the interaction between animal or plant parasites and agricultural plants. In the eighteenth century Leeuwenhoek, Réaumur, Linnaeus, Jefferson, and many others had discussed the possibilities of controlling insect pests through a greater knowledge of their life histories, and interest in this possibility steadily increased throughout the nineteenth century.[175] One can find parallel advances in phytopathology.[176] Attention here will be confined to only one aspect of this important and interesting story, that of biological control. This was the first area of ecological investigation in which American naturalists took the lead.

Agriculture was so important to the American economy that entomology inevitably became an important science. America imported many of its crops from Europe, and a suspicion developed that it might have imported insect pests as well. Kalm had written of having almost introduced accidentally the serious pest the "pease beetle" (pea weevil, *Bruchus pisorum*) into Sweden from America, and he also speculated about whether domestic flies, rats, mice, and other species were native to America or had been accidentally introduced from Europe.[177]

Since Alexander Wilson and others had emphasized the importance of birds as predators of insects, it was perhaps inevitable that someone would have the idea that the natural bird predators might better control European insects than American birds. Nicholas Pike in 1850 enlisted the assistance of others in Brooklyn for the purpose of introducing English sparrows (*Passer domesticus*) to combat geometrid caterpillars. The first effort was unsuccessful, but persistence led to success in 1853. On the basis of this achievement the federal government was persuaded in 1862 to provide financial assistance to a more comprehensive introduction of this bird, which has become far more conspicuous as a city resident than as a predator of agricultural insects.[178]

Asa Fitch (1809–79), state entomologist for New York, was not the first to speculate on the problem of introduced agricultural pests, but in his "Sixth Report on the Noxious and Other Insects of the State of New York" (1861) he went one theoretical step further than others had. While discussing the serious devastations caused by the spread of the European wheat midge (*Sitodiplosis mosellana*), he hypothesized that its spread in America was facilitated because its natural parasites had not also been imported. This idea had come to him during the disastrous harvest of 1854, and he had written for assistance to John Curtis, president of the London Entomological Society. Curtis read Fitch's letter to the society, but, not surprisingly, no one delivered the live parasites.[179]

Nevertheless, the idea did not disappear. Benjamin Dann Walsh (1808–69) was shortly thereafter to urge American government initiative in the importation of insect parasites. Walsh was born in Frome, England, and attended Cambridge University with Charles Darwin. In 1838 he married and emigrated to America, settling in Illinois. He had a casual interest in insects while in England, but it was only in the late 1850's that he began serious study of insect pests of agriculture. He was so effective in publicizing the dangers that the Illinois State Legislature in 1866–67 authorized the establishment of the post of state entomologist, to which he was appointed. He understood that Americans could not depend upon English altruism for the delivery of parasites of the wheat midge, and he urged the government to undertake the importation of its parasites. "But we should not stop here. The principle is of general application; and whenever a Noxious European Insect becomes accidentally domiciled among us, we should at once import the parasites and Cannibals that prey upon it at home."[180] Walsh did not persuade his state or any other to put this advice into practice, but neither was his message lost.

Charles Valentine Riley (1843–95), who was to become the second entomologist to the United States Department of Agriculture, received his professional training from Walsh. Riley also was from England and had an interest in insects before emigrating to Illinois. In 1868 with Walsh he founded the short-lived periodical *American Entomologist* and also, with Walsh's assistance, was appointed state entomologist to Missouri. Apparently Riley was the first person who actually distributed insect parasites from one locality to another. In 1870 he sent parasites of the weevil *Conotrachelus nenuphar* to various places in Missouri. In 1873 he did for France what the English had not bothered to do for Americans: he sent a predaceous mite, *Tyroglyphus phylloxerae*, to France to combat the *Phylloxera* grape-vine louse, which had been introduced accidentally into the French vineyards.[181] Although *Tyroglyphus* was not as effective in coltrolling *Phylloxera* as Riley had expected, he did not abandon the idea of transporting predatory insects to sites where their prey had accidentally been introduced.

He served as entomologist to the United States Department of Agriculture in 1878 and from 1881 to 1894. In 1883 he succeeded in arranging the importation from England of *Apanteles glomeratus* larvae to parasitize their natural host, the cabbage butterfly, *Pieris rapae*. These parasites were distributed in the District of Columbia, Missouri, Iowa, and Nebraska, where they became permanently established.[182]

The value of this technique for controlling insects was most conspicuously established, however, in the successful efforts against the cottony-cushion scale (*Icerya purchasi*), which had been introduced into California citrus groves about 1868. In 1887, Frazer Crawford wrote to Riley from Adelaide, Australia, reporting that *Icerya* was destroyed by a fly, now named *Cryptochaetum iceryae*. Riley did not take the claim seriously. Nevertheless, Crawford sent some of the flies to Waldemar G. Klee, California state inspector of fruit pests, who released them in 1888 in San Mateo County, where the species became established as an important predator of *Icerya*. Meanwhile, Riley sent Albert Koebele, an entomologist of the Department of Agriculture, to Australia, where he quickly discovered not only that *Icerya* was parasitized by *Cryptochaetum* but also that its eggs were eaten by the larvae of three other species of insects. Riley now became convinced that *Cryptochaetum* would provide the best control, and Koebele sent 12,000 of them to California. The most effective control, however, proved to be the Australian ladybird, *Vedalia (Rodolia cardinalis)*. The success of *Vedalia* in controlling the cottony-cushion scale was so dramatic that the state of California ever since has generously supported research in the biological control of insect pests.[183]

Concurrently with the rise of interest in the use of insect predators or parasites to control pest species was an even broader interest in using diseases to control insects. The idea was one aspect of the rise of the germ theory of disease, and two of the early pioneers of the germ theory, Agostino Bassi (1836) and Louis Pasteur (1874), were also early proponents of using pathogens against insects. In America the entomologist John Lawrence LeConte urged the development of this technique at the meeting of the American Association for the Advancement of Science in 1873. In 1879, Riley, J. H. Comstock, and J. H. Burns separately tested Herman A. Hagen's suggestion that yeast could infect and kill insects. Their results did not confirm it. In 1887, Stephen A. Forbes identified the fungal parasite *Beauveria globulifera* on the chinch bug (*Blissus leucopterus*), a serious cereal pest. The following year Otto Lugger in Minnesota attempted to infest these insects in the field with the disease, and F. H. Snow began to conduct investigations of the disease in Kansas. These studies continued for several years without definite success, but they did serve to publicize the possibility of this technique.[184]

The Emergence of the Ecological Sciences

Most ecological observations before 1900 fell in the domain of natural history, but this science was organized primarily according to the various groups of plants and animals, a system that was not optimal for theoretical developments. In 1749, Linnaeus had outlined an ecological science based upon the economy of nature, that is, the balance of nature; in 1803, Barton and around 1850, Thoreau had thought that phenology could provide the framework for one; and in 1864, Marsh had attempted to make room for ecological and conservation studies within the discipline of geography.

Ernst Haeckel (1834–1919), a noted German biologist, also realized the need for an ecological science, and in 1866, in his evolutionary treatise, *Generelle Morphologie der Organismen,* he coined the word *oekologie* for the science which he defined as "the whole science of the relations of the organism to the environment including, in a broad sense, all the 'conditions of existence.' These are partly organic, partly inorganic."[185] As a defender of Darwin's theory of evolution, Haeckel was in a better position to realize the theoretical implications for this science than were any of the pre-Darwinian exponents of an ecological science. The very structure and survival of species he understood to be tied to the relationships between organisms and their environments.

Walter Harding has shown (1965) that the prevalent belief that Thoreau used the word ecology in 1858 is incorrect. In his edition of Thoreau's correspondence (1958), the word geology in one letter was mistakenly read as ecology. J. S. Burdon-Sanderson (1893) and Stephen A. Forbes (1895) were the early expounders of the newly named science for British and American naturalists.[186]

During the second half of the nineteenth century four semidistinct ecological sciences began to develop as distinct from natural history. Listed in the chronological order of their self-conscious organization, they are oceanography, limnology, plant ecology, and animal ecology. A useful indicator of their formal organization is the earliest comprehensive treatises for each science. It is of interest to take note of the appearance of these treatises in relation to the development of the ecological sciences in America.

Oceanography. The English naturalist Edward Forbes (1815–54) in the 1840's made the first extensive studies on marine animals from the standpoint of their geographical distribution and environmental circumstances. He also wrote, before his early death, half of *The Natural History of European Seas,* which was completed by Robert Godwin-Austen (1859). Since Forbes's chapters discussed the animals and Godwin-Austen's chapters the physical features, the book as a whole constitutes the first general treatise on oceanography. Two of Forbes's papers on the stratification of coastal organisms (1843, 1844) were important for the concept of a biotic community, as was the study by the German zoologist Karl August Möbius (1825–1908) on oyster beds near Keil (1877). Haeckel was one of the early students of oceanic plankton (1862, 1890), and Victor Hensen (1835–1924), Möbius' colleague at the University of Keil, carried out important planktonic studies from 1870 to 1912.[187]

In America the first significant developments relating to biological oceanography came from Louis Agassiz. He carried out investigations on the Atlantic marine life of both North and South America, but more important for oceanography than his own discoveries was his role as teacher and organizer of the researches of his students and associates, including E. S. Morse, Alpheus Hyatt, B. Wilder, F. W. Putnam, Alpheus S. Packard, W. Strumpson, T. Lyman, H. J. Clark, A. E. Verrill, David Starr Jordan, and his son Alexander Agassiz. He organized a summer school on Penikese Island near Woods Hole, Massachusetts, in 1873. The school operated only two summers but provided the example for the ultimate establishment by Hyatt of the Marine Biological Laboratory at Woods Hole in 1888. Spenser F. Baird had already established the United States Bureau of Commercial Fisheries Laboratory at Woods Hole in 1885, having conducted research there since 1871.[188]

Limnology. The development of marine biology stimulated similar investigations in fresh water. German and Swiss naturalists were notable for leading the way, especially François Alphonse Forel (1841–1912). Kurt Lampert judged that Forel's "Introduction à l'étude de la faune profonde du lac Léman" (1869) was the beginning of scientific limnology. Forel's great three-volume *Le Léman, monographie limnologique* (published between 1892 and 1904) was the first comprehensive monograph on the subject (in which he coined the word *limnologie*). He also published the first textbook on the subject.[189]

Thoreau's studies on Walden Pond in the 1840's and 1850's have already been cited as the first significant limnological work in America, but since he was not a professional naturalist or in complete sympathy with their work, he was not in a position to establish a new science.

Stephen A. Forbes (1844–1930), on the other hand, attained a very good position for furthering limnology and animal ecology. A native of Silver Creek, Illinois, he fought in the Civil War and afterward attended Rush Medical College in Chicago. He lacked both the funds and the motivation to complete a medical degree. He then taught school and studied at the Illinois State Normal University. In 1872, when Powell resigned as curator of the Museum of the State Natural History Society at Normal, Forbes was appointed to the position. He became an instructor in zoology at the university in 1875, director of the State Laboratory of Natural History in 1877, and state entomologist in 1882. He received a doctorate from Indiana University in 1884, when he also moved with the state laboratory to the University of Illinois at Urbana.

Forbes became interested simultaneously in the conservation of wildlife and the eradication of agricultural pests. Like most conservationists, he saw the wisdom of Alexander Wilson's argument that the birds of the fields did more good for the farmer by eating insects than harm from eating his crops, but as a professor of zoology and director of the State Laboratory of Natural History he could not rest content with plausible hypothetical arguments. He wanted to build a sound understanding of the situation based upon adequate data and their theoretical interpretation. The same was true for fish preservation. Many an alarmed conservationist insisted upon the need for fish hatcheries and stocking programs, but

Forbes was convinced that preservation required a basic understanding of fishery biology.

With this orientation much of his research concerned the food of insects, birds, and fish. The natural-history literature ever since Aristotle had contained this kind of information, but Forbes appears to have been the first to make it the basis of a systematic and sustained research program. Generalities were evidently no longer sufficient, anyway, because to the argument that birds eat harmful insects some skeptics replied that birds also eat parasitic and predaceous insects that might control the pest species. To combat this claim Forbes felt the need for both data and theory. The data he obtained from stomach contents from members of the species under investigation. The steady accumulation of this data over several decades finally enabled him to write an important book on the fishes of Illinois.[190]

In 1887 he used his studies on fish feeding and populations to state his ideas on the balance of nature in a well-known and often-reprinted paper, "The Lake as a Microcosm," which seems to have been the most detailed defense of a balance of nature concept within a Darwinian theoretical framework in the nineteenth century. It also represented another important step toward the biotic-community concept in ecology.[191]

Forbes persuaded the state legislature to establish under the Illinois Natural History Survey America's first river biological field station, at Havana, on the Illinois River. Charles A. Kofoid (1865–1947), who was from Illinois and had received his doctorate from Harvard University in 1894, was director of the station from 1895 to 1901, during which time he published pioneering papers on the plankton of rivers.[192]

In the 1890's limnology also became firmly established in Wisconsin, Michigan, Indiana, and Ohio, accompanied in each state by the establishment of a lake field station. The rise of limnology in these states around the Great Lakes must have been stimulated in part by the accomplishments of Forel and Forbes and their respective associates and in part by the importance of lake resources to the states.

Edward A. Birge (1851–1950) studied zoology at Harvard University and in 1876 became instructor in natural history at the University of Wisconsin. C. Dwight Marsh (1855–1932) studied at Amherst College and in 1883 became professor of natural science at Ripon College. In the 1890's both men began publishing studies on the plankton in Wisconsin lakes. In 1904, Birge began a long series of studies on the physical characteristics of Wisconsin lakes, in which studies he was joined by Chancey Juday (1871–1944) beginning in 1908. Jacob E. Reighard (1861–1942) joined the University of Michigan faculty in 1886 and became chairman of the Department of Zoology in 1892. He worked closely with the Michigan Fish Commission and directed for it an extensive investigation in 1893 and 1894 of Lake St. Clair as an ecosystem. In 1900 he recommended that the university establish a biological station, which it did at Douglas Lake in 1909, naming him as director. In 1895, Carl H. Eigenmann (1863–1927) established a biological station for Indiana University on Lake Wawasee, and the following year David S. Kellicot established one for Ohio State University at the State Fish Hatchery on Lake Erie.[193]

Plant Ecology. Plant ecology developed into a formal science in northern Europe. It was an outgrowth of plant geography, continuing directly in the tradition established by Linnaeus, Humboldt, and Auguste Pyramus de Candolle, even utilizing at the end of the century the organization of the science established by De Candolle in 1820. The crucial works for this transition from plant geography to plant ecology were *Handbuch der Pflanzengeographie* (Stuttgart, 1890), by Oskar Drude (1852–1933), and *Plantesamfundgrundträk af den Okologiska Plantegeogre fi,* by J. Eugen B. Warming (1841–1924), (Copenhagen, 1895; German translation, 1896; English revised translation, 1909). *Pflanzengeographie auf physiologischen Grundlage,* (Jena, 1898; English translation, 1903) by Andreas W. F. Schimper (1856–1901), was also an important work in this tradition, but by the time it appeared, Americans were already busy with plant-ecology studies of their own.[194]

Some Americans were already producing similar works even before Drude's *Handbuch* appeared, because phytogeography was useful in the ongoing surveys of the plants of different regions of the country. Of particular interest in this respect is the work of John Merle Coulter (1851–1928), who grew up in Hanover, Indiana, receiving from its college his bachelor's and master's degrees and his doctorate from Indiana University. He was botanist to F. V. Hayden's geological survey of the Yellowstone country in 1872, and in 1874 he helped write *Synopsis of the Flora of Colorado.* In 1881 he was one of three authors of *Plants of Indiana,* which discussed the plants in terms of the river valleys, lake borders, prairies, and barrens. He became one of the first teachers of ecology in America.[195]

Charles Edwin Bessey (1845–1915) was from Ohio and graduated from Michigan Agricultural College in 1869. He taught botany and horticulture at Iowa Agricultural College and in 1884 accepted a professorship at the University of Nebraska. His main efforts were devoted to teaching botany, writing textbooks, and making investigations in plant phylogeny. He was also the first botanist in America to train plant ecologists. Among his students those noted for ecological studies were Conway MacMillan, Roscoe Pound, and Frederic E. Clements.[196]

Conway MacMillan (1867–1929) was born in Hillsdale, Michigan, and received his bachelor's and master's degrees from the University of Nebraska in 1885 and 1886. In 1887 he became an instructor in botany and later professor at the University of Minnesota. His *Metaspermae of the Minnesota Valley* (1892) contains a capable ecological survey of the plants of the region, following the phytogeographical tradition of Drude and other German botanists.[197]

Roscoe Pound (1870–1964), the son of a Nebraska judge, was almost lured away from his eminent career in law by the new science. He received his bachelor's, master's, and doctoral degrees from the University of Nebraska in 1888, 1889, and 1897 and also studied law at Harvard University in 1889 and 1890. He was admitted to the Nebraska bar in 1890.

Having studied botany under Bessey, he became, while a graduate student, director of the botanical survey of Nebraska. Frederic E. Clements (1874–1945) received his bachelor's, master's and doctoral degrees from Nebraska in 1894, 1896, and 1898. He also participated in the survey and published with Pound *Phytogeography of Nebraska* (1898). This was an early notable contribution to plant ecology in America, and it marked the beginning of Clements' career as one of the leading ecologists of the early twentieth century.[198]

Meanwhile, in 1896, Coulter had become head of the Department of Botany at the new University of Chicago. Warming's text convinced him of the importance of plant ecology, and he encouraged one of his students, Henry Chandler Cowles (1869–1933) to undertake doctoral research in the subject. Cowles was from Kensington, Connecticut, and had graduated from Oberlin College before going to the University of Chicago. He wrote his dissertation on the vegetative communities on the sand dunes of southern Lake Michigan, and he published a series of influential papers on the subject (1899–1901). Both Clements and Cowles did pioneering work on the description and dynamics of succession in plant communities. Cowles joined the University of Chicago faculty and was a prominent ecology professor during the first four decades of the twentieth century.[199]

Some young American botanists adopted the new perspective on their own. One of the first to do so was William Francis Ganong (1864–1944), of St. John, New Brunswick. He received his bachelor and master's degrees from the University of New Brunswick, and was at Harvard University, from 1887 until 1891, when he went to the University of Munich for his doctorate (1894). His paper "On Raised Peat-Bogs in New Brunswick" (1891) is not profound, but it does show his ecological orientation before he went to Munich and before he could have read Drude's *Handbuch.*[200]

Another who apparently developed an ecological interest on his own was John W. Harshberger (1869–1929), a lifelong resident of Philadelphia. He received his bachelor's and doctoral degrees from the University of Pennsylvania and then taught biology there throughout his career. His doctoral dissertation (1893) was a botanical and economic study of maize, but he became interested in the new science and soon published "An Ecological Study of the New Jersey Strand Flora" (1900–1902) and "An Ecological Study of the Flora of the Mountainous North Carolina" (1903). He continued throughout his career to conduct studies in both agricultural botany and plant ecology.[201]

Animal Ecology. Animal ecology was built partly upon limnology and plant ecology and partly upon pure and applied zoology. The diversity of its early years is nicely illustrated by the topical bibliographies in Charles C. Adams' *Guide to the Study of Animal Ecology* (New York, 1913). The German zoologist Karl G. Semper (1832–93) published the earliest comprehensive account of animal ecology—his Lowell Institute Lectures, *Animal Life as Affected by the Natural Conditions of Existence* (New York and London, 1881). He had become interested in animal geography and ecology during a research trip to the Philippine Islands.

Semper's book does not appear to have stimulated anyone in America to commence studies in animal ecology comparable to the influence that Forel's limnological writings and Drude and Warming's plant-ecology writings apparently exerted. America's leading investigator in animal ecology at that time was undoubtedly Stephen A. Forbes, who acknowledged a debt to Herbert Spencer's *Principles of Biology* (1867) in the formation of his own balance-of-nature concept,[202] but his ecological orientation appears to have developed out of the kinds of problems that seemed important to an economic zoologist rather than from the influence of programatic writings by others.

Since no single pattern is detectable in the animal-ecology studies of the 1880's and 1890's, a brief survey of different kinds of studies is the best way to indicate the ways in which the science was developing in America. Probably the commonest ecological observations were still being recorded in natural-history studies, which were too numerous to be cited here individually. Forbes set an example for others on food studies, and Ralph Dexter's survey of such studies in northeastern Ohio from 1879 to 1899 is probably representative of the observations being published. Few of these reports could equal the high standards of Forbes and Riley, but an exception was Leland O. Howard's study on the parasites of the white-masked tussock moth (1897). Howard (1857–1950) studied entomology under J. H. Comstock at Cornell and then served as Riley's assistant in the Division of Insects of the United States Department of Agriculture. Upon Riley's resignation in 1894, Howard succeeded him as head of the division, and in the twentieth century Howard became one of America's leading entomologists, particularly in the study of insects in relation to disease.[203]

The advance of natural-history studies of invertebrates, the germ theory of disease, and parasitism eventually led to an understanding of the role of insects and other invertebrates as vectors of certain diseases. Patrick Manson discovered in 1878 that filiariasis is caused by the worm *Wuchereria bancrofti* transmitted to man by the mosquito *Culex fatigans.* In 1893 two physicians of the United States Bureau of Animal Industry, Theobald Smith and F. L. Kilbourne, demonstrated that Texas cattle fever is caused by a protozoan, *Babesia bigemina,* which is transmitted by the tick *Boöphilus annulatus.* The work of Ronald Ross on malaria and Walter Reed on yellow fever followed in 1897 and 1900.[204] Although these studies often were viewed as within the providence of medicine or parasitology rather than of animal ecology, the demand for eradicating vector species would often lead to a demand for ecological investigation.

Darwin's theory of evolution provided an important stimulus for ecological investigation. Examples of this stimulus in American animal ecology are Alpheus S. Packard's monograph "The Cave Fauna of North America" (1888), studies on protective coloration by Abbott H. Thayer (1896) and Sylvester D. Judd (1898), and John A. Ryder's theoretical paper "A Geometrical Representation of the Relative Intensity of the Conflict between Organisms" (1892).[205] Darwin's influence is also indirectly evident in many other studies.

The direct stimulus for American insect-pollination

studies came largely from Hermann Müller's *Die Befruchtung der Blumen durch Insekten* (1873, English translation, 1883). George Valentine Riley (1892) and William Trelease (1893) published notable studies on the pollination of yucca by the yucca moth. Charles Robertson published "Flowers and Insects: Contributions to an Account of the Ecological Relations of the Entomophilous Flora and the Anthophilous Insect Fauna of the Neighborhood of Carlinville, Illinois" in 1897, and John Harvey Lovell began in the same year to study the fertilization of flowers and the color preferences of insect pollinators.[206]

The applied ecological investigations by Bashford Dean on the oyster (1890) and by Charles T. Simpson on "The Pearly Fresh-Water Mussels of the United States; Their Habits, Enemies, and Diseases; with Suggestions for Their Protection" (1898) were undoubtedly inspired by Möbius' study, as well as by the practical needs of the fisheries.[207]

C. Hart Merriam (1855–1942), who in 1888 became the first head of the Division of Economic Ornithology and Mammalogy of the United States Department of Agriculture, had a strong interest in the geographical ranges of American birds and mammals, and he and his field naturalists gathered larger amounts and more precise data on the distributions and habits of American mammals than had ever before been collected. This information was mostly published in a series of reports, beginning in 1890, entitled *North American Fauna.* In 1889 he led an expedition to the San Francisco Mountains, near the Grand Canyon in Arizona, to study the relation between temperature and geographical distribution. Since childhood he had had some awareness of Humboldt's work and an interest in this question. The data he and his associates collected on the distribution of plants and animals at different elevations enabled him to describe a series of "life zones," each of which contained characteristic species. He published these results in 1890 and attempted through the 1890's to establish the general application of his life zones for all of North America. Ever since, some naturalists in the mountainous West have found his life zones of practical convenience. Close scutiny by others, however, showed that the theoretical foundations of his scheme were never well established and cannot be sustained. Nor did his life zones have even practical utility farther eastward and northward.[208]

One of the unique American ecological studies in the nineteenth century was Murray G. Motter's "Study of the Fauna of the Grave" (1898). He examined 150 bodies that were disinterred at different stages of decomposition to determine the different species of saprophytic organisms associated with each stage. Perhaps this was a definitive study, since ecologists in the twentieth century seem not to have felt it necessary to continue this line of research.

CONCLUSIONS

This survey shows that throughout American history naturalists and others have had an interest in ecological questions. As is characteristic of a poorly organized science, many of the important questions came from without science rather than from within. This survey leads to the question why ecology was not formally organized earlier than it was, particularly in light of the suggestions for such a science from Linnaeus, Barton, and Marsh. Even after Haeckel's formal definition of it in 1866, more than two decades went by before there was a good response. Presumably naturalists found descriptive natural history—an inventory of the world—of higher priority and easier to carry out than most ecological investigations would have been. Second, the importance of ecology to the world lies in its value for resource management, and America was prodigal of its resources until they began to become scarce. Wilson, Thoreau, and Marsh were voices crying in the wilderness, and the value of ecological research depended to some extent upon how many Americans could hear them. Furthermore, the germ theory of disease undermined the medical importance of environmental studies, and, although a few people became concerned about the health hazards of pollution even in the nineteenth century, this concern was not substantial enough to raise the importance of ecology.[209] Only at the end of the century with the trend toward formal organization did ecological knowledge become well organized from a theoretical standpoint. Even so, the science was mainly descriptive rather than theoretical.

Nevertheless, Darwin's theory of evolution, the practical needs of agriculture and wildlife management, and the progress of natural history all led to the establishment of four ecological sciences—oceanography, limnology, plant ecology, and animal ecology—during the second half of the nineteenth century. Before 1900 American naturalists had already taken the lead in biological control of insect pests, and the investigation of ecological questions was already so diverse and competent that Americans were competing well with Europeans in advancing ecological understanding.

NOTES

1. Thomas S. Kuhn, *The Structure of Scientific Revolutions* (Chicago, University of Chicago Press, 1962), Chaps. 2–4.

2. Howard Mumford Jones, *O Strange New World: American Culture: The Formative Years* (New York, Viking Press, 1964), Chap. 1.

3. Raymond Phineas Stearns, *Science in the British Colonies of America* (Urbana, University of Illinois Press, 1970), 4. Hereafter cited as *Science.*

4. Thomas Harriot, *A Briefe and True Report of the New Found Land of Virginia, of the Commodities and of the Nature and Manners of the Naturall Inhabitants* . . . (Frankfurt am Main, Theodor de Bry, 1590), 31. Stearns, *Science,* 68–71.

5. John Smith, *The Generall Historie of Virginia, New-England, and the Summer Isles* . . . (London, Michael Sparkes, 1624), 25. On Massachusetts see *ibid.,* 209ff.; Stearns, *Science,* 71–74.

6. John Josselyn, *New Englands Rarities Discovered: in Birds, Beasts, Fishes, Serpents, and Plants of that Country. Together with the Physical and Chyrurgical Remedies wherewith the Natives Constantly Use to Cure Their Distempers, Wounds, and Sores* . . . (London, G. Widdowes, 1672), 43; Stearns, *Science,* 139–50.

7. Louis Armand de Lom d'Avce, Baron de Lahontan, *New Voyages to North-America* . . ., (tr. anon., 2 vols., London, H. Bonwicke, 1703); cited from reprint ed. (ed. by Reuben Gold Thwaites, 2 vols., Chicago, A. C. McClurg, 1905), I, 319–20. The first French edition (3 vols.) also appeared in 1703.

8. John Lawson, *A New Voyage to Carolina* (ed. by Hugh Talmage Lefler, Chapel Hill, University of North Carolina Press, 1967), 74, hereafter cited as *A New Voyage;* Stearns, *Science,* 305–15.

9. Lawson, *A New Voyage,* 93–94.

10. For the identification of Catesby's birds and plants see W. L. McAtee, "The North American Birds of Mark Catesby and Eleazar Albin," *J. Soc. for the Bibliog. Nat. Hist.,* Vol. 3 (January, 1957), 177–94; Frans A. Stafleu, *Taxonomic Literature: A Selective Guide to Botanical Publications with Dates, Commentaries and Types* (Utrecht, International Bureau for Plant Taxonomy and Nomenclature, 1967), 78–79; George Frederick Frick and Raymond Phineas Stearns, *Mark Catesby: The Colonial Audubon* (Urbana, University of Illinois Press, 1961), Chap. 6.

11. Mark Catesby, *The Natural History of Carolina, Florida and the Bahama Islands: Containing the Figures of Birds, Beasts, Fishes, Serpents, Insects, and Plants: Particularly, the Forest-Trees, Shrubs, and Other Plants, Not Hitherto Described, or Very Incorrectly Figured by Authors* . . . (2 vols., London, Author, 1731–43), I, ii, hereafter cited as *Natural History.*

12. *Ibid.,* iii–iv.

13. *Ibid.,* v; Frick and Stearns (*Mark Catesby,* 29) judge that he did not reach the mountains.

14. Catesby, *Natural History,* I, xxi.

15. George Basalla, "The Spread of Western Science," *Science,* Vol. 156 (May 5, 1967), 611–22. For material to illustrate his model, see Stearns, *Science,* and Brooke Hindle, *The Pursuit of Science in Revolutionary America, 1735–1789* (Chapel Hill, University of North Carolina Press, 1956), hereafter cited as *The Pursuit of Science.*

16. John Bartram, *Observations on the Inhabitants, Climate, Soil, Rivers, Productions, Animals, and Other Matters Worthy of Notice . . . in Travels from Pensilvania to Onondago, Oswego and the Lake Ontario* . . . (London, J. Whiston and B. White, 1751); Stearns, *Science,* 575–93; Ernest P. Earnest, *John and William Bartram: Botanists and Explorers* (Philadelphia, University of Pennsylvania Press, 1940), hereafter cited as *John and William Bartram.*

17. John Bartram, *Diary of a Journey Through the Carolinas, Georgia, and Florida from July 1, 1765 to April 10, 1766* (ed. by Francis Harper, American Philosophical Society Transactions, Vol. 33, Pt. 1 [Dec. 1942]), 37, hereafter cited as *Diary.* The quotation, December 27, 1765, is from a portion of the *Diary* first published in 1769.

18. Pehr Kalm, *Travels into North America; Containing Its Natural History, and a Circumstantial Account of Its Plantations and Agriculture* (trans. by John Reinhold Forster (3 vols., Warrington, William Eyres, 1770–71), III, iii–iv, hereafter cited as *Travels.* For Kalm's negative judgments see II, 194–95, 262–63; III, 7, 135. Adolph B. Benson edited a revised edition of Kalm's *Travels* (2 vols., 1937; New York, Dover, 1964), which contains important additions; however, all citations in this paper are from the first English edition. Carl Skottsberg, "Pehr Kalm, Levnadsteckning," *Levnadsteckningar över Kungl. Svenska Vetenskapsakademiens Ledamöter,* Vol. VIII, No. 139 (1951), 221–504.

19. For his general estimate of Bartram, which is favorable, see Kalm, *Travels,* I, 112–14.

20. *Ibid.,* 65–66.

21. *Ibid.,* 75.

22. *Ibid.,* 162.

23. *Ibid.,* II, 90.

24. *Ibid.,* I, 142–43.

25. *Ibid.,* 311–20.

26. John Livingston Lowes, *The Road to Xanadu: A Study in the Ways of the Imagination* (Boston, Houghton Mifflin, 1927). Nathan B. Fagin, *William Bartram: Interpreter of the American Landscape* (Baltimore, Johns Hopkins Press, 1933), Part 3.

27. William Bartram, *Travels Through North & South Carolina, Georgia, East & West Florida, the Cherokee Country, the Extensive Territories of the Muscogulges, or Creek Confederacy, and the Country of the Chactaws; containing an Account of the Soil and Natural Productions of Those Regions* . . . (Philadelphia, James Johnson, 1791), 7–8, hereafter cited as *Travels.* The authoritative reprint, with commentary, is by Francis Harper (ed.), *The Travels of William Bartram, Naturalist's Edition* (New Haven, Yale University Press, 1958), 4–5. See also Earnest, *John and William Bartram;* Fagin, *William Bartram,* Parts 1, 2.

28. Bartram, *Travels* (1st ed.), xxvii–xxxi; (Harper ed.), lviii–lix.

29. *Ibid.* (1st ed.), 118; (Harper ed.), 75.

30. *Ibid.* (1st ed.), 8; (Harper ed.), 5.

31. *Airs, Waters, and Places,* in *Hippocrates* (trans. by W. H. S. Jones, Cambridge, Mass., Harvard University Press, 1923), I, 71.

32. William E. Knowles Middleton, *The History of the Barometer* (Baltimore, Johns Hopkins Press, 1964); William E. Knowles Middleton, *A History of the Thermometer and Its Use in Meteorology* (Baltimore, Johns Hopkins Press, 1966); William E. Knowles Middleton, *Invention of the Meteorological Instruments* (Baltimore, Johns Hopkins Press, 1969).

33. Genevieve Miller, "'Airs, Waters, and Places' in History," *J. Hist. Med. and Allied Sciences,* Vol. XVII (1962), 129–40; Kenneth Dewhurst, *Dr. Thomas Sydenham (1624–1689)* (Berkeley and Los Angeles, University of California Press, 1966), 65–67; Gordon Manley, "The Weather and Diseases; Some 18th Century Contributions to Observational Meteorology," *Notes and Records of the Roy. Soc. London,* Vol. IX (1952), 300–307; James H. Cassedy, "Meteorology and Medicine in Colonial America: Beginnings of the Experimental Approach," *J. Hist. Med. and Allied Sciences,* Vol. XXIV (1969), 193–204, especially 194–96, hereafter cited as "Meteorology and Medicine."

34. John Clayton, "A Letter from Mr. John Clayton Rector of Crofton at Wakefield in Yorkshire to the Royal Society, May 12, 1688, giving an Account of several Observables in Virginia, and in his Voyage thither, more particularly concerning the Air," *Phil. Trans. Roy. Soc. London,* Vol. XVII (1693), 781–89; reprinted in Edmund Berkeley and Dorothy Smith Berkeley (eds.), *The Reverend John Clayton, a Parson with a Scientific Mind: His Scientific Writings and Other Related Papers, Edited, with a Short Biographical Sketch* (Charlottesville, University Press of Virginia, 1965), 45, hereafter cited as "Concerning the Air," *Clayton;* Stearns, *Science,* 183–95.

35. Cadwallader Colden, "An Account of the Climate and Diseases of New-York," *Amer. Med. and Philos. Register,* Vol. I (1811), 304–10; Saul Jarcho, "Cadwallader Colden as a Student of Infectious Disease," *Bull. Hist. Med.,* Vol. XXIX (1955), 99–115, especially 102, hereafter cited as "Colden as a Student." Colden had discussed the nature of smallpox and other diseases in 1716–19 in letters published by Saul Jarcho, "The Correspondence of Cadwallader Colden and Hugh Graham on Infectious Fevers," *Bull. Hist. Med.,* Vol. XXX (1956), 195–211. See also Hindle, *The Pursuit of Science,* 39–48 et passim; Stearns, *Science,* pp. 559–75; Cassedy, "Meteorology and Medicine," *J. Hist. Med. and Allied Sciences,* Vol. XXIV (1969), 197–98.

36. First published anonymously in *New-York Weekly Postboy,* December 26, 1743–January 9, 1744; reprinted in *Amer. Med. and Philos. Register,* Vol. I (1811), 310–30. Jarcho, in "Colden as a Student" (*Bull. Hist. Med.,* Vol. XXIX [1955], 102–103), observed that about half of this essay was paraphrased from Giovanni Maria Lancisi's *De Noxiis Paludum Effluviis* (1717).

37. Letter to Dr. John Mitchell, Urbanna, Va., November 7, 1745, Jarcho, "Colden as a Student," *Bull. Hist. Med.,* Vol. XXIX (1955), 105.

38. James Thomas Flexner, *Doctors on Horseback: Pioneers of American Medicine* (New York, Viking, 1937; Dover reprint, 1968), 93–94.

39. "Extract of a Letter from Cadwallader Colden, Esq., to Dr. Fothergill concerning the Throat Distemper, Oct. 1, 1753," *Med. Observations and Inquiries,* Vol. I (1757), 211–29. Jarcho, "Colden as a Student," *Bull. Hist. Med.,* Vol. XXIX (1955), 106–109.

40. [Cadwallader Colden], "The Cure of Cancers. From an Eminent Physician at New-York," *Gentleman's Magazine,* Vol. XXI (1751), 305–308; on scurvy see letters to Dr. John Mitchell (ca. 1745) and Pehr Kalm (1751), Urbanna, Va., Jarcho, "Colden as a Student," *Bull. Hist. Med.,* Vol. XXIX (1955), 109–10.

41. John Lining, "Extracts of Two Letters . . . giving an Account of Statical Experiments made several times in a Day upon himself, for one whole Year, accompanied with Meteorological Observations; to which are subjoined Six General Tables, deduced from the whole Year's Course," *Phil. Trans. Roy. Soc. London,* Vol. XLII (1743), 491–509, especially quotation on 492. Joseph Ivor Waring, *A History of Medicine in South Carolina, 1670–1825* (Charleston, South Carolina Medical Association, 1964), 254–60, hereafter cited as *History.*

42. Santorio, *Medicina Statica: Being the Aphorisms of Sanctorius, Translated into English, with Large Explanations. To which is added Dr. Keil's Medicina Statica Britannia* . . . (London, 1723).

43. John Lining, "A Letter from . . . Charles-Town, South Carolina . . . Concerning the Quantity of Rain Fallen There from January 1738, to December 1752," *Phil. Trans. Roy. Soc. London,* Vol. XLVIII (1753), 284–85 and table.

44. John Lining, "A Description of the American Yellow Fever . . .," *Essays and Observations, Physical and Literary,* Vol. II (1756), 370–95. This report was reprinted at Philadelphia in 1799. Robert Croom Aldredge, "Weather Observers and Observations at Charleston, South Carolina, 1670–1871," *Year Book of the City of Charleston for the Year 1940,* pp. 190–257, see especially 204–18, hereafter cited as "Weather Observers"; Everett Mendelsohn, "John Lining and His Contribution to Early American Science," *Isis,* Vol. LI (1960), 278–92. Cassedy, "Meteorology and Medicine," *J. Hist. Med. and Allied Sciences,* Vol. XXIV (1969), 199–201.

45. Kalm, *Travels,* I, 361–68, especially quotation on 368; II, 120.

46. *Ibid.,* I, 371–72; see also 369–71, 376–78.

47. *Ibid.,* II, 253–54.

48. Lionel Chalmers, *Account of the Weather and Diseases of South Carolina* (2 vols., London, 1776), I, charts at 42 and at end of vol., hereafter cited as *Account;* Waring, *History,* 188–97 and portrait; Aldredge, "Weather Observers," *Year Book of the City of Charleston for the Year 1940,* 219–22.

49. Chalmers, *Account,* I, 6.

50. *Ibid.,* 33–34.

51. *Ibid.,* 150–51.

52. Charles-Edward Amory Winslow, *The Conquest of Epidemic Disease: A Chapter in the History of Ideas* (Princeton, Princeton University Press, 1943), 193, hereafter cited as *Conquest.*

53. Benjamin Rush, *An Account of the Bilious Remitting Yellow Fever as It Appeared in the City of Philadelphia in the Year 1793* (Philadelphia, 1794). The conclusion by the College of Physicians is quoted by Winslow, *Conquest,* 198.

54. Karl Popper, *Conjectures and Refutations; the Growth of Scientific Knowledge* (New York, Basic Books, 1962).

55. Quoted from Benjamin Spector, "Noah Webster, His Contribution to American Medical Thought and Progress," in Noah Webster, *Letters on Yellow Fever Addressed to Dr. William Currie* (first published 1797; ed. by Benjamin Spector, Baltimore, Johns Hopkins Press, 1947), 11, hereafter cited as "Webster," *Letters.* See also Charles-Edward Amory Winslow, "The Epidemiology of Noah Webster," *Conn. Acad. Arts and Sciences,* Vol. XXXII (1934), 21–109; see 45 showing the date of the questionnaire as October 31, 1795.

56. In a letter to Benjamin Rush, 4 Dec. 1798 Webster complained that the theory of imported disease "is a serious attack on the commercial interests of the country." Quoted from Spector, "Webster," in Webster, *Letters,* 15.

57. Noah Webster, *Brief History of Epidemic and Pestilential Diseases, with the Principal Phenomena of the Physical World, Which Precede and Accompany Them, and Observations Deduced from the Facts Stated* (2 vols., 1799), II, Secs. 15–16; I, x.

58. Samuel Noah Kramer, *The Sumerians: Their History, Culture, and Character* (Chicago, University of Chicago Press, 1964), 340–42 for the almanac and 105–109 for a discussion of it.

59. Kalm, *Travels,* II, 76, 90–91.

60. *Ibid.,* 318–52.

61. *Ibid.,* I, 105–12.

62. *Ibid.,* II, 167–68.

63. Kalm, "Beskrifining pa de vilda Dufvar I Norra America," *Kongl. Vetenskaps-Akademiens Handlingar,* Vol. XX (1759), 275–95, cited from S. M. Gronberger (trans.), "A Description of the Wild Pigeons Which Visit the Southern English Colonies in North America, During Certain Years, in Incredible Multitudes," *Auk,* Vol. XXVIII (1911), 53–66, especially 61–64, hereafter cited as "A Description of the Wild Pigeons."

64. Benjamin Stillingfleet, *Miscellaneous Tracts relating to Natural History, Husbandry, and Physick, To which is added: The Calendar of Flora* (2d ed., London, R. & J. Dodsley, 1762), 229–327. The bibliography of Linnaean works is complex; see B. H. Soulsby, *A Catalogue of the Works of Linnaeus* . . . (2d ed., London, British Museum, 1933).

65. Alexander McAdie, "A Colonial Weather Service," *Popular Science Monthly,* Vol. XLV (1894), 331–37; Edward T. Martin, *Thomas Jefferson: Scientist* (New York, Henry Schuman, 1952), Chap. 5.

66. Thomas Jefferson, *Garden Book, 1766–1824, with Relevant Extracts from His Other Writings* (ed. by Edwin Morris Betts, Philadelphia, American Philosophical Society, Memoir 22, 1944), 4, hereafter cited as *Garden Book.*

67. *Ibid.,* 129; see also Thomas Jefferson, *Notes on the State of Virginia* (ed. by William Peden, Chapel Hill, University of North Carolina Press, 1955), 75, hereafter cited as *Notes.*

68. Quoted in *Garden Book,* 151.

69. *Ibid.,* 161–62.

70. *Ibid.,* 578–79.

71. The plan was briefly described by Samuel Vaughan in a letter to Humphry Marshall dated May 22, 1786 quoted in William Darlington (ed.), *Memorials of John Bartram and Humphry Marshall, with Notices of Their Contemporaries* (Philadelphia, 1849), 558; facsimile (ed., with introduction and indices, by Joseph Ewan, New York, Hafner, 1967); Hindle, *The Pursuit of Science,* 332 et passim; *National Cyclopaedia of American Biography* (1891), I, 257.

72. Samuel Williams, *Natural and Civil History of Vermont* (1794; enlarged ed., 1809), 44–56, 78, 112–13 (all citations from 1st ed.), hereafter cited as *Vermont.* Bartram, *Travels* (1st ed., 286–302; (Harper ed.), 179–91. Benjamin Smith Barton, *Fragments of the Natural History of Pennsylvania . . .* (Philadelphia, Author, 1799). Witmer Stone, "Bird Migration Records of William Bartram, 1802–1822," *Auk,* Vol. XXX (1913), 325–58 and 3 plates.

73. Kalm, *Travels,* I, 58, 102–103 (on degeneration); II, 194–95 (on American farming).

74. *Ibid.,* I, 104, 390; II, 189; III, 8–9.

75. Comte Georges Louis Leclerc de Buffon, "The Natural History of Man," in *Natural History, General and Particular* (trans. by William Smellie, 9 vols., Edinburgh, William Creech, 1780), III, 181, 164–65, hereafter cited as *Natural History.*

76. *Ibid.,* V, 129; Clarence J. Glacken, *Traces on the Rhodian Shore: Nature and Culture in Western Thought from Ancient Times to the End of the Eighteenth Century* (Berkeley and Los Angeles, University of California Press, 1967), 663–85, hereafter cited as *The Rhodian Shore.*

77. Antonello Gerbi, *The Dispute of the New World: the History of a Polemic, 1750–1900* (trans. by Jeremy Moyle, Pittsburgh, University of Pittsburgh Press, 1973); Gilbert Chinard, "Eighteenth Century Theories of America as a Human Habitat," *Amer. Philos. Soc. Proc.,* Vol. XCI (1947), 27–57, hereafter cited as "Eighteenth Century Theories."

78. Jefferson, *Notes,* 48; he cited Buffon, *Histoire naturelle,* VIII, 134; Martin, *Thomas Jefferson: Scientist,* Chaps. 6–7.

79. Williams, *Vermont,* 105–107.

80. Benjamin Franklin, *Writings* (Memorial ed.), XVIII, 170 on the Paris party; "Letter to Mr. Nairne, of London, on Hygrometers," *Amer. Philos. Soc. Trans.,* o.s. II (1786), 51–56 on climate data. Both citations from Chinard, "Eighteenth Century Theories," *Amer. Philos. Soc. Proc.* Vol. XCI (1947), 43.

81. John Clayton, "Concerning the Air," in Berkeley and Berkeley (eds.), *Clayton,* 48.

82. Reprinted in *ibid.,* 80–81.

83. John Woodward, "Some Thoughts and Experiments Concerning Vegetation," *Phil. Trans. Roy. Soc. London,* Vol. XXI (1699), 193–227, especially 208–209. Woodward's speculations on the evolution of the environment were explained in *An Essay Towards a Natural History of the Earth* (1695), which has been discussed by Glacken, *The Rhodian Shore,* 409–11.

84. Kalm, *Travels,* II, 119, 127–30; III, 258.

85. Hugh Williamson, "An Attempt to Account for the Change of Climate, Which Has Been Observed in the Middle Colonies in North America," *Amer. Philos. Soc. Trans.,* Vol. I (1771), 272–80; Whitfield J. Bell, Jr., *Early American Science: Needs and Opportunities for Study* (Williamsburg, Institute of Early American History and Culture, 1955), 77–78, hereafter cited as *Early American Science.*

86. Jefferson, *Notes,* 80.

87. On the history of photosynthesis studies in the eighteenth century see Leonard K. Nash, *Plants and the Atmosphere,* Harvard Case Histories in Experimental Science, No. 5 (ed. by James B. Conant, Cambridge, Mass., Harvard University Press, 1966); Howard S. Reed, "Jan Ingenhousz, Plant Physiologist, with a History of the Discovery of Photosynthesis, *Chronica Botanica,* XI, Nos. 5–6 (1949).

88. Marquis François Jean de Chastellux, *Travels in North America in the Years 1780, 1781, and 1782* trans. by George Grieve, rev. and ed. by Howard C. Rice, Jr., 2 vols., Chapel Hill, University of North Carolina Press, 1963, II, 395–96.

89. Benjamin Rush, "An Enquiry into the Cause of the Increase of Bilious and Intermitting Fevers in Pennsylvania, with Hints for Preventing Them," *Amer. Philos. Soc. Trans.,* Vol. II (1786), 206–12; Thomas Wright, "On the Mode Most Easily and Effectually Practicable of Drying Up the Marshes of the Maritime Parts of North America," *Amer. Philos. Soc. Trans.,* Vol. IV (1797), 243–46; William Currie, "An Enquiry into the Course of the Insalubrity of Flat and Marshy Situations," *Amer. Philos. Soc. Trans.,* Vol. IV (1797), 127–42; Adam Seybert, "Experiments and Observations on the Atmosphere of Marshes," *Trans. Amer. Philos. Soc.,* Vol. IV (1797), 262–71; all cited from Gilbert Chinard, "The American Philosophical Society and the Early History of Forestry in America," *Amer. Philos. Soc. Proc.,* Vol. LXXXIX (1945), 444–88, especially 454–55.

90. Williams, *Vermont,* 50, 74–76, 60–61, 54.

91. Noah Webster, "On the Supposed Change of Temperature in Modern Winters," in *A Collection of Papers on Political, Literary and Moral Subjects* (New York, 1843; facsimile ed., New York, Burt Franklin, 1968), 119–62; Glacken, *The Rhodian Shore,* 560–61, 660–63.

92. Herodotus, *History,* Book III, Chaps. 108–109; Book II, Chap. 68; Plato, *Protagoras,* 320d–321b; Plato *Timaeus,* 30c–d; Carl Linnaeus, *Specimen Academicum de Oeconomia Naturae* (I. J. Biberg, respondent, Upsala, 1744); Frank N. Egerton, "Changing Concepts of the Balance of Nature," *Quart. Rev. Biol.,* Vol. XLVIII (1973), 322–50, hereafter cited as "The Balance of Nature."

93. Kalm, *Travels,* I, 291, 116–17, 143–45, 289–92, 343–44, 353–54; II, 59–60, 72, 196, 200; III, 296–97; "A Description of the Wild Pigeons," *Auk,* Vol. XXVIII (1911), 59.

94. Buffon, "The Wild Animals," in *Natural History,* IV, 73; Kalm, *Travels,* II, 59–60.

95. Kalm, *Travels,* II, 135.

96. *Ibid.,* I, 371–72.

97. *Ibid.,* II, 3.

98. Franklin's "Observations Concerning the Increase of Mankind, Peopling of Countries" was written in 1751 and first appeared as an appendix in William Clarke, *Observations on the Late and Present Conduct of the French, with Regard to Their Encroachments upon the British Colonies in North America* (Boston, 1755); reprinted in *The Writings of Benjamin Franklin* (ed. by Albert Henry Smyth, 10 vols., New York, Macmillan, 1905–1907), III, 63–73. See also Conway Zirkle, "Benjamin Franklin, Thomas

Malthus and the United States Census," *Isis,* Vol. XLVIII (1957), 58–62.

99. Kalm, *Travels,* I, 294.

100. Benjamin Franklin, *Works* (ed. by John Bigelow) (New York: G. P. Putnam's Sons, 1904), VI, 334.

101. Williams, *Vermont,* 110–11, 84ff.

102. Jefferson, *Notes,* 53–54; Williams, *Vermont,* 103.

103. Paul Russell Cutright, *Lewis and Clark: Pioneering Naturalists* (Urbana, University of Illinois, 1969), 56. The first edition of the report of the Lewis and Clark Expedition was edited by Nicholas Biddle. His text was later the basis for Elliott Coues (ed.), *History of the Expedition Under the Command of Lewis and Clark . . . A New Edition, . . . with Copious Critical Commentary . . .* (4 vols., New York, Francis P. Harper, 1893; reprinted, 3 vols., New York, Dover, 1965).

104. Humboldt's influence on Darwin is discussed in some detail in Frank N. Egerton, "Humboldt, Darwin, and Population," *J. Hist. Biology,* Vol. III (1970), 325–60.

105. Alexander von Humboldt and Aimé Bonpland, *Essai sur la Géographie des Plantes; accompagné d'un Tableau physique des Régions équinoxiales, Fondé sur des Mesures Exécutées, depuis le Dixième Degré de Latitude boréale jusqu'au Dixième Degré de Latitude australe, pendant les Années 1799, 1800, 1801, 1802 et 1803* (Paris, Levrault, Schoell, 1807), 14–18.

106. *Ibid.,* 154–55; Benjamin Smith Barton, "Specimen of a Geographical View of the Trees and Shrubs, and Many of the Herbaceous Plants of North America, Between the Latitudes of Seventy-one and Twenty-five" (a set of galley proofs is in the Yale University Library, New Haven, Conn.).

107. Charles Pickering, "On the Geographical Distribution of Plants," *Amer. Philos. Soc. Trans.,* n.s., Vol. III (1830), 274–84 and map.

108. Charles Pickering, *Geographical Distribution of Animals and Plants,* Vol. XIX in *U.S. Exploring Expedition* (Boston, Little, Brown, 1854); "On the Geographical Distribution of Species," *Amer. Acad. Arts and Sci. Proc.,* 4 (1859), 192–94; "Relative to the Geographical Distribution of Species," *ibid.,* 5 (1860), 81–82. On Pickering see Asa Gray, *Scientific Papers* (ed. by Charles S. Sargent, 2 vols., Boston, 1889), II, 406–10.

109. Henry R. Schoolcraft, *Narrative Journal of Travels, Through the Northwestern Regions of the United States Extending from Detroit Through the Great Chain of American Lakes, to the Sources of the Mississippi River . . .* (Albany, N.Y., E. & E. Hosford, 1821), 71 on phenology, 161 on the Huron plain. For a guide to the literature of his and other American expeditions see Max Meisel, *A Bibliography of American Natural History: the Pioneer Century, 1769–1865* (3 vols., New York, Premier, 1924–29; reprinted, New York, Hafner, 1967).

110. Chase S. Osborn and Stellanova Osborn, *Schoolcraft—Longfellow—Hiawatha* (Lancaster, Pa., Jacques Cattell, 1942), contains Schoolcraft's biography and bibliography.

111. Louis Agassiz, *Lake Superior: Its Physical Character, Vegetation, and Animals, Compared with Those of Other and Similar Regions . . .* (Boston, Gould, Kendall and Lincoln, 1850), 153; Frank N. Egerton, "Louis Agassiz and Biogeography," to be published in a volume commemorating the centennial of Woods Hole Marine Biological Laboratory (ed. by Harold L. Burstyn); Edward Lurie, *Louis Agassiz: A Life in Science* (Chicago, University of Chicago Press, 1960).

112. Asa Gray, review of Philip Franz von Siebold, *Flora Japonica,* in *Amer. J. Sci.,* Vol. XXXIX (1840), 175–76; "Analogy Between the Flora of Japan and That of the United States," *Amer. J. Sci.,* n.s., Vol. II (1846), 135–36; "Diagnostic Characters of New Species of Phaenogamous Plants, Collected in Japan by Charles Wright, Botanist of the United States North Pacific Exploring Expedition. With Observations upon the Relations of the Japanese Flora to That of North America, and of Other Parts of the Northern Temperate Zone," *Mem. Amer. Acad. Arts and Sci.,* Vol. VI (1859), 377–452; Asa Gray and Joseph Dalton Hooker, "The Vegetation of the Rocky Mountain Region and a Comparison with That of Other Parts of the World," *Bull. U.S. Geol. and Geogr. Survey of the Territories,* Vol. VI (1881), 1–77; Asa Gray, *Darwiniana* (1st ed., 1876; reprinted, Cambridge, Mass., Harvard University Press, 1963), Art. 4 and 5. A. Hunter Dupree, *Asa Gray, 1810–1888* (Cambridge, Mass., Harvard University Press, 1959).

113. Leviticus 13–14; Lucretius 6. 1090–1137; Varro 1. 12; Vitruvius 1. 4. 11. Columella 1. 4; Palmer Howard Futcher, "Notes on Insect Contagion," *Bull. Hist. Medicine,* Vol. IV (1936), 536–58, esp. 544–45; Winslow, *Conquest,* Chap. 5.

114. Jerome Cardan, *De Rerum Varietate* (Basel, 1557); Charles Singer and Dorothea Singer, "The Development of the Doctrine of Contagium Vivum, 1500–1750," *XVIIth International Medical Congress, Section XXIII* (London, 1914), 187–207; Winslow, *Conquest,* Chaps. 7–8; Giovanni Maria Lancisi, "On the Noxious Exhalations of Marshes" (trans. by Samuel Latham Mitchill), *Medical Repository,* Vol. XIII (1809–10), 9–18, 126–35, 237–45, 326–30; n.s., Vol. IV (1818), 201–12, 322–32, 442–67. More recently William C. Gorgas and Fielding H. Garrison have translated the most significant passages in "Ronald Ross and the Prevention of Malarial Fever," *Sci. Monthly,* Vol. III (1916), 133–50, esp. 135–36. Their translation is also quoted by Futcher, "Notes on Insect Contagion," *Bull. Hist. Med.* Vol. IV (1936), 547–48.

115. John Crawford published "Remarks on Quarantine" and "Dr. Crawford's Theory and an Application of It to the Treatment of Disease," *Observer,* Vols. I–II (1807), and "Observations on the Seats and Causes of Disease," *Baltimore Medical and Physical Recorder,* Vol. I (1809), 40–52, 81–92, 206–21. See also Raymond N. Doetsch, "John Crawford and His Contribution to the Doctrine of *Contagium Vivum,*" *Bacteriological Reviews,* Vol. XXVIII (1964), 87–96.

116. Daniel Drake, *A Practical Treatise on the History, Prevention, and Treatment of Epidemic Cholera, Designed for Both the Profession and the People* (Cincinnati, Corey & Fairbank, 1832); Raymond N. Doetsch, "Daniel Drake's Aetiological Views," *Medical History,* Vol. IX (1965), 365–73.

117. Josiah Clark Nott, "Yellow Fever Contrasted with Bilious Fever—Reasons for Believing It a Disease Sui Generis—Its Mode of Propagation—Remote Cause—Probable Insect or Animalcular Origin," *New Orleans Med. Surg. J.,* Vol. IV (1848), 563–601; Josiah Clark Nott, "Report on Yellow Fever," *Memorial Record,* Vol. VI (1871), 451–59; Emmett B. Carmichael, "Josiah Clark Nott," *Bull. Hist. Medicine,* 22 (1948), 249–62 and portrait; William Leland Holt, "Josiah Clark Nott of Mobile, an American Prophet of Scientific Method" [based on researches of Bayard Holmes], *Medical Life,* Vol. XXXV (1928), 487–504; Robert Wilson, "Dr. J. C. Nott and the Transmission of Yellow Fever," *Annals of Medical Hist.,* n.s. Vol. III (1931), 515–20.

118. John Kearsley Mitchell, *On the Cryptogamous Origin of Malarious and Epidemic Fevers* (Philadelphia: Lea & Blanchard, 1849), reprinted in his *Five Essays* (ed. by S. Weir Mitchell, Philadelphia, J. P. Lippincott, 1859), 13–140; Raymond N. Doetsch, "Mitchell on the Cause of Fevers," *Bull. Hist. Med.,* Vol. XXXVIII (1964), 241–59.

119. Erwin H. Ackerknecht, "Anticontagionism Between 1821 and 1867," *Bull. Hist. Med.,* Vol. XXII (1948), 562–93. Phyllis Allen [later Richmond], "Etiological Theory in America Prior to the Civil War," *J. Hist. Med. and Allied Sci.,* Vol. II (1947), 489–520; Phyllis Allen [Richmond], "Some Variant Theories in

Opposition to the Germ Theory of Disease," *J. Hist. Med. and Allied Sci.*, Vol. IX (1954), 290–303; Phyllis Allen [Richmond], "American Attitudes Toward the Germ Theory of Disease (1860–1880)," *J. Hist. Med. and Allied Sci.*, Vol. IX (1954), 428–54; Charles Rosenberg, "The Causes of Cholera: Aspects of Etiological Thought in Nineteenth Century America," *Bull. Hist. Med.*, Vol. XXXIV (1960), 331–54.

120. Constantin François Chasseboeuf, Comte de Volney, *A View of the Soil and Climate of the United States of America . . .* trans. by C. B. Brown, Philadelphia, J. Conrad, 1804), Chap. 10 (facsimile ed., Intro. by George W. White, New York, Hafner, 1968), hereafter cited as *Soil and Climate.*

121. Joseph Lovell, *Regulations for the Medical Department, by Order D. Parker, Adj. & Insp. Gen.* (September, 1818, 31 pp.); Quoted from Edgar Erskine Hume, "The Foundations of American Meteorology by the United States Army," *Bull. Inst. Hist. Med.*, Vol. VIII (1940), 202–38, esp. 206, 208, hereafter cited as "Foundations."

122. *Meteorological Register for the Years 1822, 1823, 1824, & 1825 from Observations Made by the Surgeons of the Army at the Military Posts of the United States* (prepared under the direction of Joseph Lovell, M.D., Surgeon General of the United States Army, Washington, D.C., Edward de Krafft, 1826).

123. *Meteorological Register for the Years 1826, 1827, 1828, 1829, and 1830; from Observations Made by the Surgeons of the Army and Others at the Military Posts of the United States* (prepared under the direction of Thomas Lawson, M.D., Surgeon General of the United Sates Army, Philadelphia, Haswell, Barrington, and Haswell, 1840); Samuel Forry, "Statistical Researches Elucidating the Climate of the United States and Its Relation with Diseases of Malarial Origin; Based on the Records of the Medical Department and Adjutant General's Office," *Amer. J. Med. Sci.*, n.s., Vol. I (1841), 13–46; Samuel Forry, *The Climate of the United States and Its Endemic Influences: Based Chiefly on the Records of the Medical Department and Adjutant General's Office, United States Army* (New York, J. & H. G. Langley, 1842), 244, cites "Cowan's Additions to Louis on Phthises."

124. Hume, "Foundations," *Bull. Inst. Hist. Med.*, Vol. VIII (1940), 214–38. Donald R. Whitnah, *A History of the United States Weather Bureau* (Urbana, University of Illinois Press, 1961), Chaps. 1–2.

125. Published in *Philadelphia Medical and Physical J.*, Vol. III (1808), 85–90; reprinted by Henry D. Shapiro and Zane L. Miller (eds.), *Physician to the West: Selected Writings of Daniel Drake on Science & Society* (Lexington, University of Kentucky Press, 1970), 1–4, hereafter cited as *Physician*; Emmet Field Horine, *Daniel Drake (1785–1852): Pioneer Physician of the Midwest* (Philadelphia, University of Pennsylvania Press, 1961).

126. Shapiro and Miller, *Physician*, p. 5. They reprinted the entire *Notices*, pp. 6–56.

127. Selections are reprinted by Shapiro and Miller (eds.), *Physician*, 66–124. Adolph E. Waller, because of his unfamiliarity with the history of ecology before Drake, has overstated Drake's originality in this book ("Daniel Drake as a Pioneer in Modern Ecology," *Ohio State Archeological and Historical Quarterly*, Vol. LVI [1947], 262–73).

128. Drake's cholera book is cited above, n. 116. Drake's evidence for *contagium vivum* [in *A Systematic Treatise, Historical, Etiological, and Practical, on the Principal Diseases of the Interior Valley of North America, as They Appear in the Caucasian, African, Indian, and Esquimaux Varieties of Its Population* (2 vols., 1850–54), hereafter cited as *A Systematic Treatise*] occurs in Book 2, Part I, Chap. II, which, confusingly, is included near the end of Vol. I and also near the beginning of Vol. II. This chapter is included in the extracts from *A Systematic Treatise* in Shapiro and Miller (eds.), *Physician*, 366–79. The chapter is also reprinted by Norman D. Levine (ed.), *Malaria in the Interior Valley of North America: a Selection from a Systematic Treatise . . . by Daniel Drake* (Urbana, University of Illinois Press, 1964), hereafter cited as *Malaria.*

129. Drake, *A Systematic Treatise*, I, 455–59 on weather, 623–36 on distributions of trees, mammals, birds (these pages are not reprinted by either Shapiro and Miller [eds.], *Physician*, or Levine [ed.], *Malaria*).

130. *Ibid.*, 341.

131. For example: Manning Simons, "Report on the Climatology and Epidemics of South Carolina," *Amer. Med. Assoc. Trans.*, Vol. XXIII (1872), 275–331; Billy M. Jones, *Health Seekers in the Southwest, 1817–1900* (Norman, University of Oklahoma Press, 1967); John E. Baur, *The Health Seekers of Southern California, 1870–1900* (San Marino, Calif., Huntington Library, 1959).

132. Benjamin Smith Barton, *Elements of Botany: or Outlines of the Natural History of Vegetables* (3 parts, Philadelphia, Author, 1803), Part I, 300; Francis W. Pennell, "Benjamin Smith Barton as Naturalist," *Amer. Philos. Soc. Proc.*, Vol. LXXXVI (1942), 108–22; William Martin Smallwood and Mabel Sarah Coon Smallwood, *Natural History and the American Mind* (New York, Columbia University Press, 1941), 289–93. See also above, n. 72.

133. Jacob Bigelow, *Florula Bostoniensis; a Collection of Plants of Boston and Its Environs, with Their Generic and Specific Characters, Synonyms, Descriptions, Places of Birth, and Time of Flowering* (Boston, 1814, 268 pp.; 2d ed., 1824, 423 pp.; 3d ed., 1840, 468 pp.), hereafter cited as *Florula Bostoniensis*; Volney, *Soil and Climate*, 120–21. On Muhlenberg see Bell, *Early American Science*, 67–68; see also Constantine S. Rafinesque, "A Journal of the Progress of Vegetation near Philadelphia, Between the 20th of February and the 20th of May, 1816, with Occasional Zoological Remarks," *Amer. J. Sci.*, Vol. I (1818), 77–82.

134. *Amer. Acad. Arts and Sci. Mem.*, Vol. IV (1818), 77–85; George E. Ellis, "Memoir of Jacob Bigelow, M.D., L.L.D.," *Mass. Hist. Soc. Proc.*, Vol. XVII (1880), 383–467 and portrait.

135. Jacob Porter, "Floral Calendar for Plainfield, Massachusetts, 1818," *Amer. J. Sci.*, Vol. I (1818), 254–55; Stephen W. Williams, "Floral Calendar Kept at Deerfield, Massachusetts, with Miscellanious Remarks," *Amer. J. Sci.*, Vol. I (1819), 359–73. Jefferson to Bigelow, April 11, 1818, in Jefferson, *Garden Book* (ed. by Betts), 578–79.

136. Quoted from Alfred J. Henry, "Early Individual Observers in the United States," *U.S. Dept. Agriculture, Weather Bureau Bull.*, Vol. XI (1895), 291–302, esp. 300; William M. Meigs, *Life of Josiah Meigs, by His Great Grandson* (Philadelphia, J. P. Murphy, 1887).

137. Franklin B. Hough, *Historical and Statistical Record of the University of the State of New York During the Century from 1784 to 1884* (Albany, N.Y., Weed, Parsons, 1885), 767; Leo Stoller, "A Note on Thoreau's Place in the History of Phenology," *Isis*, Vol. XLVII (1956), 172–81, esp. 177, hereafter cited as "Thoreau's Place."

138. Franklin B. Hough, *Results of a Series of Meteorological Observations Made in Obedience to Instructions from the Regents of the University at Sundry Academies in the State of New York, from 1826 to 1850 Inclusive, Compiled from the Original Returns of the Annual Reports of the Regents of the University* (Albany, N.Y., Weed, Parsons, 1855); reviewed in *Amer. J. Sci.*, Vol. XXI (1856), 149.

139. [Joseph Henry], "Smithsonian Institution—Registry of Periodical Phenomena," *Amer. J. Sci.*, Vol. XII (1851), 293–95; Franklin B. Hough, "Observations upon Periodical Phenomena

in Plants and Animals from 1851 to 1859, with Tables of Opening and Closing of Lakes, Rivers, Harbors, &c.," 36th Cong., 1st Sess., *House Ex. Doc. 55* (Washington, D.C., 1864); Romeyn B. Hough on Hough in Howard A. Kelly and Walter L. Burrage, *American Medical Biographies* (Baltimore, Norman, Remington, 1920), 562–63; Edna L. Jacobsen, "Franklin B. Hough, a Pioneer in Scientific Forestry in America," *New York History*, Vol. XV (1934), 310–25.

140. Charles Peirce, *A Meteorological Account of the Weather in Philadelphia, from January 1, 1790, to January 1, 1847, Including Fifty-Seven Years; with an Appendix, Containing a Great Variety of Interesting Information* . . . (Philadelphia, 1847).

141. Aldo Leopold and Sara Elizabeth Jones, "A Phenological Record of Sauk and Dane Counties, Wisconsin, 1935–1945," *Ecological Monographs*, Vol. XVII (1947), 81–122, esp. 83. Their claim has been repeated often, as by Philip Whitford and Kathryn Whitford in "Thoreau: Pioneer Ecologist and Conservationist," *Scientific Monthly*, Vol. LXXIII (1951), 291–96, esp. 292.

142. Nina Baym, "Thoreau's View of Science," *J. Hist. Ideas*, Vol. XXVI (1965), 221–34, esp. 224, 226.

143. *Ibid.*, 225. On Howitt see George C. Boase in *Dict. Nat. Biog.*; Wendell Glick, "Three New Early Manuscripts by Thoreau," *Huntington Lib. Quart.*, Vol. XV (1951), 59–71, esp. 61–68.

144. Henry David Thoreau, "Natural History of Massachusetts," *Dial*, Vol. III (July, 1842), 19–40, esp. 23, 40.

145. Various memoranda are in Henry David Thoreau, *The Journals* (ed. by Bradford Torrey and Francis H. Allen, 14 vols., Boston, Houghton Mifflin, 1906, New York, Dover, 1962), III, 364, 377, 438; IV, 8, 15, 66; IX, 158. These memoranda and the chart (in Henry and Albert A. Berg Collection, New York Public Library) are discussed by Stoller in "Thoreau's Place," *Isis*, Vol. XLVII (1956), 173–74.

146. Baym, "Thoreau's View of Science," *J. Hist. Ideas*, Vol. XXVI (1965), 227.

147. On Bartram's influences upon literary authors, see Lowes, *Road to Xanadu*, and Fagin, *William Bartram*, Part 3; Joseph Ewan (ed.), *William Bartram: Botanical and Zoological Drawings, 1756–1788, Reproduced from the Fothergill Album in the British Museum (Natural History)*, (Philadelphia, American Philosophical Society, 1968).

148. Hans Huth, *Nature and the American: Three Centuries of Changing Attitudes* (Berkeley and Los Angeles: University of California Press, 1957); Robert Henry Welker, *Birds and Men: American Birds in Science, Art, Literature, and Conservation, 1800–1900* (Cambridge, Mass., Harvard University Press, 1955); Philip Marshall Hicks, *The Development of the Natural History Essay in American Literature* (Philadelphia, 1924); Arthur A. Ekirch, Jr., *Man and Nature in America* (New York, Columbia University, 1963); Roderick Nash, *Wilderness and the American Mind* (New Haven, Yale University, 1967); *The American Environment: Readings in the History of Conservation* (Reading, Mass., Addison-Wesley, 1968); Donald Fleming, "Roots of the New Conservation Movement," *Perspectives in Amer. Hist.*, Vol. VI (1972), 7–91; Frank Graham, Jr., *Man's Dominion: The Story of Conservation in America* (New York, M. Evans, 1971); Henry Clepper (ed.), *Origins of American Conservation* (New York, Ronald Press, 1966); Henry Clepper (ed.), *Leaders of American Conservation* (New York, Ronald Press, 1971).

149. Van Wyck Brooks, *The World of Washington Irving* (New York, E. P. Dutton, 1944), 87. Alexander Wilson, "The Foresters, Descriptive of a Pedestrian Journey to the Falls of Niagara, in the Autumn of 1804," *The Port Folio*, n.s., Vols. I–III (1809–10), and four later separate editions; *The Poems and Literary Prose . . .*, (ed. by Alexander B. Grosart, 2 vols., Paisley, Scotland, 1876); Robert Cantwell, *Alexander Wilson: Naturalist and Pioneer* (Philadelphia, J. B. Lippincott, 1961); Emerson Stringham, *Alexander Wilson: A Founder of Scientific Ornithology* (Kerrville, Texas, Author, 1958); James Southall Wilson, *Alexander Wilson, Poet-Naturalist: A Study of His Life with Selected Poems* (New York, Neale, 1906); Gordon Wilson, "Alexander Wilson, Poet-Essayist-Ornithologist" (unpublished Ph.D. dissertation, Indiana University, 1930).

150. Alexander Wilson, *American Ornithology; or, the Natural History of the Birds of the United States* . . . (9 vols., Philadelphia, Bradford and Inskeep, 1808–14), II, 72, on kingbird; I, 27, on Baltimore oriole and 59–60 on blue bird; II, 29, on hummingbird and 79 on wood pewee; IV, 96, on bald eagle; V, 24–26 on Osprey. On his bird census, see *ibid.*, IV, v–x, and Frank L. Burns, "Alexander Wilson in Bird Census Work," *Wilson Bulletin*, Vol. XIX (1907), 100–102.

151. Ralph Waldo Emerson, *The Early Lectures, 1833–1836* (ed. by Stephen E. Whicher and Robert E. Spiller, Cambridge, Mass., Harvard University Press, 1959), 10; Ralph Waldo Emerson, *Journals* (ed. by Edward W. Emerson and Waldo E. Forbes, Boston, Houghton Mifflin, 1909–14), V, 285.

152. Thoreau's readings are well documented by John A. Christie, *Thoreau as World Traveler* (New York, Columbia University Press, 1965); Walter Harding, *Thoreau's Library* (Charlottesville, University of Virginia Press, 1957); Kenneth W. Cameron (ed.), *Thoreau's Fact Book*, 2 vols. (Hartford, Conn., Transcendental Books, 1966).

153. Baym, "Thoreau's View of Science," *J. Hist. Ideas*, Vol. XXVI (1965), 221–34; Raymond Adams, "Thoreau's Science," *Scientific Monthly*, Vol. LX (1945), 379–82; Charles R. Metzger, "Thoreau on Science," *Annals of Sci.*, Vol. XII (1956), 206–11; Whitford and Whitford, "Thoreau: Pioneer Ecologist and Conservationist," *Scientific Monthly*, Vol. LXXIII (1951), 291–96; James McIntosh, *Thoreau as Romantic Naturalist: His Shifting Stance Toward Nature* (Ithaca, N.Y., Cornell University Press, 1974).

154. Henry David Thoreau, *A Week on the Concord and Merrimack Rivers* (1849), cited from 1906 edition (Boston, Houghton Mifflin, 1906), 29.

155. Edward S. Deevey, Jr., "A Re-examination of Thoreau's *Walden*," *Quart. Rev. Biol.*, 17 (1942), 1–11.

156. Henry David Thoreau, "The Succession of Forest Trees," *Excursions* (1863, cited from reprint, Boston, Houghton, Mifflin, 1893), 225–50. Thoreau read this essay before the Middlesex Agricultural Society in Concord in September, 1860; Kalm, *Travels*, I, 75, 162, 312; Wilson, *American Ornithology*, I, 17; Kathryn Whitford, "Thoreau and the Woodlots of Concord," *New England Quart.*, Vol. XXIII (1950), 291–306; Leo Stoller, *After Walden: Thoreau's Changing Views on Economic Man* (Stanford, Calif., Stanford University Press, 1957), 72–88. For a survey of studies on plant succession before 1860 see Frederic E. Clements, *Plant Succession: An Analysis of the Development of Vegetation* (Washington, D.C., Carnegie Institution, 1916), 8–17.

157. Thoreau, *The Journals*, XIV, 94–95, 125–27, 130–33, 145–46, 203–205, 211.

158. David Lowenthal, *George Perkins Marsh: Versatile Vermonter* (New York, Columbia University Press, 1958), 251–53, hereafter cited as *Marsh*; David Lowenthal (ed.), "Introduction," xvii–xviii, in George Perkins Marsh, *Man and Nature; or, Physical Geography as Modified by Human Action* (ed. by David Lowenthal, Cambridge, Mass., Harvard University Press, 1965); *ibid.*, 102–105, hereafter cited as *Man and Nature*.

159. Lewis Mumford, *The Brown Decades: A Study of the Arts in America, 1865–1895* (New York, 1931), 78; Lowenthal,

Marsh, 246.

160. Lowenthal (ed.), "Introduction," in Marsh, *Man and Nature*, xiii.

161. Marsh, *Man and Nature*, 53.

162. *Ibid.*, 247; Darwin, *On the Origin of Species by Means of Natural Selection, or the Preservation of Favoured Races in the Struggle for Life* (London, John Murray, 1859), Chaps. 3–4, 10; Egerton, "The Balance of Nature," *Quart. Rev. Biol.*, Vol. XLVIII (1973), 341–42.

163. Marsh, *Man and Nature*, 36–37, 76–78, 82–87, 99–102, 105–108.

164. Lowenthal (ed.), "Introduction," xxi–xxiii, xxvii–xxviii, in *ibid.*; George Perkins Marsh, *The Earth as Modified by Human Action* (New York, Scribner, Armstrong, 1874; facsimile ed., New York, Arno, 1970).

165. Letters of E. P. Evans to Marsh, November 24, 1870, January 20, 1871, letter of Marsh to Evans, December 5, 1870, Marsh Collection, University of Vermont. Lowenthal (ed.), n. 51, in Marsh, *Man and Nature*, 49–50.

166. Lowenthal (ed.), "Introduction," xxii, in *ibid.*; Clepper (ed.), *Leaders of American Conservation*, 172–73, 217–18; Herbert A. Smith, "The Early Forestry Movement in the United States," *Agricultural History*, Vol. XII (1938), 326–46, esp. 326–31.

167. On Forbes see below. On the Division of Economic Ornithology and Mammalogy see Jenks Cameron, *The Bureau of Biological Survey: Its History, Activities and Organization* (Baltimore: Johns Hopkins University Press, 1929; New York: Arno, 1974), hereafter cited as *Biological Survey*; Keir B. Sterling, *Last of the Naturalists: The Career of C. Hart Merriam* (New York, Arno, 1974), Chaps. 3–4, 7.

168. A. Starker Leopold, "The Conservation of Wildlife," in *A Century of Progress in the Natural Sciences, 1853–1953* (ed. by Edward L. Kessel, San Francisco, California Academy of Science, 1955), 795–807; Norman G. Benson (ed.), *A Century of Fisheries in North America* (Washington, D.C., American Fisheries Society, 1970); Frank G. Roe, *The North American Buffalo: A Critical Study of the Species in Its Wild State* (2d ed., Toronto, University of Toronto Press, 1970); A. W. Schorger, *The Passenger Pigeon* (Madison, University of Wisconsin Press, 1955); James C. Greenway, Jr., *Extinct and Vanishing Birds of the World* (2d ed., New York, Dover, 1967), 35–48, 304–11, 322–27.

169. James S. Lippincott, "Geography of Plants," *Agricultural Report of the United States Commissioner of Patents for 1863* (Washington, D.C., Government Printing Office, 1864), 464–525.

170. John Wesley Powell, *Report on the Lands of the Arid Region of the United States, with a More Detailed Account of the Lands of Utah* (ed. by Wallace Stegner, Cambridge, Mass., Harvard University Press, 1962), 25, hereafter cited as *Report*; William Culp Darrah, *Powell of the Colorado* (Princeton, Princeton University Press, 1951); Wallace Stegner, *Beyond the Hundredth Meridian: John Wesley Powell and the Second Opening of the West* (Boston, Houghton Mifflin, 1954).

171. Caleb Atwater, "On the Prairies and Barrens of the West," *Amer. J. Sci.*, Vol. I (1818), 116–25; R. W. Wells, "On the Origin of Prairies," *Amer. J. Sci.*, Vol. I (1819), 331–37.

172. Powell, *Report*, 30–31.

173. *Ibid.*, 114.

174. Stegner (ed.), "Editor's Introduction," xxiv, in Powell, *Report.*

175. F. S. Bodenheimer, *Materialien zur Geschichte der Entomologie* (2 vols., Berlin, W. Junk, 1928–29); Leland O. Howard, *A History of Applied Entomology (Somewhat Anecdotal)* (Smithsonian Miscellaneous Collection, Vol. LXXXIV, 1930), 207–12; Harry B. Weiss, "Thomas Jefferson and Economic Entomology," *J. Economic Entomology*, Vol. XXXVII (1944), 836–41.

176. John A. Stevenson, "The Beginnings of Plant Pathology in North America," in *Plant Pathology: Problems and Progress, 1908–1958* (ed. by C. S. Holton, G. W. Fischer, R. W. Fulton, Helen Hart, and S. E. A. McCallan, Madison, University of Wisconsin Press, 1959), 14–23. Despite the limited scope indicated by the title of the volume, Stevenson surveys the whole history. E. C. Large, *The Advance of the Fungi* (London, Jonathan Cape, 1940; New York, Dover, 1962).

177. Kalm, *Travels*, I, 177; II, 46–48; III, 60.

178. Cameron, *Biological Survey*, 12–13.

179. Fitch, "Sixth Report on the Noxious and Other Insects of the State of New York," *New York State Agricultural Society Trans.*, Vol. XX (for 1860, pub. 1861), 745–868, esp. 824; Richard L. Doutt, "The Historical Development of Biological Control," see *Biological Control of Insect Pests and Weeds* (ed. by Paul DeBach, New York, Reinhold, 1964), 26–27, hereafter cited as "Control"; Arnold Mallis, *American Entomologists* (New Brunswick, N.J., Rutgers University Press, 1971), 37–43 et passim; Samuel Rezneck on Fitch, *Dictionary of Scientific Biography* (1972), V, 11–12.

180. Walsh, "Imported Insects; the Gooseberry Sawfly," *Practical Entomologist*, I (September 29, 1866), 117–24; Doutt, "Control," 27–31; Mallis, *American Entomologists*, 43–48 et passim.

181. Large, *Advance of the Fungi*, Chap. 11; Mallis, *American Entomologists*, 69–79 et passim; Alpheus S. Packard, "Charles Valentine Riley," *Science*, n.s., Vol. II (1895), 745–51; G. B. Goode, "Charles Valentine Riley (1843–95)," *Science*, n.s., Vol. III (1896), 217–25.

182. Riley, "Parasitic and Predaceous Insects in Applied Entomology," *Insect Life*, Vol. VI (1893), 130–41.

183. Doutt, "Control," 31–38.

184. Edward A. Steinhaus, "Microbial Control—The Emergence of an Idea: A Brief History of Insect Pathology Through the Nineteenth Century," *Hilgardia*, Vol. XXVI (1956), 107–60.

185. Ernst Haeckel, *Generelle Morphologie der Organisme. Allgemeine Grundzüge der organischen Formen-Wissenschaft, mechanisch begründet durch die von Charles Darwin reformirte Descendenz-Theorie* (2 vols., Berlin, Reimer, 1866), II, 286–87; trans. in Robert C. Stauffer, "Haeckel, Darwin, and Ecology," *Quart. Rev. Biol.*, Vol. XXXII (1957), 138–44, esp. 140; Georg Uschmann on Haeckel, *Dictionary of Scientific Biography* (1972), VI, 6–11.

186. Walter Harding, "Thoreau and 'Ecology': Correction," *Science*, Vol. CXLIX (1965), 707; J. S. Burdon-Sanderson, "Biology in Relation to Other Natural Sciences," *Nature*, Vol. XLVIII (1893), 464–72, reprinted in *Smithsonian Institution Annual Report for 1893* (Washington, D.C., 1894), 435–63, esp. 439; Stephen A. Forbes, "On Contagious Disease in the Chinch-Bug (*Blissus leucopterus* Say)," *Illinois Dept. Agriculture Trans.*, Vol. XXXII (1896), 16–176, esp. 16–18.

187. Edward Forbes, "Report on the Molluscs and Radiata of the Aegean Sea, and on Their Distribution Considered as Bearing on Geology," *British Assoc. Advancement of Sci. Report*, Vol. XIII (1843), 130–93, esp. 173; Edward Forbes, "On the Light Thrown on Geology by Submarine Researches," *Edinburgh New Philos. J.*, Vol. XXXVI (1844), 318–27; Frank N. Egerton on Forbes in *Dictionary of Scientific Biography* (1972), V, 66–68; Karl August Möbius, *Die Auster und die Austernwirtschaft* (Berlin, 1877); trans. by H. J. Rice, "The Oyster and Oyster-Culture," *U.S. Commission of Fish and Fisheries Report for 1880* (Washington, D.C., 1880), 683–751; Hans Querner on Möbius, *Dictionary of Scien-*

tific Biography, IX (1974), 431–432; John Lussenhop, "Victor Hensen and the Development of Sampling Methods in Ecology," *J. Hist. Biology*, Vol. VII (1974), 319–37, hereafter cited as "Hensen"; Stauffer, "Haeckel, Darwin, and Ecology," *Quart. Rev. Biol.*, Vol. XXXII (1957), 141–42.

188. Edward Lurie, *Nature and the American Mind: Louis Agassiz and the Culture of Science* (New York, Science History, 1974), 37ff.; Ralph W. Dexter, "From Penikese to the Marine Biological Laboratory at Woods Hole—the Role of Agassiz's Students," *Essex Institute Historical Collections*, Vol. CX (1974), 151–61; Frank R. Lillie, *The Woods Hole Marine Biological Laboratory* (Chicago, University of Chicago Press, 1944), Chaps. 2–3; Paul S. Galtsoff, *The Story of the Bureau of Commercial Fisheries Biological Laboratory, Woods Hole, Massachusetts* (U.S. Fish and Wildlife Service, Bureau of Commercial Fisheries, Circular 415, 1962); Donald J. Zinn, "Study of Marine Life," *Dictionary of American History* (new ed., New York, Charles Scribner's Sons, 1976).

189. Kurt Lampert, *Das Leben der Binnengewässer* (2d ed., Leipzig, Tauchnitz, 1910), 13; Frank N. Egerton, "The Scientific Contributions of François Alphonse Forel, the Founder of Limnology," *Schweizerische Zeitschrift für Hydrologie*, Vol. XXIV (1962), 181–99; W. C. Allee, "Ecological Background and Growth Before 1900," in W. C. Allee, Alfred E. Emerson, Orlando Park, Thomas Park, and Karl P. Schmidt, *Principles of Animal Ecology* (Philadelphia, W. B. Saunders, 1949), 13–43, esp. 40–42.

190. Stephen A. Forbes and Robert E. Richardson, *The Fishes of Illinois* (Urbana, Illinois State Laboratory of Natural History, 1908, 2d ed., 1920); Leland O. Howard, "Stephen Alfred Forbes, 1844–1930," *National Academy of Sciences Biographical Memoirs,* Vol. XIV (1932), 3–25; H. C. Oesterling, "Bibliography of Stephen Alfred Forbes," *National Academy of Sciences Biographical Memoirs*, Vol. XIV (1932), 26–54; Harlow B. Mills, "Stephen Alfred Forbes," *Systematic Zoology,* Vol. XIII (1964), 208–14; Harlow B. Mills, "From 1858 to 1958," *Illinois Natural History Survey Bull.* (special issue, *A Century of Biological Research)*, Vol. XXVII (1958), 85–103, esp. 94–98; Thomas G. Scott, "Wildlife Research," *Illinois Natural History Survey Bull.*, Vol. XXVII (1958), 179–201, esp. 183–84; Mary P. Winsor on Forbes, *Dictionary of Scientific Biography* (1972), V, 69–71.

191. Stephen A. Forbes, "The Lake as a Microcosm," *Peoria Scientific Assoc. Bull.*, 1887, 77–87; Egerton, Balance of Nature," *Quart. Rev. Biol.*, Vol. XLVIII (1973), 342–43.

192. Gerald E. Gunning, "Illinois," in *Limnology in North America* (ed. by David G. Frey, Madison, University of Wisconsin Press, 1966), 163–89, esp. 166–71, hereafter cited as *Limnology*; George W. Bennett, "Aquatic Biology," *Illinois Nat. Hist. Survey Bull.*, Vol. XXVII (1958), 163–78, esp. 163–68; Pierce C. Mullen on Kofoid in *Dictionary of Scientific Biography* (1973) VII, 447. Lussenhop, "Hensen," *J. Hist. Biol.*, Vol. VII (1974), 334.

193. G. C. Sellery, *E. A. Birge, a Memoir*, with "An Appraisal of Birge the Limnologist, an Explorer of Lakes," by C. H. Mortimer (Madison, University of Wisconsin Press, 1956); Florence W. Marsh, "Professor C. Dwight Marsh and His Investigation of Lakes," *Wisconsin Acad. Sci. Arts Letters Trans.*, Vol. XXXI (1938), 535–43; David G. Frey, "Wisconsin: the Birge-Juday Era," in Frey, *Limnology*, 3–54; David C. Chandler, "Michigan," in *ibid.*, 95–115, esp. 98–99; Shelby D. Gerking, "Central States," in *ibid.*, 239–68, esp. 239.

194. Auguste Pyramus de Candolle, "Géographie botanique," *Dictionnaire des science naturelles* (ed. by F. G. Levrault (1820), XVIII, 359–422. William J. Hooker published a modified translation, "Geography Considered in Relation to the Distribution of Plants," in Hugh Murray, *An Encyclopaedia of Geography: Comprising a Complete Description of the Earth . . .* (London, Longman, 1834), 227–46. See also A. G. Tansley, "The Early History of Modern Plant Ecology in Britain," *J. Ecology,* Vol. XXXV (1947), 130–37, esp. 130; Howard S. Reed, "A Brief History of Ecological Work in Botany," *Plant World*, Vol. VIII (1905), 163–70, 198–208, esp. 198–201.

195. Andrew Denny Rodgers III, *John Merle Coulter: Missionary in Science* (Princeton, Princeton University Press, 1944); William Trelease, "John Merle Coulter, 1851–1928," *Nat. Acad. Sci. Biog. Mem.*, Vol. XIV (1932), 97–108 and portrait, J. C. Arthur, "Bibliography of John Merle Coulter," *Nat. Acad. Sci. Biog. Mem.*, Vol. XIV (1932), 109–23.

196. Raymond J. Pool, "A Brief Sketch of the Life and Work of Charles Edwin Bessey," *Amer. J. Botany*, Vol. II (1915), 505–18, and portrait; Joseph Ewan on Bessey, *Dictionary of Scientific Biography* (1970), II, 102–104; Richard A. Overfield, "Charles E. Bessey: The Impact of the 'New' Botany on American Agriculture, 1880–1910," *Technology and Culture,* Vol. XVI (1975), 162–81.

197. Conway MacMillan, *The Metaspermae of the Minnesota Valley. A List of the Higher Seed-Producing Plants Indigenous to the Drainage-Basin of the Minnesota River* (Minneapolis, Harrison and Smith, 1892); Conway MacMillan, "Observations on the Distribution of Plants Along Shore at Lake of the Woods," *Minnesota Botanical Studies*, Vol. I (1897), 949–1023 and pls. 70–81; Harry Baker Humphrey, *Makers of North American Botany* (New York, Ronald, 1961), 159–60, hereafter cited as *Makers*.

198. Raymond J. Pool, "A Memorial, Frederic Edward Clements," *Ecology*, Vol. XXXV (1954), 109–12; Roscoe Pound, "Frederic E. Clements as I Knew Him," *Ecology*, Vol. XXXV (1954), 112–13; J. Phillips, "A Tribute to Frederic E. Clements and His Concepts in Ecology," *Ecology*, Vol. XXXV (1954), 114–15; David Wigdor, *Roscoe Pound: Philosopher of Law* (Westport, Conn., and London, Greenwood, 1974), Chap. 3.

199. Cowles, "The Ecological Relations of Vegetation on the Sand Dunes of Lake Michigan," *Botanical Gazette*, Vol. XXVII (1899), 95–117, 167–202, 281–308, 361–91; "The Physiographic Ecology of Chicago and Vicinity," *Botanical Gazette*, Vol. XXXI (1901), 73–108, 145–82; "The Plant Societies of Chicago and Vicinity," *Bull. Geographical Soc. Chicago*, Vol. II (1901), 1–76; William S. Cooper, "Henry Chandler Cowles," *Ecology*, Vol. XVI (1935), 281–83 and portrait.

200. William Francis Ganong, "On Raised Peat-bogs in New Brunswick," *Botanical Gazette*, Vol. XVI (1891), 123–26; "The Vegetation of the Bay of Fundy and Diked Marshes: an Ecological Study," *Botanical Gazette*, Vol. XXXVI (1903), 161–86, 280–302, 349–67, 429–55; Humphrey, *Makers*, 91–92.

201. John W. Harshberger, "An Ecological Study of the New Jersey Strand Flora," *Acad. Nat. Sci. Philadelphia Proc.*, 1900, 623–71, 1902, 642–69; John W. Harshberger, "An Ecological Study of the Flora of Mountainous North Carolina," *Botanical Gazette*, Vol. XXXVI (1903), 241–58, 368–83; John W. Harshberger, *Life and Work of John Harshberger, Ph.D.: An Autobiography* (Philadelphia, 1928); G. E. Nichols, "Obituary Notice: John William Harshberger, 1869–1929," *Ecology*, Vol. XI (1930), 443–44.

202. Stephen A. Forbes, "On Some Interactions of Organisms," *Illinois State Lab. Nat. Hist. Bull.*, Vol. I (1880), 3–17, esp. 6; "The Food of Birds," *Illinois State Lab. Nat. Hist. Bull.*, Vol. I (1880), 80–148, esp. 85; Herbert Spencer, *The Principles of Biology* (2 vols., London, 1865–67; New York, D. Appleton, 1866–67), II, 397–478.

203. Ralph Dexter, "Birds and Insects in Relation to Horticulture in Northeastern Ohio, 1879–1899," *Biologist*, Vol. XLIV (1961), 1–6; Leland O. Howard, "A Study in Insect Parasitism: a

Consideration of the Parasites of the White-Masked Tussock Moth, with an Account of Their Habits and Interrelations, and with Descriptions of New Species," *U.S. Dept. Ag., Technical Series*, Vol. V (1897), 5–57; Leland O. Howard, *Fighting the Insects: The Story of an Entomologist, Telling of the Life and Experiences of the Writer* (New York, Macmillan, 1933); Mallis, *American Entomologists*, 79–86 et passim.

204. Winslow, *Conquest*, Chap. 17; Jean Théodoridès, "Les grandes Étapes de la Parasitologie," *Clio Medica*, Vol. I (1966), 129–45, 185–208, esp. 196–204; William B. Herms, *Medical Entomology, With Special Reference to the Health and Well-being of Man and Animals* (3d ed., New York, Macmillan, 1939), Chap. 1.

205. Alpheus S. Packard, "The Cave Fauna of North America, with Remarks on the Anatomy of the Brain and Origin of the Blind Species," *National Acad. Sci. Mem.*, Vol. IV (1888), 3–156 and 27 plates; T. D. A. Cockerell, "Biographical Memoir of Alpheus Spring Packard, 1839–1905," *Nat. Acad. Sci. Biog. Mem.*, Vol. IX (1920), 181–236; Abbott H. Thayer, "The Law Which Underlies Protective Coloration," *Auk*, Vol. XIII (1896), 124–29, 318–20; S. D. Judd, "The Efficiency of Some Protective Adaptations in Securing Insects from Birds," *American Naturalist*, Vol. XXXIII (1899), 461–84; Mary Fuertes Boynton, "Abbott Thayer and Natural History," *Osiris*, Vol. X (1952), 542–55; Sterling, *Merriam*, 332–35; Hugh Cott, *Adaptive Coloration in Animals* (London, Methuen, 1957); J. A. Ryder, "A Geometrical Representation of the Relative Intensity of the Conflict Between Organisms," *American Naturalist*, Vol. XXVI (1892), 923–29.

206. George Valentine Riley, "The Yucca Moth and Yucca Pollination," *Missouri Botanical Garden Annual Report,* Vol. III (1892), 99–158; William Trelease, "Further Studies of Yuccas and Their Pollination," *Missouri Botanical Garden Annual Report,* Vol. IV (1893), 181–226; Charles Robertson, "Flowers and Insects: Contributions to an Account of the Ecological Relations of the Entomophilous Flora and the Anthophilous Insect Fauna of the Neighborhood of Carlinville, Illinois," *Acad. Sci. St. Louis Trans.*, Vol. VII (1897), 151–79; Frank N. Egerton on Lovell, *Dictionary of Scientific Biography* (1973), VIII, 518; On Trelease see Humphrey, *Makers of North American Botany*, 251–53, and Joseph Ewan, *Dictionary of Scientific Biography* (1976), XIII, 456.

207. Bashford Dean, "The physical and Biological Characteristics of the Natural Oyster Grounds of South Carolina," *U.S. Fish Commission Bull.*, Vol. X (1890), 335–61; Charles T. Simpson, "The Pearly Fresh-water Mussels of the United States; Their Habits, Enemies, and Diseases; with Suggestions for Their Protection," *U.S. Fish Commission Bull.*, Vol. XVIII (1898), 279–88.

208. C. Hart Merriam, "Results of a Biological Survey of the San Francisco Mountain Region and Desert of the Little Colorado, Arizona," *North American Fauna*, Vol. III (1890), 1–136, 14 plates and 5 maps. This and other writings by Merriam on the subject have been reprinted in facsimile, *Selected Works of Clinton Hart Merriam* (ed. by Keir B. Sterling, New York, Arno, 1974); Sterling, *Merriam*, Chap. 6.

209. But see Robert Clarke, *Ellen Swallow: The Woman Who Founded Ecology* (Chicago, Follett, 1973).

ECOLOGY SINCE 1900

Robert P. McIntosh

Reprinted from *Issues and Ideas in America,* edited by Benjamin J. Taylor and Thurman J. White, published and copyrighted by the University of Oklahoma Press, Norman, Oklahoma.

17

ECOLOGY SINCE 1900

By Robert P. McIntosh*

In 1938 the distinguished philosopher of science Rudolph Carnap recognized a logical need for a branch of biology dealing with the interactions of individual organisms, groups of organisms, and their environment (Allee et al., 1949). He apparently was not aware that this logical need had been felt by biologists for over half a century and that the already thriving young science of ecology was filling the logical niche he recognized. Carnap's lapse was particularly striking because ecology in the United States was, in considerable degree, a product of the Middle West, notably his own institution, the University of Chicago.

THE RISE OF SELF-CONSCIOUS ECOLOGY, 1900–20

Ecology began to be formalized as a discipline in Europe in the late 1800's, when biology was becoming well established as a professional and scientific area discrete from medicine and was still largely dominated by taxonomy, anatomy, morphology, physiology, and controversies over the new theory of natural selection. The same logic that struck Carnap must have been apparent to European biologists, for in 1866, E. H. Haeckel, the leading proponent of Darwinian thought in Germany, christened the science "oekologie" (Egerton, 1976). Stauffer (1957) asserted that the logical justification for the introduction of the term was a notable consequence of Darwin's thought. Lynn White (1968) attributed the "crystallization of the novel concept of ecology" to the recognition that scientific knowledge means technological power over nature that developed principally in the mid-nineteenth century. Haeckel was not so much the founder of a science as he was the first to name and define a logically consistent area of biology bringing together an extremely heterogeneous mix of natural history, physiology, hydrobiology, biogeography, and evolutionary concerns of nineteenth-century biology. He raised a standard around which biologists of similar mind could gather, thus providing a focus for a previously inchoate area of science (Egerton, 1976).

Among the seminal European ecological works that exerted major influence on the genesis of ecology in the United States were E. Warming's *Oecological Plant Geography* (1895), A. F. W. Schimper's *Plant Geography on a Physiological Basis* (1898), both of which stressed the relation of plants to environment (Goodland, 1975), and F. A. Forel's (1892) monograph on Lake Leman introducing the term *limnology* (Egerton, 1962). Prominent among American biologists of the nineties were J. M. Coulter and C. E. Bessey, professors of botany at the University of Chicago and the University of Nebraska, respectively; E. A. Birge, zoologist of the University of Wisconsin; and S. A. Forbes, entomologist associated with the University of Illinois, the Illinois State Laboratory of Natural History, and its successor, the Illinois Natural History Survey. In 1887, Forbes wrote *The Lake as a Microcosm,* which later came to be recognized as a classic of ecology. The work of these four biologists forms something of a watershed in the history of ecology in the United States. Although not themselves trained as ecologists, they became interested in the new science of ecology and were influential in the careers of many of the biologists who established ecology as a science in America.

"Oekologie" was formally considered at the Madison Botanical Congress of 1893, where the "O" was dropped and the

*Ecology has expanded so rapidly in recent years that by far the greatest amount of ecological work has been done in the past twenty-five years. Thus a history of ecology since 1900 is soon involved in what may better be called current events. Entry into this area is fraught with dangers of omission and commission because of the vast amount of material, much of it not yet assimilated into a clear conceptual framework. My efforts to boil down ecology in the United States to an essence suitable for publication in a short article were greatly aided by the kindness of several colleagues in reading the commenting on preliminary drafts of the manuscript. I express my appreciation to Carl von Ende, Grant Cottam, Robert Whittaker, Orie Loucks, and Frank Egerton. I am also grateful for the interest and comments of Paul B. Sears, a premier practitioner and expositor of ecology, whose lucid writing did much to bring ecology to the attention of scientists and the lay public. I do not intimate that all of the above persons necessarily agree with my selection or interpretations or that they share any onus for the inevitable omissions or errors. The real appreciation for any history of ecology goes to those whose work established a body of ecological concepts and facts long before ecology became a familiar word in the household, councils of government, and funding agencies. It seems a nice example of scientific preadaptation that ecology was there when it was needed.

Anglicized spelling, "ecology," adopted, a source later of confusion and controversy. The early development of ecology as a named and recognized science is commonly attributed to botanists. Charles Elton (1933), a prominent British animal ecologist, commented, "Animal ecology began as a science by following rather closely the lines laid down by the earlier work of plant ecologists"; and the American animal ecologists, R. N. Chapman (1931), and Allee and his associates (1949), also acknowledged the early leadership of botanists.

At the University of Chicago, H. C. Cowles transferred his interests from geology with T. C. Chamberlin, to botany with J. M. Coulter and received his doctoral degree in 1898 for studies entitled "Ecological Relations of the Sand Dune Flora of Northern Indiana." This, and his subsequent studies of the vegetation and physiographic ecology of the Chicago region, constituted the pioneer studies in America of succession—the process of change in ecological systems—that preoccupied ecologists in subsequent decades and remains one of the central concepts of modern ecology. Perhaps Cowles's major contribution was his impact on the first generation of ecology students.

In the same year, 1898, F. E. Clements completed his doctoral work with Bessey, a distinguished mycologist, at the University of Nebraska. Clements had previously published on fungi, but his doctoral dissertation was a study of the phytogeography of Nebraska, later published with Roscoe Pound, who abandoned ecology for a distinguished career in law. By virtue of his prolific research and writing (in 1901 he presented five of the twelve papers on the program of the meeting of the Botanical Society), Clements became the major codifier and theorist of early-twentieth-century plant ecology, and in 1905 he produced the first American book on ecology, *Research Methods in Ecology.* In this volume he evidenced the tendency that was to earn for ecology the pejorative definition "that part of biology which has been totally abandoned to terminology." Its glossary contained the classical definition of *geotome* (complete with Greek derivation): "An instrument for obtaining soil samples"—that is, a shovel. The bibliography of Clements' book suggested the newness of ecology in America, for it referred mostly to European works, and only two references (by Cowles) used the word "ecology."

Ecology, one of the newest branches of biological science, slightly antedating the formalization of genetics, was taken up rather quickly, particularly in botanical circles. By 1902 it was incorporated without comment in a high-school botany book, although still suspect as a bastard version of physiology in some circles and unknown in others. Even Haeckel described ecology as a part of physiology, an unfortunate classification (Haeckel, 1898). Indeed, ecology was still so new that in 1902 a letter to *Science* from a well-known editor was critical of the word being used without explanation when it did not appear in the standard dictionaries. A number of respondents pointed out that the word was in the dictionaries as *oecology,* and an extended discussion ensued concerning its origin, etymology, and proper spelling. William Morton Wheeler (a distinguished entomologist and later the first vice-president of the Ecological Society of America) complained that botanists had usurped the word, had eliminated the *o* improperly, and were distorting the science. He urged that zoologists adopt the word "ethology" instead (Wheeler, 1902). There is in the exchange a suggestion of the split between plant and animal ecologists that pervaded ecology from its beginnings and only recently shows signs of closing.

Those who find recent ecology difficult to comprehend may be relieved to know that this has long been so. Cowles (1904), attempting to review the work in ecology for 1903, said: "It is almost impossible to do such a task for the field of ecology since the field of ecology is chaos. Ecologists are not agreed even as to fundamental principles or motive; indeed no one at this time, . . . is prepared to define or delimit ecology." Early ecologists apparently did agree on the central role of evolution, for Cowles also said, "If ecology has a place in modern biology, certainly one of its great tasks is to unravel the mysteries of adaptation." In the same issue of *Science,* W. F. Ganong (1904), an early Canadian ecologist, listed "seven cardinal principles of ecology," all concerned with adaptation. This central concern with evolution, which was widespread in early ecology, was not as intensely pursued as it might have been, for many years later John Harper (1967) complained that ecology had abandoned evolution to genetics, although evolution is predominantly an ecological subject. Allee et al. (1949), in their review of the history of ecology, also noted the tendency of some ecologists to "veer away from an evolutionary viewpoint," a tendency clearly reversed in recent textbooks of ecology. The diffuseness of ecology that often drew critcism of its publications was also noted early. Ganong (1904) commented:

Ecological publications in American are too often characterized by a vast prolixity in comparison with their real additions to knowledge, by a pretentiousness of statement and terminology unjustified by their real merits, and by a weakness of logic deserving the disrespect they receive. The subject suffers, I fear, from a phase of the "get-rich-quick spirit."

Although the real rise of the get-rich-quick spirit in ecology was many years in the future, with the rise, in the 1960's, of the popularity and status of ecology in an era of large-scale funding and public concern for the environment, perhaps the earliest funding of ecology in America appeared in 1903, as both Cowles and Ganong noted. The Carnegie Institution supported a proposal to establish a desert laboratory on a hill overlooking Tucson in the then munificent sum of eight thousand dollars, which in those days built a substantial building (which continues to serve its purpose) and funded its first year's operation. The institution also supported early investigations of transpiration of plants by B. E. Livingston (inventor of one of the earliest instruments used by ecologists, the atmometer, a clay bulb for measuring water loss to the atmosphere) and other significant studies in physiological ecology. Forest Shreve, a major student of desert ecology, was on the staff of the Carnegie Institution from 1908 to 1945 and Clements, the premier American plant ecologist, from 1917 to 1946.

Already well into his long career as zoologist, student of

zooplankton, and university administrator, E. A. Birge made his debut as an aquatic ecologist, or limnologist, in 1895 (Mortimer, 1956; Frey, 1963). Like Forbes, he had earlier been interested in taxonomy of aquatic organisms *(Cladocera)*. His distinguished career as a limnologist began with studies of plankton, but this led him into significant work in the physics and chemistry of lakes in the very first formative years of limnology. The Wisconsin Geological and Natural History Survey, the counterpart of Forbes's department in Illinois, afforded him a similarly strategic position to influence the development of aquatic ecology. In 1900, Chancey Juday joined the survey, and the partnership of Birge and Juday, which was substantially to lay the foundations of limnology in America, was formed. Birge established in detail the physical phenomenon of lake stratification and mixing resulting from change in temperature and density of water.

In the long collaboration of Birge and Juday (their joint publications exceeded the total of those published separately) they considered the "lake as a microcosm," to borrow Forbes's apt phrase. The processes of photosyntehsis, respiration, and decay and the attendant changes in dissolved gases were primary concerns of this synergistic pair, who also developed the earliest assays of biological productivity of lakes and studied the key factors controlling it. It may well be that the view of the lake as a whole, as a complex of physical and biological processes, articulated by Forbes and developed by Birge and Juday, is the reason that the moderate concept of the ecosystem and its functioning leans so heavily on aquatic studies. Chapman (1931) commented: "These relationships and processes are to be found everywhere in nature, though they may be less susceptible to analysis in other environments (than aquatic). The student of terrestrial synecology may do well to study the results of the aquatic synecologists."

Around and shortly after the turn of the century the pioneers of formal animal ecology in America also appeared. Charles C. Adams and Victor Shelford were both influenced by Cowles and Davenport at the University of Chicago. Adams (1917) said that he gave the first course in general animal ecology at Chicago in 1902, and he subsequently worked with S. A. Forbes at the Illinois Natural History Survey. Shelford completed his doctoral thesis in 1907 on "Tiger Beetles of Sand Dunes," describing the parallels of beetle populations and vegetational succession shown earlier by Cowles. In 1913, Adams published the first text on animal ecology, *Guide to the Study of Animal Ecology,* and Shelford published one of the monuments of ecology, *Animal Communities in Temperate America,* which was somewhat mistitled, for it stressed the relations of plants and animals in the Chicago region. Shelford emphasized the community concept clearly developed, before ecology was named, by the geographer-explorer Alexander von Humboldt and the marine biologist Karl August Möbius, which became the essence of ecology. According to Shelford:

Ecology is the science of communities. A study of the relations of a single species to the environment conceived without reference to communities and, in the end, unrelated to the natural phenomena of its habitat and community associated is not properly included in the field of ecology.

Shelford underplayed the study of single-species populations, their properties, and relations to the environment *(autecology),* which actually came to be a major focus of ecology along with the central concern with the community and its properties. The relative contemporariness of ecology is seen in the fact that Shelford, one of its founders in America, published his last book in 1963.

It is perhaps unfair to mention only a few of the personalities in the early development of ecology in America, for others are surely worthy of note. C. H. Merriam (1894) had propounded his famous (if inadequate) doctrine of the distribution of animals related to temperature based on his extensive studies of mammals. H. A. Gleason, later to abandon ecology for taxonomy, was formulating his "individualistic concept," which would make him the first apostate to the overwhelming Clementsian credo of the climax association as an organism (McIntosh, 1975). Joseph Harshberger was pursuing his extensive surveys leading to the *Phytogeographic Survey of North America* (1911), the first summary description of the major communities of North America. Forrest Shreve was well into his studies of the American deserts, J. E. Weaver was beginning the work that would make him the foremost student of the American grassland, while E. N. Transeau was beginning to formulate his ideas of vegetational distribution leading to his well-known phrase "the Prairie Peninsula," describing the eastward extension of grassland into the deciduous forest. E. W. Hilgard was developing the idea of plants as indicators of soils, and C. MacMillan, one of the first to use the new word "ecology," was studying Minnesota's bogs.

The early part of the century also saw the rise of the professional areas of applied ecology, forestry, range management, wildlife management, and fisheries management; the related sciences of soils and meteorology; and the great tradition of conservation in America with its notable successors to John Wesley Powell and George Perkins Marsh. Carrying on the interlocking traditions of natural history, love of the out-of-doors and the aesthetic, almost mystical, appreciation of nature earlier seen in American poets, writers, and artists such as Walt Whitman, Henry David Thoreau, William Cullen Bryant, and Thomas Cole, were John Burroughs, John Muir, and the key figures of the American conservation movement, Gifford Pinchot and Theodore Roosevelt. The common cultural heritage of ecology and conservation and their joint development in the early 1900's were the beginning of a continued and fruitful interaction.

By 1915 a body of scientists had risen in America who regarded themselves as ecologists, although their areas of interest in biology ranged widely over plants, insects, aquatic organisms, and terrestrial vertebrates; and their approaches to these diverse taxonomic groups and different habitats were, similarly, wide-ranging, from the ecological physiology and morphology of the several taxa to their biogeographical distributions and aggregation in communities. Ecologists were notably concerned with the phenomena of succession and population and environmental changes. Following in the midlands tradition, R. H. Wolcott, professor of zoology in the University of Nebraska, wrote to Shelford, at the University of Illinois, suggesting the formation of a society of ecologists. Shelford consulted with Cowles, who agreed, and in December, 1915, the Ecological Society of America was founded, with Shelford as its first president. At the end of the year there were 307 members (Shelford, 1938).

Ecologists, then as now, were a heterogeneous group. The second president of the new society was Ellsworth Huntington, a pioneer of human ecology and an early proponent of the profound impact of the environment on human beings. Unfortunately, this early hybrid of general and human ecology was not viable; and at intervals since, particularly in recent years, there have been earnest, if only partly successful, efforts to merge human sociology and ecology in a science of human ecology.

The early American ecologists were not unduly modest. Adams (1917) saw the United States as the world leader in the "new Natural history," as he called ecology. He commented in a very modern vein, ". . . we look upon science as a tool to aid us in securing better human living in the broadest and best sense." He and Barrington Moore (1920), the first editor of the journal *Ecology,* which was established in 1920 by the Ecological Society, asserted that ecology represented the synthetic phase of biology with the function of bringing together myriads of isolated facts. Moore, along with Adams and Forbes, the latter two occupying strategic positions as directors of the state natural history surveys of New York and Illinois, respectively, stressed the importance of ecology as the basis for applied ecology and natural-resource management, a consideration that persisted throughout the development of ecology as a science.

By 1920 self-conscious ecology was reasonably well established and recognized as an academic discipline in America, although it was hardly known to the general public. It had accumulated extensive surveys and descriptions of communities, both aquatic and terrestrial (mostly qualitative though primitive quantitative methods were developing), extended the tradition of "relations physiology" as *autecology* (the relation of the organism to the physical environment), produced a limited number of specifically ecological books and references, and developed in some detail one major concept—succession. It had also established its own professional society and publication outlet and had a number of notable spokesmen in the biological community. Ecology was still notoriously heterogeneous. As one of them, Adams (1913), stated, "At present ecology is a science with its facts out of all proportion to their organization or integration." Adams (1917) also noted that the hostile opponents had mellowed with time, agnostics had been convinced and the younger generation accepted ecology, as a matter of course, like physiology.

ECOLOGY, 1920–50

From 1920 to about 1950 ecology developed rapidly. Far from the synthetic ideal of its early proponents, it proliferated its own set of special interests and remained an eclectic science that frustrates writers of ecological history. Gleason (1936) correctly described ecology as "extraordinarily polymorphic." The distinction between plant and animal ecology persisted, and there was little commonality between aquatic and terrestrial aspects of ecology. It was not uncommon for authors dealing with either plants or animals to apply the word "ecology" to their particular interests without bothering with the limiting descriptor plant or animal (Nichols, 1928). Thus Gleason (1936) considered *Twenty-five Years of Ecology 1910–1935* without reference to animals, and Victor Shelford (1929) published *Laboratory and Field Ecology* as a methods book for animal ecology, although he did include references to plants as food for animals.

The first formal textbooks of ecology that might establish a "normal" science of ecology in the sense of Kuhn (1962) were Pearse's *Animal Ecology* (1926), W. B. McDougall's *Plant Ecology* (1927), Weaver and Clements' *Plant Ecology* (1929), R. N. Chapman's *Animal Ecology* (1931), and R. S. Welch's *Limnology* (1935). Pearse, for example, considered the major subdivisions of animal ecology to be physical and chemical factors, biological factors, succession, animals of the ocean, fresh-water animals, terrestrial animals, relations of animals to plants, relations of animals to color, intraspecific relations, and economics aspects of ecology. Weaver and Clements included chapters on methods of studying vegetation, plant succession, units of vegetation, causes of succession, migration, ecesis (establishment) and aggregation, competition and invasion, soils, reaction (effects of plants on environment) and stabilization, coaction (effects of organisms on each other) and conservation, a series of chapters on the environment—humidity, temperature, light, water, plants as indicators of the environment, and a description of the climax-vegetation formations of North America.

Major European influents were Charles Elton's *The Ecology of Animals* (1927) and R. Hesse's *Tiergeographie auf oekologischer Grundlage* (1924), which, translated into English as *Ecological Animal Geography* (1937), had major impact on American as well as European ecologists. It emphasized the importance of ecological aspects of animal geography, much as Warming's earlier (1895) *Oecological Plant Geography* had done for plant ecology, and, according to its translators, Allee and Schmidt, marked a new phase in the development of animal ecology. This phase emphasized the geographical distribution of animals and communities, the structural features of communities, and the feeding relations and quantitative makeup of the animal community. The nature of animal aggregations was pursued extensively, notably by Allee in a series of articles and books (1927, 1931, 1938) that emphasized studies of experimental populations and of social organization and cooperation among animals.

A significant development early in this middle period was the rise of paleoecology. Paleoecology, based largely on the study of the layers of fossil pollens (palynology) preserved in peat bogs or lake sediments, began in Scandinavia about 1916 and was rapidly adopted in America. Clements, the encyclopedist of American plant ecology, published *Methods and Principles of Paleo-ecology* (1924) and pollen studies by C. L. Fenton, Paul B. Sears, J. Potzger, Henry Hansen, and others in the 1930's and 1940's traced the history of vegetation and associated climates as they had changed with successive advances and recessions of the Pleistocene ice sheets. These studies made clear the significance of history in any consideration of the distribution of plant and animal communities and, along with the concept of succession, required that ecology develop a dynamic view of communities. Communities continued as major objects of study for ecologists and, in 1935, were rechristened and incorporated in an expanded concept "ecosystems" by Sir Arthur Tansley, a prominent British plant ecologist. *Ecosystem,* defined as the complex of living organisms (the community) and its nonliving environment, was, however, to remain in the wings until called forth by dramatic post–World War II developments in ecology. Although the actual study and analysis of the entire ecosystem was, until recently, only a gleam in the intellectual eye of ecologists, the "holistic" ideal was clearly evident much earlier. According to W. F. Taylor (1936), the key words for ecology were integration, correlation, coordination, synthesis, holistic, emergent, and relationship, all of which are prominent in the lexicon of recent ecologists. The extension of this ideal of ecology into the realm of the human environment, with its economic and political overtones, was foreshadowed in Taylor's quotation from Henry Wallace, then secretary of agriculture and left-of-center political leader, "There is as much need today for a declaration of interdependence as there was for a Declaration of Independence in 1776."

A particularly significant event that occurred about 1920 was the flourishing of population ecology, stemming from demography, which was primarily associated with animal ecology. It was notable in that it represented the first attempt to use a mathematical and theoretical approach in ecology, and the tradition continues in current ecology at an accelerated pace. Treatments of populations (demography) have a distinguished lineage most familiarly from Thomas Malthus' famous *Essay on Population,* first published in 1798, and the famed logistic equation, the most ubiquitous mathematical model of population growth, was devised in the early nineteenth century (Verhulst, 1838). Like many another promising scientific idea, the logistic equation lay fallow until rediscovered and extensively exploited in the 1920's and 1930's by several authors in Europe and, in America, notably by Raymond Pearl (Pearl and Reed, 1920; Pearl, 1925). Although Allee and his associates (1949) commented that "theoretical population ecology has not advanced to a great degree in terms of its impact on ecological thinking," the background was established for subsequent expansion of analytic population studies. Theoretical, experimental, and field studies of populations were to become a major part of ecology and, along with the study of communities, constitute those aspects of biological phenomena which are most distinctively ecology. Animal-population studies explored the theoretical form of growth of populations, the effects of population density on growth and development of individual organisms, reproduction and mortality, and the factors controlling population growth and community organization. Pearl's contribution was salient in that it represented the impact of the new biometric, or statistical, methods, which were then being developed and incorporated into ecological theory and practice. One of the cornerstones of modern ecology is the concept of a population as a group of individuals with a set of attributes to be considered in terms of the population, not of the individuals (for example, mortality rates); and the gradual recognition of the power of statistics to describe and, more particularly, to assess probable error in the numerical estimates of population characteristics and measures of association and correlation between species was necessary for the population concept to develop.

An important corollary of the logistic equation and mathematical theory in ecology derived from the classical work in theoretical and experimental population ecology by G. F. Gause in Europe, and in America the extended studies of Thomas Park with flour beetle *(Tribolium)* populations were a significant extension of this work. These and similar studies led to the formulation of "Gause law," or the "competitive exclusion principle" (Hardin, 1962), which became a cornerstone of ecological thought. The general thesis of the competitive exclusion principle is that two species that compete for a common and limiting resource cannot coexist and one will be eliminated. Thomas Park corroborated Gause's findings, showing that one of two competing species normally becomes extinct. From this kind of study, and observations in the field, it was assumed that species that coexisted in nature inhabited somewhat different niches, the niche being the organism's place in the community. Recent formulations of competitive exclusion agree that two species, to survive in the same stable community, must have different controls on their populations, but not necessarily resource limitation. Few aspects of ecological theory or experimentation have a more involved, or even convoluted, history than the study of populations, the phenomenon of competition and their offspring, the *niche.* The niche concept, first specifically named by Joseph Grinnell in the United States in 1917 and built into community interpretation by Charles Elton, blushed substantially unseen until the 1950's, when it was seized upon and moved into place with the hope that it would be a major tenet of modern ecological theory. The famous individualistic concept of Gleason (1917, 1926) was also an early statement of the species niche.

Although theoretical-mathematical aspects of ecology were largely confined to considerations of populations, ecologists in America were aware of the need for quantitative approaches to communities. The earliest attempts to apply rigorous quantitative sampling and statistical methods to field populations were made by a European marine biologist, V. Hensen, about 1880 (Lussenhop, 1974). Aquatic sampling was extended to fresh waters in Illinois about 1897; and Clements and Pound, as early as 1898, used sample areas called "quadrats" in terrestrial studies. They noted the difficulty of estimating numbers of individuals and urged the need for actual counts. W. L. McAtee (1907) censused animal populations in four square feet, and in the same year Forbes (1907) introduced the idea of a coefficient of association designed to show the frequency with which one species was found with another. Forbes (1909) also published a survey of bird populations in an E-W transect 150 feet wide across the entire state of Illinois (191.86 miles) and analyzed the relative abundance of bird species in the different habitats, mostly corn fields, that were encountered. Early proponents of statistics in ecology assumed that populations were uniformly (evenly) distributed in nature; later, randomness was accepted as a possibility. Although ecology has been widely criticized as being largely a descriptive and qualitative science, there has been a continuity of interest in quantitative approaches from the pioneers in the field to the present. Statistical sophistication on the part of ecologists grew slowly, however, and as late as 1935 even as astute an individual as Gleason could assert that plant species were distributed at random, and therefore, he said, it was possible to determine densities (number of individuals) from frequency (the number of samples in which a species occurred). It would be unfair to allow this uncharacteristic lapse of judgment to obscure the fact that Gleason was one of the major proponents of the use of mathematical methods in ecology. He was in the 1920's an early student of species-area curves (the increase in number of species with increasing sample size), and he noted that attempts to determine number of species by empirically developed equations had failed. In this area he anticipated the recent interest in species diversity. Gleason correctly predicted the continuing development of quantitative and mathematical methods in ecology, although, as late as 1944, Stanley Cain (1944), another proponent of quantitative methods in ecology, worried, with considerable justice as it now appears, that ecological systems might be difficult, or even too complex, to treat mathematically. It is only in the past decade that the validity of Cain's apprehension is being tested. Quantitative ecologists, particularly plant ecologists and marine ecologists, were extensively concerned with sampling methods and statistical analysis of the resulting data. These aspects of quantitative ecology, which were largely used in field studies of population and communities, were carried on quite independently of the theoretical mathematical concerns of population growth and population interactions that were developing concurrently.

Ecologists, plant and animal, aquatic and terrestrial, were expanding their horizons from studies of communities as lists of species to assessments of numerical abundance and measures of relations among species. In this they followed in the wake of applied areas of ecology, such as fisheries, which had been concerned with numerical assessment, at least of commercial species, for many years. Differences in techniques of sampling and in ease of catching various kinds of organisms obscured the general similarity of the means and ends of quantitative approaches to ecology. Animal ecologists generally were assured that they could recognize individuals for sampling; plant ecologists worried about recognition of individuals. Plant ecologists and marine benthic ecologists, dealing with relatively stationary organisms, were most confident of their estimates, while animal ecologists generally experienced more difficulty in arriving at accurate estimates of abundance. The quantitative interests of ecologists focused on a number of central concerns:

1. Single-species distribution and the dispersion of individuals that complicated sampling and statistical analyses by being nonrandom or distributed in patterns (usually clumped) for varied reasons, contrary to early assumptions of uniform or random distribution.
2. The number of species and their relative abundances in an area or community.
3. Association or correlations between species that were suggestive of common reactions to the environment or of interactions among species.
4. Measures of similarity between areas or communities.
5. Assessment of the pattern or homogeneity of an area. Homogeneity had been widely assumed to be a criterion of a natural ecological community. In fact, the nature and scale of pattern or heterogeneity under a series of new terms (heterogeneously diverse, patchiness, grain) continues as a major concern of sampling technique and of recent theoretical ecology.

Ecology, like biology generally, proved relatively difficult to fit into a numerical or mathematical framework, and the model of the so-called hard sciences like physics was not readily followed (Haskell, 1940). This was, in substantial part, due to the difficulty of isolating single ecological variables for study and to the difficulty of recognizing, or at least arriving at a consensus on, the objects for ecological study, essentially the population and the community (the aggregation of populations) and their interaction with the environment. Much of the first fifty years of self-conscious ecology in America was spent in describing, largely qualitatively, the complex mantle of environments, vegetation, and animal communities that blended in subtle and ill-understood ways over the landscape and were rapidly being modified or eliminated by the activities of man and his animals even as ecologists sought to study them. The composition, structure, distribution, successional status, classification, and terminology of the plant community were a pervasive concern of American plant ecologists (phytosociologists). The theoretical concepts of Clements dominated American plant ecology, largely isolating it from the several European schools and producing a provincial tone lamented by F. E. Egler (1951), then as now a perceptive, if acerbic, critic of the needs and failings of ecology.

Ecology was, to a large degree, a descriptive and empirical science in 1920. As late as 1939, Allee and Thomas Park wrote, "The statement is frequently made that ecology deals mainly with facts which are organized around relatively few principles." Recognizing this limitation, they attempted to summarize distinctly ecological principles from the literature. In this endeavor they followed Ganong, who in 1904 had recognized seven ecological principles, all dealing with adaptation; and they anticipated K. E. F. Watt, who in 1973 recognized an unwieldy thirty-eight principles, starting with the laws of thermodynamics. Allee and Park recognized about nine sets of principles, which they said dealt with "interrelations between an organism or one or more groups of organisms and its or their environment." They serve to characterize ecology at approximately the midpoint in its formal existence in the United States:

1. *Law of the minimum,* or, more generally, the capacity of an organism to tolerate the minimum or maximum extreme of environment. Familiarly, some organisms have narrow, others broad, ranges of tolerance.

2. *Adaptation,* which was so prominent in Ganong's principles (1904), appeared also as a fundamental principle of Allee and Park, who cited as an example the familiar "Bergman's rule," which suggested an adaptive relation of body size of related warm-blooded animals to temperature.

3. *Community,* at that time often thought of as the major, or even the only, ecological concern, particularly by some plant ecologists.

4. *Succession,* the process of maturation of the community towards a "climax," in those days usually thought of as a regional formation recognized by its dominant organisms, substantially self perpetuating under the control of the regional climate according to the ideas of Clements.

5. *Population growth,* particularly the quantitative aspects summarized in the logistic curve.

6. *Cooperation* and *competition* among organisms suggested a whole set of ecological principles centering about evolution. Allee and Park noted that the factors of natural selection are definitely ecological and laid claim to evolutionary dynamics as a province of ecology, but they noted that this was a field that ecologists had avoided.

7. *Niche separation,* a principle which was based on the observation that related species occupy separate niches and thus are in less direct competition. This principle has been seized on by recent ecological theorists and extensively exploited.

8. *Geographic distribution.* Among these they noted the tendency for tropical species to have fewer individuals per species than temperate species, an aspect of another central concern of current ecological thought under the name diversity.

9. *Emigration* and *dispersal* of organisms.

The 1939 statement of ecological principles was very loosely structured, and Allee and Park recognized its inadequacy. It serves to illustrate the very tentative thinking of two leading animal ecologists, who said, "We believe that focusing attention on a theoretical framework will lead to more important work in ecology."

Another landmark of ecology in midcourse was *Plant and Animal Communities* (Just, 1939), which was the result of a 1938 conference described as "the first ambitious attempt to arrange a general public stocktaking of ecology" (Allee, 1939). It marked, according to Allee, a change from the earlier evangelical tone of ecologists to "frank scepticism about every aspect of ecology except the value of the subject itself." The general stocktaking included the following titles and authors:

Plant Associations on Land, H. S. Conard
Littoral Marine Communities, G. E. McGinitie
Fresh-Water Communities, F. E. Eggleston
The Biome, J. R. Carpenter
The Individual Concept of the Plant Association, H. A. Gleason
The Unistratal Concept of Plant Communities (The Unions), Theodore Lippma
The Climax and its Complexities, S. A. Cain
Social Coordination and the Superorganism, A. E. Emerson
On the Analysis of Social Organization Among Vertebrates with Special Reference to Birds, N. Tinbergen
Analytical Population Studies in Relation to General Ecology, T. Park

The list of authors included many of the ecological luminaries of the day, but Allee noted that the conference was dominated by the ideas of those who were not present, such as Clements, Shelford, J. Braun-Blaunquet, Juday, Elton, and K. Lorenz. The reviewer, a distinguished animal ecologist, commented "The papers on plant sociology seemed at times to be fuller of words than meat," which suggested that the traditional plant-animal ecology schism still persisted; and he noted that the concept of the *biome* (biocommunities considering both plants and animals) received lip service only. He also deplored the continuing problem of proliferating terminology, which continues unabated to the present, and called for more effective means of collecting, analyzing, and interpreting data.

The symposium was distinctive in that in included plant and animal, terrestrial and aquatic, fresh-water and marine aspects of ecology. Aquatic ecology had developed along somewhat distinct lines from terrestrial ecology, and in 1936 the Limnological Society of America had been founded with 221 members and Juday as its first president. In 1948 it became the American Society of Limnology and Oceanography. Fitting all the taxonomic and habitat interests into one ecological bed was not notably successful, and the several aspects of ecology continued to develop in parallel. Since most limnologists were trained as zoologists, there was more interaction between terrestrial animal ecology and limnology. Plant ecology, especially *phytosociology* (the study of plant communities), largely dominated by the views and terminology of Clements, went substantially its own way (Egler, 1951). Sears (1956) provided an apt phrase, "the ecology of ecologists," to describe the environmental reasons why ecologists and ecology continued to be divided into parochial interests.

A striking aspect of the 1938 conference was the inclusion of the famous animal behaviorist N. Tinbergen, and the reference to Lorenz as one of the notable absentees. The chronology of the development of animal behavior parallels that of ecology, and in the United States outstanding work, such as that of Margaret Morse Nice (1937) on the song sparrow, emphasized the significance to ecology of aspects of animal behavior such as territoriality and social hierarchy, concepts which were being developed in this period.

Another important publication of 1939 was *Bio-ecology,* by Clements and Shelford, then among the major figures in American ecology. As the hyphenated title implies, *Bio-ecology* was an attempt to achieve the synthetic ideal ecologists often talked about but rarely practiced. It incorporated animal, plant, and aquatic ecology but recognized that the desirable addition of human ecology would be delayed until the feeling of the need for synthesis became more general. *Bio-ecology* may in some ways mark a turning point in ecology. In a review, G. E. Hutchinson (1940), one of the titans of modern ecology, noted the failures of the book and pointed the way ecology should go and, indeed, has gone (Hutchinson, himself, usually in the lead). Hutchinson criticized the book's emphasis on classification and terminology. He specifically deplored (1) its failure to use statistics and mathematics in dealing with ecological problems, (2) the insistence of *Bio-ecology* on the community, and (3) a complete lack of the biogeochemical approach, which integrates the living community with the nonliving environment. At this time Hutchinson was beginning his famous studies on Linsley Pond and developing his interests in limnology (Hutchinson, 1957b), biogeochemistry, and theoretical ecology, which have had major influence on subsequent developments in ecology by virtue not only of his own enormous research output but of his impact on a large number of the leading figures in modern ecology (Edmondson, 1971; Kohn, 1971; Riley, 1971). Some of the needs of ecology urged by Hutchinson in 1940 had been foreshadowed in the world of E. N. Transeau, who, as early as 1926, worked out a reasonably detailed energy budget of a cornfield; and by 1940, Juday had worked out an energy budget of a lake, and Raymond Lindeman (1941, 1942) had elaborated in detail the subject of tropic (nutritional) relationships and developed the framework of the metabolism of biological communities.

The latter part of this middle period in the growth of ecology was marked by the appearance of a number of significant publications. Plant ecology, previously dominated by Weaver and Clements' premier textbook, *Plant Ecology*, was subdivided into *Plants and Environment: A Textbook of Plant Autecology*, by R. F. Daubenmire (1947), and H. J. Oosting's *The Study of Plant Communities* (1948), emphasizing *synecology* (plant communities). Another milestone was the appearance of the first encyclopedic attempt to review animal ecology, *Principles of Animal Ecology* (1949), which was written, appropriately enough, by five distinguished second-generation ecologists from Chicago: Allee, A. E. Emerson, Thomas Park, Orlando Park, and K. P. Schmidt. The major sections of this volume were: (1) The history of ecology (to 1940), (2) the environment; (3) populations, (4) the community, and (5) ecology and evolution. The authors' major interest was in the documentation of the general ecological principles that Allee and Thomas Park had discussed earlier. However, principles are nearly buried in the wealth of detail amassed by these ecologists, although they disavowed detail for the sake of detail. Nevertheless, the explicit exposition of ecological concepts, which had been developing in the notoriously diffuse subject ecology had become, served to summarize the state of animal ecology about 1949 and to anticipate the dramatic developments of the next two decades. The delineation in the book of the essence of trophic ecology, the quantitative treatment of population and energy budgets, the emphasis on biomass and the re-emphasis on natural selection and evolution as ecological subjects point to what came to be called "the wave of the future" in the full flush of post–World War II ecology.

Ecologists in the consolidation period of ecology continued the strong impetus toward an important role for ecology in the management and conservation of natural resources that had been given it by Forbes, Clements, and Adams and that was the basis for the establishment of a committee on the preservation of natural conditions in 1917 as one of the first acts of the newly formed Ecological Society of America. The reciprocal effects of ecology and various professional areas of management were represented in H. L. Stoddard's *The Bobwhite Quail* (1932), Aldo Leopold's *Game Management* (1933), H. J. Lutz and R. F. Chandler's *Forest Soils* (1946), J. Kittredge's *Forest Influences* (1948), and other similar books. The early-warning system of the current ecological crisis, which unfortunately had little influence, was seen in Paul B. Sears's *Deserts on the March* (1935), William Vogt's *Road to Survival* (1948), Fairfield Osburn's *Our Plundered Planet* (1948), and Aldo Leopold's *Sand County Almanac* (1949). The last, in the 1960's, would become the bible of the new converts to the land ethic.

It is not possible in the short compass of this paper to suggest the important impacts on ecology of contemporary developments in other sciences, such as meteorology, geology, soils, and chemistry, as well as other facets of biology. Preeminently a field science, ecology depends on many sources of information about the physical environment, as well as living organisms, which it hopes to integrate into its effort to understand the complex of populations, communities, and environment that is the distinctive goal of ecology.

By midcentury ecology was established widely, if not universally, in the academic world of the United States as a science with its own concepts, techniques, and objects of study—the population and community. It had its own societies, publication outlets, basic specialized literature, influential textbooks, and an embarrassment of terminology for which it was roundly criticized. Ecology had progressed and improved on the work of its founders, whose active professional lives continued up to, and even past, the half-century mark, but it had not much transcended them. There had been few if any of what in modern scientific jargon are called "breakthroughs." Perhaps the diffuse subject matter of ecology does not lend itself to the quantum jump of insight or the revolutionary new interpretation that has marked the dramatic turning point of other scientific disciplines. Or, as so often is the case, the promising but unconventional idea is ignored. This was true of Gleason's "individualistic concept" of the species and of the association that he developed in three papers between 1917 and 1939 and were almost totally ignored in the standard ecological texts and references before 1950. Late in the 1940's Gleason's concept was resurrected, in the early 1950's it was supported by a substantial body of quantitative evidence, and since it has become a standard idea in most ecological textbooks and references, often being taken as almost axiomatic (McIntosh, 1967, 1975).

Ecology had in its ranks critics and prophets who tried to interpret it for other ecologists, to extend its insights to wider horizons, and to lead it into the role of public service and utility that had been envisioned by its founders, particularly Adams, Forbes, and Clements. Up to 1950, however, ecology was little known, and its worth as a basis for management of natural resources was poorly exploited. Notable among those ecologists of the pre–World War II generation striving to bring its name and ideas out of the academic shadow were Frank Egler, Paul B. Sears, and Aldo Leopold. Egler worked both sides of the street, introducing abstract and formal aspects of philosophy into interpretations of vegetation study (Egler, 1942) and, on the other hand, urging the empirical use of ecological ideas upon practical men of affairs.

Paul Sears was the master of the lucid essay, intriguing specialist and nonspecialist alike, in seeking a wider audience for ecological ideas inside and outside the scientific community. He was one of the few ecologists whose writing gave a glimpse of what ecology was and, moreover, what it ought to be. It was Sears who, as late as 1944, wondered about the factor that had retarded the development of ecology and suggested as a partial answer the failure of ecology to recognize its ancestors in biology whose work pointed toward "organized elaborations of the energy and material cycle." He called attention then to the early work in energetics and nutrient dynamics, and in 1949 he stressed the absence of information on "unifying dynamic principles that may apply to the community process." He noted the relevance of the laws of thermodynamics, the concept of entropy, complex channels of energy flow, and number of species in community development as some of the things ecologists should attend to.

Aldo Leopold, apart from being the founder of game management as an extension of ecology, extrapolated ecological ideas and placed them in the framework of the human conscience. He took the ordinary stuff of the farmer, woodsman, naturalist, and scientist and spun it into the web of ethics and aesthetics (Leopold, 1949).

Ecology had not yet achieved the public acclaim accorded its sister science, genetics, as Sears (1949) noted, nor had it become the synthetic science that its advocates in 1950, as in 1920, wished it to be.

THE RISE OF THE NEW ECOLOGY, 1950–76

The immediate post–World War II period provides a convenient and reasonably logical break in the continuum of the history of ecology. Many ecologists who were to bring to fruition concepts of ecology born in its first half century in America and add new ones were trained in the postwar flush in higher education and on the G.I. Bill. The war period had seen the development of some radically new techniques in other scientific disciplines that had filtered into ecology, and the concept of the relation of science to society had begun to undergo significant changes, which were suggested in Vannevar Bush's *Science the Endless Frontier* (1946). It can as truly be said of ecology of 1950 as Allee and his associates (1949) had said of ecology of 1900: "Scientific attention in general was focused on nonecological phases of biology, and the science of ecology, now well and firmly rooted, could continue to develop outside the distorting influences often accompanying high popularity." In the early postwar period scientific and popular attention was drawn largely to the dramatic developments in molecular biology, especially molecular genetics, and ecology progressed quietly, undisturbed by public adulation. It was a not unmixed blessing to ecology later, in the 1960's, to be thrust onto center stage with the rise of the environmental movement, and the consequent distorting influences attendant upon high popularity, foreseen by Allee and others (1949) became apparent.

As Adams had described ecology in 1913 as "the new natural history," so ecologists by the 1960's were referring to the "new ecology," a phrase suggesting that ecology had undergone a sharp break with its prewar "classical" past (Odum, 1964; Darnell, 1970). This was, in fact, the case. Although much ecological work in the postwar era continued and expanded the earlier traditions, there were a number of significant departures. In the space remaining, the emphasis will be placed on what seem to be distinctive aspects of the break with traditional ecology. It must, unfortunately, omit references to very extensive and excellent work in ecology that extended and improved on the information base and insights of "classical" ecology. I have selected some aspects of ecology for emphasis because of their notoriety as well as because of their solid accomplishment, since some of the most recent developments in ecology are in a state of flux, and their impact on the future of ecology may be problematical. Some of the impacts are more cultural or sociological than scientific.

One of the striking changes in the postwar years was that ecology emerged as "big science" compared with its early traditions (McIntosh, 1974b). Most pre–World War II ecologists were university faculty members working as individuals with a few students and minimal funding. The advent of large-scale federal funding and an era when "Grant Swinger" became an apocryphal figure in science greatly changed the expectations of ecologists, as well as those of other scientists. The comparatively ample funds available helped spread ecology inside and outside the academic arena and carried ecological studies extensively into arctic and tropical regions. The National Science Foundation (NSF), established in 1950, began funding ecological work in 1952 and by 1956 was supporting many awards in the newer thrusts of ecology, such as trophic structure, productivity, energetics, and nutrient cycling of ecosystems. Foundation-financed training programs helped support the new generation of ecologists. About the same time the United States Atomic Energy Commission (AEC) became aware of the health and environmental hazards of radiation. As early as 1955 ecological studies were underway at several AEC installations, and the ecology program at Oak Ridge National Laboratory was begun. It expanded rapidly after 1958, and under the able leadership of Stanley Auerbach it became the largest single ecological-research enterprise in the United States, pioneering in the development of radiation ecology and other aspects of the new ecology.

A consequence of large-scale funding was a new approach to ecological research made possible by the conjunction of new and expensive instrumentation and techniques, the money to exploit the possibilities of these, many more ecologists, and logistical support to do the work. The classical ecologist was described as working with apparatus which, "if not just a ruler and a piece of string, was seldom much more complex than a thermometer and a rain gauge" (Darnell, 1970). The new ecologists, as time went on, confidently looked for the tools of the trade, to include radioisotopes and counting equipment, gas analyzers, chromatographs, respirometers, spectrophotometers, neutron probes, colorimeters, biotelemetry, automatic recording devices, computers, controlled-growth chambers, and biotrons. An old-fashioned car-window ecology was in some instances replaced by airplanes or helicopters. This and other elements of a richer diet for ecologists sometimes led to a rather smug air about the limitations of earlier ecologists. It would be well for the new breed of ecologists to ponder the words of Thomas H. Huxley: "It is easy to sneer at our ancestors, . . . but it is much more profitable to try to discover why they, who were really not one whit less sensible than our own excellent selves, should have been led to entertain views which strike us as absurd" (Huxley, 1903).

In addition to the availability of more complex instrumentation and an extended capacity to collect data, ecology in the last twenty-five years has seen a great increase in the use of quantitative techniques to analyze data. The earlier, somewhat grudging, acceptance of quantitative methods gave way to an almost unbridled enthusiasm for applying numbers, sometimes calculated to unwarranted numbers of significant digits. Nevertheless, ecologists have devoted much attention to sampling methods and quantitative analyses of populations and communities of diverse kinds of organisms, from redwood forests to minute fresh-water or marine plankton. Earlier, often naïve approaches have given way to increasingly sophisticated and rigorous use of sampling and statistical methods stimulated by the appearance of books that made quantitative methods more readily available to ecologists (Greig-Smith, 1957; Pielou, 1969; Patil et al., 1971).

The basic problems of many aspects of ecology, however, remained: securing reasonably accurate estimates of numbers of species and individuals (abundance), assessing their distribution in space (pattern), and determining their associations or correlations with other organisms and the environment. The static view of populations or communities increasingly gave way to studies of population and community dynamics as plant ecologists followed the lead of animal ecologists, and both studied such characteristics of populations as age structure and birth or mortality rates and their effects on population growth and distribution or, reciprocally, the effects of abundance on population characteristics or interspecific interactions such as predator-prey relations, parasitism, or competition.

Quantitatively minded plant ecologists improved their sampling methods in the postwar era, introducing point or distance techniques that increased the efficiency of sampling (Cottam and Curtis, 1949; Goodall, 1952). The traditional preoccupation of plant ecologists with community classification, reviewed by Whittaker (1962), was continued, and numerical methods of ordination and classification were introduced in the United States that rekindled the long-term dispute concerning Gleason's "individualistic concept," which argued that the vegetational community varied continuously and that therefore classifications were necessarily arbitrary. Studies in the 1950's of vegetation by J. T. Curtis in Wisconsin (Curtis, 1959) and R. H. Whittaker (Whittaker, 1967) converged in supporting Gleason's contention, and the continuum, or gradient, concept of the plant, and in some cases the animal, community has been widely if not universally accepted as a useful concept (McIntosh, 1967). Simple empirical methods of numerical analysis were rapidly supplanted by more elaborate mathematical techniques, especially principal components and discriminant analysis, which are still being explored. However, recent studies have raised questions about the utility of these presumably more powerful methods because they presume linear relations, and organisms and environment are not linearly related (Austin and Noy-Meir, 1971; Gauch, Chase, and Whittaker, 1974; Beals, 1973). The interst of animal ecologists in the community has been rekindled in recent years, and they are also using gradient analysis and numerical techniques to examine the animal community (Terborgh, 1970; Green, 1974).

During most of its history ecology held to the holistic approach, that is, the study of the entire complex of nature as a unit. In practice ecologists usually studied single species or single environmental variables, taxonomic groups, particular functions or specific relations between organisms, for example, parasitism or competition predation. The study of the whole complex of organisms and environment, the ecosystem, had been deferred, probably because most prewar ecologists agreed with Egler (1942), who said of the ecosystem, "Although the concept is intellectually more acceptable than any of the preceding, the natural phenomena appear at present too complicated for study and development as a separate science." Among the most far-reaching changes in postwar ecology was a determined effort to produce a science of ecosystems—"ecosystem ecology." The ecosystem was said to be to the ecologist what the cell is to the molecular biologist (Odum, 1964). The rise of the ecosystem concept in ecology was paralleled by a reduction in the traditional distinction between plant ecology and animal ecology and by the development of research teams, bringing together experts in various aspects of ecology and other fields necessary to deal with the complexities of the ecosystem.

This change of emphasis was stimulated in ecological education by the appearance of several textbooks of general ecology in the early 1950's (Odum, 1953; Woodbury, 1954; Clark, 1954). These, particularly the "green book" of E. P. Odum and H. T. Odum, were influential in leading the way to courses in general ecology in addition to the traditional courses in plant ecology, animal ecology, and limnology. Perhaps the most significant emphasis in Odums' popular book was its stress on basic principles of ecology and on the ecosystem as a functional entity. These had been called for in prewar ecology by Hutchinson, and Sears, who had urged the need for a biogeochemical and metabolic approach to the ecosystem. The ground for these ideas had been laid by the work of Juday (1940) and Lindeman (1941, 1942) and in the monumental animal ecology text by Allee and others (1949), but they were introduced to a new generation of ecologists substantially by the Odums' book and their effective campaign for the functional view of the ecosystem in their other writing and research.

The "new ecology"—functional ecology, trophic ecology, ecosystem ecology, to cite a few of its aliases—had a number of distinctive emphases. Beginning with the study of aquatic ecosystems, it emphasized productivity, and ecologists devised new methods to measure the rate of formation or organic material (primary productivity) and the accumulated amount of living material (biomass) in plants and its transfer to other components (consumers) of the trophic (food) web, such as herbivores and carnivores and, eventually, from all of these to decomposers (microorganisms) that returned the materials back to the inorganic state, making them again available to plants, thus completing the biogeochemical cycle. These studies were aided by some of the new techniques developed during the war, for example the use of radioisotopes and new instrumentation that had become available after 1950. Associated with productivity were studies of ecological energetics and nutrient cycling.

The quantitative measurements of essential nutrients in various trophic levels, their pathway or cycle through the ecosystem, and, in some cases, loss from the ecosystem were significant developments in the 1950's and early 1960's. Widespread recognition of the vital importance of biogeochemical cycles to human welfare and concern about losses of essential plant nutrients or unfortunate concentrations of toxic substances stimulated scientific interest in mineral cycles. Recognition of widespread distribution of radioactive iodine and strontium, excessive phosphorus from sewage outfalls and detergents and its deleterious effects on lakes and streams, or the occurrence of toxic levels of nitrogen from fertilizer runoff brought nutrient cycles into the public domain.

Ecosystem energetics seized on the calorie as a common denominator to assess the function of ecosystems, and studies of ecological thermodynamics proliferated in various guises. E. P. Odum (1968) asserted, "Ecoenergetics is the core of ecosystem analysis." Studies of the energy dynamics of ecosystems are an important aspect of ecosystem ecology. The ultimate energy source for the earth's ecosystems, the sun, provides a nearly constant amount of energy, but its availability and use varies greatly over the earth's surface. Energy ultimately limits the productivity of ecosystems, and some ecologists (Gates, 1968; Gates and Schmerl, 1975) have urged the application of physical, micrometeorological, and mathematical techniques in ecology to produce detailed energy budgets that would allow the workings of the ecosystem to be understood.

H. T. Odum (1971), another notable proponent of ecological energetics, urges an engineering approach to ecology. In his view all phenomena are based on energy flows and can be expressed in the common denominator of energy and the language of electrical circuitry to allow a quantitatively integrated approach to ecology and to human affairs in general.

Other ecologists have reservations and question the faith in the calorie as the common denominator of ecology (Goldman, 1968; Paine, 1971). The ecosystem, like all other systems, obeys the first and second laws of thermodynamics, but it is not as yet clear that it does not have idiosyncracies that introduce complications. Some organisms simply do not eat or cannot digest perfectly good calories in the wrong package. A deer does poorly on balsam needles, while a moose thrives on them; or, as Goldman (1968) put it, "Animals are not bomb calorimeters."

A development that was to have a dramatic effect on ecology was the International Biological Program (IBP). It grew from the widespread concern in the 1950's about rapidly intensifying resource-related and environmental problems facing mankind. These problems, recognized by earlier ecologists, were brought into sharper focus by rapidly increasing human populations, incipient food shortages, and increasingly obvious environmental degradation. The IBP was initiated an an international undertaking around 1960, and the United States National Committee for the IBP was appointed in the National Academy of Sciences in 1965. A somewhat distinctive American twist was given to the American IBP at a meeting at Williamstown, Massachusetts, in 1966 (U.S. National Committee for the IBP, 1974). The proposals emanating from this meeting led to the analysis-of-ecosystems program, a valiant and unprecedented effort to develop large-scale, relatively well-funded, closely-linked, cooperative studies of whole ecosystems. The operational phase of the IBP began in 1967 with a group of Integrated Research Programs (IRP) as its primary concern. Perhaps the major impact of the IBP program was that it drastically revised the way ecology was done, and it produced a managerial class in ecology. A new era was clearly indicated when an ecological article described the deployment by a manager of a staff of mathematical modelers in an ecological research program (Pielou, 1972). These programs have not only supported but directed the course of ecology, and it is not entirely clear where they may lead. The travail of breaking the trail in the wilderness of management of large-scale ecological research is effectively described by Van Dyne (1972), the program director of the earliest and largest of the United States IBP ecosystem studies.

The IBP had multiple goals, including: (1) understanding the interactions of the many components of complex ecological systems, (2) exploiting this understanding to increase biological productivity, (3) increasing predictive capacity of the effects of environmental impacts, (4) enhancing human capacity to manage natural resources, and (5) advancing the knowledge of human genetic, physiological, and behavioral adaptations. It had made substantial progress in advancing understanding of ecosystems, particularly the physical and biological processes operating in them: photosynthesis and productivity, water and mineral-nutrient cycling, competition, predation, and, particularly, decomposition processes and the role of detritus in the ecosystem. It has certainly changed the way in which ecology is pursued, and some have proposed that a substantial restructuring in the education of ecologists is in order to train people to cope with the demands of the new ecology (Van Dyne, 1969a; Watt, 1966;

Council on Environmental Quality, 1974). Although the IBP ended as a nationally coordinated program in June, 1974, the substance of its results has yet to be fully appreciated. A synthesis of the results is to be published in a series of a dozen book-length monographs. A continuing impact will likely be felt in large-scale and long-term ecosystem studies and in applications of ecological research to management of natural resources and environmental concerns.

Modeling

One of the major aspects of the new ecology, particularly of the IBP, was a greatly increased emphasis on modeling. A striking departure from most of ecological tradition, except for population ecology, was the introduction of substantial numbers of mathematically sophisticated persons, in some cases physicists, mathematicians, or engineers, either as individuals or as members of teams. It was their function and interest to make models amenable to analytic or digital treatment that would transfer the myriad variables and properties of populations, processes, or even of entire ecosystems into an abstract image that, it was hoped, would fit or elucidate the "real world" of ecology. Models were designed to be dynamic and to possess in varying degrees: (1) realism—to correspond accurately to the "real-world" ecosystem, (2) precision—to predict quantitative changes, (3) generality—to be widely applicable, and (4) resolution—to show detail. It is not possible to achieve all these goals in a single model, and the recent trend in ecology has been to emphasize limited models when more ambitious dreams of whole ecosystem models proved difficult to realize. Nevertheless, the new breed of modelers in ecology believe that more general models will be useful in ecology as they have in other sciences. Robert May (1974), a prominent proponent of the potential of models in ecology, stated:

In ecology, I think it is true that tactical models . . . applied to specific individual problems of resource and environmental management have been more fruitful than has general theory and they are likely to remain so in the near future. But in the long run, once the "perfect crystals" of ecology are established, it is likely that a future ecological engineering will draw upon the entire spectrum of theoretical models, from the very abstract to the very particular, just as the more conventional (and more mature) branches of science and engineering do today.

May was likely speaking metaphorically, but the contrast of a "perfect crystal" model with the entire tradition and realities of ecology brings the ecologist up sharply. Crystals and living cells or organisms are commonly considered as two distinct classes (Blackburn, 1973), and the rigidity of crystal structure, even as a metaphorical model, is difficult to perceive in the organization of an ecosystem.

In the attempt to develop the perfect-crystal model for ecology, myriads of models, large and small, have appeared, and there is even some dispute over what constitutes a model. It harks back to an earlier day, when Allee and his associates (1949) warned: ". . . for all our emphasis on the need of ecological principles it must be emphasized again that in the formulating of principles, as in testing and extending them, evidence is basic." Some ecologists, concerned about the proliferation of models, suggest the need for more testing of models (Pielou, 1972), Schoener (1972) warns of a "constipating accumulation of models," and there is some suggestion that old models never die (McIntosh, 1962). A good example of the resilience of a model is the "broken-stick" model of Robert MacArthur (1957) which purported to represent the way relative abundance of individuals was distributed among the species of a community. It generated great interest among animal ecologists and was tried in a variety of communities, fitting some (mostly birds) but not others. After Pielou (1966) pointed out a mathematical deficiency, MacArthur (1966) described his intellectual offspring "an obsolete approach to community ecology which should be allowed to die a natural death." Far from being allowed a decent burial, the model was widely used until Hairston (1969) reviewed its inadequacies and asserted that "no biological significance can be attributed to the fact that a collection does or does not show a fit to the broken-stick model, and its usefulness in any ecological context is challenged." Nevertheless, as recently as 1974 the broken-stick model was in use, not even bent, much less broken, by these several attacks. It was still appealed to as an adequate model that showed consistent patterns in real communities, all without acknowledgment of the many criticisms of the model or even its author's request that it be allowed to die. Mathematical models of ecological processes and ecosystems proliferate, and articles and books about models in ecology appear with disquieting regularity, all suggesting that modelers have not yet finished with ecology and ecologists have not finished with models (Kadlec, 1974; Jeffers, 1972; May, 1973; Smith, 1974). It is not clear how much longer it will be true that "ecology is still a branch of science in which it is usually better to rely on the judgement of an experienced practitioner than on predictions of a theorist" (Smith, 1974).

Ecosystems Analysis

The extreme complexity and diversity of ecological systems, which had long made ecology a notably prolix subject, proved a challenge to a new generation of nontraditional ecologists. Stimulated by recent demands on ecology to provide more effective bases for resource management and attacking environmental problems, many ecologists have seized upon systems analysis as the wave of the future (Odum, 1971). Certainly, systems analysis washed over ecology during the last decade and thoroughly saturated some aspects of it, but whether the tide is still rising or ebbing is not clear. One of the reasons that the World War II era is an apt break in the history of ecology is that operations research and systems analysis were spawned during the war. Subsequently, systems analysis flourished in the physical sciences and engineering, and, more recently, was introduced into ecology. The systems approach has been one of the most heralded of the postwar entries into ecology, and Spurr (1964) commented on the etymological connection with the ecosystem: "And how should such a system be studied if not by systems analysis and the computer"? E. P. Odum (1971) noted several pioneers of systems analysis (G. Van Dyne, J. Olson, B. Patten, K. Watt, C. Holling and H. T. Odum) who were "revolutionizing the field of ecology." Systems analysis involves some formal methods that were introduced to the ecologist, for example, linear and nonlinear programming, game theory, decision theory, queuing theory, and extensive use of mathematics and computers. The methods of systems analysis were developed to apply to man-dominated, "complex aggregates of simpler components that interact to produce a desired result" and in which "the components are under design control" (Bode, 1967). Their application to biological systems that have a "mind" of their own may call for new developments in systems analysis (Reichle and Auerbach, 1972), much as Johnson (1970) suggested that description of living things would require modification of information theory. It is, however, a thesis of the new ecology that the time is ripe for advances in ecology through systems analysis, and the wave of the future rolls on, ignoring any commands to the sea to stand still uttered by ecological counterparts of King Canute.

One of the difficulties of the neophyte in systems ecology is to find out just what it is. A system as defined as "a set of objects together with relations between the objects and between their attributes" (Young, 1956). Systems ecology is pre-eminently the application of advanced mathematical techniques and computer technology to the analysis of ecosystems, but it is not a lineal descendent of traditional quantitative ecology or even cousin to other quantitative and mathematical approaches that have been more recently introduced into ecology. According to Rosen (1972), systems theory "amounts to a profound revolution in science—a revolution which will transform human thought as deeply as did the earlier ones of Galileo and Newton." Rosen noted further that systems analysis is in the eye of the beholder and that it is different things to different people, an observation that is amply born out in a review of the literature on systems analysis in ecology.

The first linking of ecology with systems theory seems to have been a publication of a Spanish marine biologist and theoretical ecologist, R. Margaleff (1958). It was not until the mid-1960's, however, that systems analysis, as a more-or-less explicit approach, became a force in ecology. Strikingly, systems analysis, the mostly extravagantly heralded development in the history of ecology, is poorly represented in the ecological research literature. It is difficult to identify its beginnings because, like modeling, it came in so many guises and largely appeared outside the conventional ecological research publications in *Memo Reports of the IBP Biomes* and several recent books (Watt, 1966, 1968; Van Dyne, 1969; Patten, 1971).

The techniques and concepts of systems analysis derive substantially from engineering (Bode, 1967), and the future linking of ecology and engineering is commonly indicated. Odum (1971) comments on systems ecologists who are "providing a vital link with engineering where systems analysis procedures have been in use for some time." Patten (1971) noted that systems ecology is a hybrid of biology and engineering and predicted that if ecologists "open the door ever so slightly" a shotgun marriage, if not a rape. In either case he anticipates an inevitable kinship between engineers and ecologists helped by demonstrating how ecology can be cast into *"their terms"* (italics added). Levins (1968) viewed at least one approach to systems ecology as a "school of ecological engineering," and engineers, notably electrical engineers, have entered into ecology through systems analysis. Some ecologists have embraced systems analysis wholeheartedly and see in it the wave of the future for ecology; others are less confident that systems ecologists can stay afloat. Deevey (1968), for example, said ". . . it is not certain that the systems approach, powerful though it is, can be made general enough to handle the many weak interactions involved in community organization." Slobodkin (1968), however, predicted that systems analysis "will be a major area of intellectual concern 20 years from now." Given the inertia of scientific developments, this seems a safe prediction.

Quite apart from the emphasis on ecosystem ecology, but with the similar effect of breaking down the distinction between plant and animal ecology, a significant development in postwar ecology was a renewal of interest in the plant-animal interface. The importance of the interaction of plant and animal in community control was argued by Hairston, Smith, and Slobodkin (1960), followed by an extended interchange exploring the question. Harper (1969) examined the role of herbivores in modifying or controlling vegetation. Jansen (1970) placed major importance on the role of animals in controlling plant populations in the tropical forest. Studies of food preferences, nutritional quality, and factors that influence herbivory are causing a reassessment of the role of secondary substances in plant metabolism. These had long been considered waste products, but in the light of the renewed interest in chemical interactions of plants and animals this evaluation is being reinterpreted (Whittaker and Feeney, 1971; Dethier, 1970). Chemical interactions of plants (allelopathy) had been known since 1881, but the years since World War II have produced effective demonstrations of its importance in certain communities (Rice, 1964; Del Moral and Muller, 1970), and it may play an important role in community organization and evolution (Muller, 1970).

Theoretical Ecology

One of the most frequently encountered criticisms of ecology throughout its early years and up to the present was the extensive collection of data in the absence of organizing principles or theory. Yet in one aspect of ecology—populations—there was a long history of mathematical treatment that was available to begin a theoretical framework for ecology. These analytic mathematical treatments of populations did not have substantial influence on ecology generally until recently (Allee et al., 1949; Cole, 1954; Watt, 1962). Among the noteworthy changes of the last decade are increased calls for and interest in mathematical and theoretical approaches to ecology; and in animal ecology at least there has been an extension of the early mathematical theory of populations to the community.

There had been one premature effort (Haskell, 1940) to provide for ecology a mathematical systemization of *environment, organism,* and *habitat*—terms that about sum up ecology. Its author was critical of the structure of ecological concepts and terminology and asserted, "It is no more possible to make present ecological theory produce accurate predictions than to make a wild cherry tree produce fancy dessert cherries." He called somewhat patronizingly for a drastic linguistic transformation of ecology, saying, "For ecology to become an exact science, . . . all its basic notions will have to be redefined and stated in such ways as to be fairly functionally constant and amiable to progressive regularization." In a prescient discussion anticipating recent theoretical interests in ecology (for example, ordination and niche), he said that the only kind of geometry that would encompass the environment-organism relation is an *N*-dimensional geometry, and he referred to "hyperbodies" as spaces representing the effective regions of various processes of an organism. He also posed the concept of entropy as "the first contribution to a unified ecological theory." Although Haskell's paper left no ripple in ecology in the 1940's, redefinition of old concepts (such as niches) in terms of set theory, the concepts of entropy, and multidimensional or *N* hyperspaces all came to the fore in the 1950's and 1960's in a dramatic resurgence of the search for a theoretical ecology.

The earliest postwar entrant into ecological theory was information theory in the work of R. Margalef (1958). The concepts of information, entropy, and cybernetics were briefly intriguing to ecologists, and Patten (1959) proposed a general theory of ecosystem dynamics elucidated by information theory. Information theory has not been generally productive in biology (Johnson, 1970), and it has largely disappeared from ecology, leaving behind a residue in the form of the Shannon-Weaver equation, which has been widely applied to quantify anything that came to hand, particularly species diversity and niche breadth.

Ecologists had long been intrigued by the variation in the number of species in certain areas and by the variation in the number of individuals per species. Plant ecologists from Gleason on had been interested in species-area curves or species-individual curves. The relation between number of species and number of individuals per species in a community, under the rubric diversity, has become an active area of theoretical ecology. Diversity was regarded as a fundamental attribute of a community, has been variously related to productivity, niche, successional development, and particularly stability, and was the subject of a wide-ranging but inconclusive symposium in 1969 (Brookhaven Symposia in Biology, 1969). The conventional wisdom of ecology held that diversity enhanced ecosystem stability by increasing the number of links in the ecological web. This idea became almost axiomatic to some biologists despite indications that diversity of trees in relatively stable or climax forests was less than that in seral or changing forests, and recently May (1974) has argued against the tradition on theoretical-mathematical grounds.

A more fruitful effort in theoretical ecology, at least bibliographically, was the formalization of the niche concept in terms of set theory by G. E. Hutchinson (1957a). The concept of the niche as an *N*-dimensional hyperspace was avidly seized on, especially by animal ecologists, has gone through a series of modifications (Maguire, 1973; Whittaker, Root, and Levin, 1972; Vandemeer, 1972), and is incorporated in ecological theory (Levins, 1968). Darnell (1970) said that nomenclature in ecology "has subsided into a few useful handles for the most basic concepts." This is somewhat wishful thinking, for new terms spring up as frequently as ever. Niche alone has already generated a number of terms peculiar to its discourse (among them fundamental, realized, included, partial, actual, virtual, and envelopes) in addition to the confusion about the meaning of the term niche itself (Whittaker, Levin, and Root, 1973). Notwithstanding, there has developed a considerable literature on niche mensuration by use of several numerical indices (Levins, 1968; Colwell and Futuyma, 1971), which have given rise to a number of theoretical generalizations. Niche has been said to suggest a "predictive practical ecology" (Lewandowsky, 1972) that would substantially fulfill the dream of a theoretical ecology. What is intriguing is the spurious accuracy to which niche measures are sometimes determined to four significant digits in violation of elementary canons of quantitation.

The 1960's was the decade in which theoretical-mathematical ecology burgeoned. The IBP was seen as a step toward developing a new field of theoretical ecology, allowing predictions about natural and man-made ecological changes (U.S. National Committee for IBP, 1974). R. Lewontin (1970) commented that ecology was being transformed from a descriptive science to a quantitative and theoretical one. He attributed the change to a union of mathematics and evolution and said that the theoretical framework for ecology was *"the concept of the vector field in N-dimensional space,"* citing particle physics as a model for ecology. *Science* (Kolata, 1974) published a commentary, "Theoretical Ecology: Beginnings of a Predictive Science," which reviewed the thrust of theoretical ecology produced by the late Robert MacArthur, his associates, and his students. It seems to be a misfortune of ecology that several of its most productive and imaginative personalities (such as Robert MacArthur, Raymond Lindemand, and John Curtis) died prematurely following major contributions to ecology. MacArthur's work is perhaps the outstanding example of a mathematical-theoretical approach to ecology, and he initiated or stimulated theoretical ecological work in species diversity, island biogeography, community organization, niche structure, *r*- and *k*-selection, and the effect of pattern (coarse and fine grain) on community organization. Whether this thrust of theoretical ecology is a major breakthrough of modern ecology (Fretwell, 1975) or "a public relations web," as one ecologist suggested (Mitchell, 1974), remains to be seen.

The revolutionizing of a science is not accomplished without breaking a few eggs, and the infusion of mathematics and new theoretical approaches was no exception. During the 1960's mathematically and theoretically oriented persons complained bitterly that they were unable to get their work published in the usual ecological journals (Ecology Study Committee, 1965). In 1969 Wilson expressed a fear of an "unholy alliance between the population model builders and molecular biologists to exclude systematists and descriptive ecologists." In 1974, Van Valen and Pitelka voiced a complaint that mathematical ecologists, having entered the ecological establishment, were pursuing a policy of intellectual censorship against ecologists who studied real organisms. The suggestion of a schism or a divisive power struggle in ecology with some sacred cows endangered is not unique in the history of science, but it is hoped that this can be avoided.

Mathematical and theoretical ecology is in full cry in the 1970's, and, like many active scientific areas, there are differences of opinion, some honest, some not so honest, and even basic differences in philosophy. Lewontin (1968) called for a reductionist approach to ecology and evolution and deplored "biologists who reject the analytic method and insist that the problems of ecology and evolution are so complex that they cannot be treated except by holistic statements." Lane, Levins, and Lauff (1974) take a diametrically opposed position, saying:

In ecology there has been too much reliance on the assumption that once small portions of the system are studied, the whole can be reconstituted from the part. . . . While reductionist approaches have always been more popular in all areas of science, we believe it is possible and indeed desirable to develop holistic approaches for ecosystem analysis.

There are those who look on the physicists, mathematicians, and engineers recently entering ecology and bringing new types and levels of mathematical approaches with them as carpetbaggers. Others are more hospitable, but many questions are raised. Slobodkin, an active student of population mathematical and theoretical ecology, commented that "the normal criteria of scientific quality which we use as biologists, are not the same as those of the physicists and mathematician" (Slobodkin, 1962). Ten years later, when the full force of mathematical ecology was impinging on ecology, Slobodkin (1974) offered a list of ten caveats to mathematicians involved in population biology and ecology. Among the more trenchant of these are the second and the fifth:

2. I wish they would not build theories that involve extremely precise exponential functions. . . . The ecological world is a sloppy place.

5. I wish mathematical ecologists would stop rediscovering the wheel. . . . All would-be mathematical modelers should read the extant works on mathematical ecology, and three other books, at a minimum, on descriptive ecology.

This particular criticism of theoretical ecology had been voiced earlier in a review of a book (one of a prominent series on theoretical ecology) presenting a theory of forest structure and succession:

Much of this type of canopy analysis has been done before in a much more sophisticated manner; reading this book is therefore rather like discovering a tribe lost to civilization that has quite independently discovered a primitive form of the internal combustion engine. Does one praise the originality or sympathize with the ignorance? [*Harper, 1972*].

Slobodkin's sixth caveat is:

6. I wish they would stop developing biological nonsense with a mathematical certainty.

One of the major criticisms of mathematical-theoretical approaches in ecology is that they commonly rest on simplifying assumptions, often unstated, that make them tractable mathematically but nonsense biologically. The problem is basically the same one raised by Stanley Cain (1944): whether the complexities of ecology are amenable to mathematical treatment. If not, it will not be for want of trying, and there is a danger that ecology may be cut or stretched to fit not a bed of Procrustes but a rectangular matrix. Theorists have been accused of building unfalsifiable hypotheses into untestable theories and of cooking up the ecological omelet by their own recipe. A more optimistic hope is that new mathematics may be called forth by the challenge of ecological phenomena:

There is every reason to hope that original and creative mathematics will be developed by mathematicians who will become well versed in biology as a discipline, rather than attempting to force biological phenomena into a mold created by hydrodynamics, economics, physics or what have you [*Slobodkin, 1974*].

Perhaps the best advice for bruised egos and stepped-on toes comes from one of the most active of mathematical-theoretical ecologists, Robert May (1974), who counsels the need for a full spectrum of theoretical activity from abstract to specific *"with sympathetic handling"* (italics added). May, however, repeats the familiar comment that ecology is "immature," by which he means it is not mathematical in the manner of physics. This may be a questionable generalization, as Slobodkin's comments suggest.

Whatever the case for the new ecology or any of the elements of the new look in ecology—modeling, systems analysis, theoretical ecology—the relatively ugly duckling of biological science of the 1920's and the 1950's has emerged with a new vitality and a new authority. With the influx of very large sums of money many additional hands and minds and a new visibility and public image, ecology has fulfilled some of the ideals claimed by its early practitioners. It has increased understanding of the functioning of both natural and human-controlled ecosystems, has begun really effective work on the role of decomposer organisms in the maintenance of cycles of nitrogen and other vital nutrients, has made substantial progress in the study of population mechanisms (for example, competition, predation) and their role in the organization of communities, and has developed an impressive range of techniques for ordering large amounts of data to analyze communities and niche relationships of species in communities. There has been significant increase in the knowledge of the physiological properties of organisms as these relate to ecological attributes and response to environmental stress and of the genetic mechanisms involved in these properties. Ecology has returned to the emphasis of its founders in their concern for the centrality of evolutionary biology with the emergence of a comparative approach to species and even community evolution. The new look in ecology is seen in the very existence of a document entitled *The Role of Ecology in the Federal Government,* which begins with the unequivocal assertion, "Ecology as a science has come of age" (Council on Environmental Quality and Federal Council for Science and Technology, 1974).

The Environmental Movement

In 1950 ecology was fairly well established in the academic world, although it was still peripheral in many biology departments. Outside of academic and limited professional circles ecology was little known, and its potential importance for human well-being was not recognized, although it had been a tenet of ecologists from its very beginnings in America. Perhaps the really significant development in ecology was its emergence from obscurity in the shadow of molecular biology into the limelight. This clearly was a consequence of the rapidly mounting concern of the American public about the quality of its environment. The awakening was stimulated by early concerns about the effects of radioactive fallout and by Rachel Carson's book on the hazards of pesticides, *Silent Spring* (1962), which succeeded where earlier cries of alarm had failed to arouse much interest (Fleming, 1972).

Once aroused, concern with the environment grew exponentially, in proper fashion, and myriads of articles and books in the popular press belabored the point that the world's ecosystems were in serious trouble. There were ample expressions of concern about radiation, the spread of industrial and urban decay, the population explosion, and attendant consequences, including the food crisis and the ultimate reality of starvation. Ecology was dubbed the subversive science and was itself in danger of subversion and incorporation into various ideologies associated with the environmental movement. As Allee and others had warned (1949), high prosperity and the accompanying notoriety had distorting influences as well as stimulating a rapid and turbulent growth of ecology. If the ecology of 1903 as seen by Cowles was chaos, that of 1970 was *Chaos chaos,* making it an amoeba, which is not an inapt term, for ecology grew and spread in many directions, sometimes propelled by its inner intellectual motivations, at other times simply drawn into a vortex of environmental questions with which it was ill-prepared to deal. Instant ecologists like Barry Commoner came to the fore to fill the rhetorical gap not adequately filled by working ecologists. The environment movement threatened to turn ecology from science to ideology, and the long-held dogmas of ecology could, without undue distortion, be linked with those of Zen (Barash, 1973). The terms "revolution" and "explosion" have frequently been applied to developments in ecology and environmental concerns in recent years, and perhaps the greatest burden placed on ecology was the revolution in expectations about its role in the development and management of natural resources and maintenance of the quality of the human environment (Slobodkin, 1968; McIntosh, 1974a; Council on Environmental Quality and Federal Council for Science and Technology, 1974).

Ecology had always been closely aligned with the traditional conservation movement as its scientific arm. The environmental movement is basically an enormous expansion of the conservation movement with many enlarged, scientific, economic, historical and political consequences. In effect, it has become the central issue of modern society. Thus it is not surprising to find a resurgence of interest in human ecology, which was a primary interest in early ecology but was too indigestible for the limited conceptual structure of ecology to embrace. In spite of abortive efforts in the 1930's and 1950's, it is only in the 1970's that there seems to be a determined effort to develop a general ecology incorporating the complex of human affairs, including urban ecosystems.

Ecologists, and the Ecological Society of America from its inception, were involved in alerting America to concerns of the environment and what has come to be called the "environmental crisis." As individuals ecologists have devoted much time and effort to providing professional expertise in questions of waste disposal; resource management; the impact of human use of the environment; erosion effects and control; the effects of strip mining and revegetation of mined areas; the effects and role of fire, herbicides, pesticides, and pollutants generally in ecosystems; and innumerable other environmental concerns. Ecologists have increasingly become involved in matters that are beyond their ecological training and are called upon to interact with other scientists, engineers, lawyers, economists, and other social scientists, as well as diverse governmental agencies at local, state, federal, and even international levels.

Because of his expertise the ecologist often finds that he can no longer stick to his last but must frequently enter the public arena. The Ecological Society of America has recently proposed adoption of a code of ethics for ecologists, and a more conspicuous person on the environmental scene is the ecologist as a practitioner applying his ecological knowledge to practical problems. Only in the past few years have ecologists been hired by consulting firms, banks, and utilities. The result has been a substantial commercial enterprise of consulting ecology. The principal stimulus was the National Environmental Policy Act (NEPA) of 1970, which calls for environmental impacts for major federally financed developments to provide a statement of environmental consequences. Ecologists necessarily cooperate with other scientists and nonscientists in preparing these statements.

An institutional approach to these same ends is the Institute of Eology (TIE), founded in 1971 by ecologists. TIE is an international consortium devoted to ecological research to meet human needs and, particularly, to incorporation of ecological analyses in policy formation and education of the public on environmental issues. It deal with problems beyond the scope of individuals, single institutions, or even of ecologists as specialists, and its structure incorporates individuals and organizations not restricted to professional ecologists. TIE is the latest and, in some respects, the most encompassing effort to make ecological science relevant in human affairs.

Perhaps as a consequence of the increasing role of ecology in environmental affairs, there has appeared in the last decade a rash of interdisciplinary programs, institutes, and other academic aggregations focusing on environmental and ecological concerns, embracing, at least rhetorically, the essential precepts of ecology as a way of dealing with the environment and, as Slobodkin (1969) notes, attracting the attention of academic vice-presidents. Not the least significant indication of the new visibility of ecology is that the august National Academy of Sciences (NAS) has seen fit to increase its representation of ecologists from two (G. E. Hutchinson and Alfred Emerson) to a modest number. The lack of representation of ecologists in the NAS was noted by W. F. Taylor (1936), who counseled ambitious ecologists, "If you want to get elected to the National Academy, . . . select an old established science and stay right in the middle of it!" It may be a slight stretch to note that the distinguished animal behaviorist N. Tinbergen, who was among the authors in the first ambitious attempt to assess ecology in 1939, in 1973 shared (with K. Lorenz and K. von Frisch) the Nobel Prize in Biology and Medicine. This marks a dramatic change in eligibility for that award recently lusted after largely by molecular biologists. A new era is also evidenced in the institution in 1972 of the Tyler Ecology Award. It is the largest monetary award for science. In 1974 it was shared by G. E. Hutchinson with A. Haagen-Smit and M. F. Strong. In 1975 it was bestowed on Ruth Patrick, a leading aquatic ecologist.

The essence of the modern environmental movement is that it is now clear at all levels of human concern—scientific, sociopolitical, and religious—that man's fate, on earth at least, is intimately and ultimately integrated with the natural and managed ecosystems of the earth, their biogeochemical processes, nutrient cycling, transformation of energy, and productive capacity. While all sciences contribute to the factual describing of the universe in which man functions, it is essentially biology, particularly ecology, that integrates the information available into an understanding of the ecosystems of which man is a part, which he in part creates, which he can readily destroy, and the rules of which he must live by however he manipulates them. Aldo Leopold (1949) expressed this eloquently in his land ethic, and there has been ecological feedback into philosophy and ethics:

Ecology provides a model to philosophy and to the other human sciences of a new way of viewing the interrelationships between the phenomena of nature. Central to its perspective is the idea of ecosystem analysis and the concepts of the balance of Nature. . . . The answer to the value question then from an ecological point of view is this: human values are founded in objectively determinable ecological relations within Nature [*Colwell, 1970*].

It is fortunate that this passage was written not by an ecologist but by a philosopher, for ecologists have been accused of having visions of grandeur, and so broad a claim would simply confirm that accusation.

A philosophical overtone in ecological thought is not entirely new, nor is an ecological overtone in philosophical thought; but a concept that allows nature or scientific consideration of nature to inform philosophy contradicts a long-argued precept of philosophy. Clearly, early American writers and philosophers like Thoreau and Emerson were influenced by nature in their philosophical discourse. Ecologists have often ventured into aspects of value and ethics but have rarely argued from ecological grounds to aesthetic or ethical conclusions. Something of a departure from this is seen in the writings of Aldo Leopold (1949) and, from the philosophical camp, in *Man's Responsibility for Nature* (Passmore, 1973). What is most interesting is to see interpretations by philosophers such as that quoted above and more guardedly discussed in the pages of *Ethics,* by Rolston (1975). Rolston poses a new and difficult axis to the ecologist's *N*-dimensional space where "the topography is largely uncharted; to cross it will require the daring and caution of a community of scientists and ethicists who can together map both the ecosystem and the ethical grammar appropriate for it."

Perhaps the greatest of challenges to the new ecologist is on the horizon of Charles Darwin's and Aldo Leopold's vision.

BIBLIOGRAPHY

Adams, C. C. 1913. Guide to the study of animal ecology. MacMillan Co., New York, 183pp.

Adams, C. C. 1917. The new natural history-ecology. Am. Mus. J. 7:491–94.

Allee, W. C. 1927. Animal aggregations. Quart. Rev. Biol. 2:367–98.

Allee, W. C. 1931. Animal aggregations. A study of general sociology. Univ. Chicago Press, Chicago, 431pp.

Allee, W. C. 1938. The social life of animals. Norton, New York, 293pp.

Allee, W. C. 1939. An ecological audit. Ecology. 20:418–21.

Allee, W. C., and T. Park. 1939. Concerning ecological principles. Science. 89:166–69.

Allee, W. C., A. E. Emerson, O. Park, T. Park, and K. P. Schmidt. 1949. Principles of animal ecology. W. B. Saunders Co., Phila., 837pp.

Austin, M. P., and Noy-Meir. 1971. The problem of nonlinearity in ordination: experiments with two gradient models. J. Ecol. 59:763–73.

Barash, D. P. 1973. The ecologist as Zen master. Am. Mid. Nat. 89:375–92.

Beals, E. W. 1973. Ordination: mathematical elegance and ecological naivete. J. Ecol. 61:23–36.

Blackburn, T. R. 1973. Information and the ecology of scholars. Science. 181:1141–46.

Bode, H. W. 1967. The systems approach. pp. 73–94. In Applied science and technological progress. Report to the Committee on Science and Astronautics by the National Academy of Sciences, Washington, D.C.

Brookhaven Symposia in Biology. 1969. Diversity and stability in ecological system. No. 22. Brookhaven Nat'l. Lab., Upton, New York, 264pp.

Bush, V. 1945. Science the endless frontier. U.S. Govt. Printing Office, Washington, D.C., 184pp.

Cain, S. A. 1944. Foundations of plant geography. Harper and Brothers, New York, 556pp.

Carson, R. 1962. Silent spring. Houghton Mifflin Co., Boston, 368pp.

Chapman, R. N. 1931. Animal Ecology. McGraw-Hill Book Company, Inc., New York, 463pp.

Clarke, G. L. 1954. Elements of ecology. John Wiley and Sons Inc., New York, 534pp.

Clements, F. E. 1905. Research methods in ecology. Univ. Pub. Co., Lincoln, Nebr., 334pp.

Clements, F. E. 1924. Methods and principles of paleo-ecology. Year Book. Carnegie Inst. Wash. 32.

Clements, F. E., and R. Pound. 1898–1902. A method of determining the abundance of secondary species. Minnesota Bot. Stud. 2:19–24.

Clements, F. E., and V. Shelford. 1939. Bio-ecology. J. Wiley and Sons Inc., New York, 425pp.

Cole, L. C. 1954. The population consequences of life history phenomena. Quart. Rev. Biol. 29:103–37.

Colwell, R. K., and D. J. Futuyma. 1971. On the measurement of niche breadth and overlap. Ecology. 52:567–76.

Colwell, T. B., Jr. 1970. Some implications of the ecological revolution for the construction of value. pp. 245–58. In E. Lazlo and J. B. Wilbur (ed.). Human values and natural science. Gordon and Breach Sci. Publ., New York.

Cottam, G., and J. T. Curtis. 1949. A method for making rapid surveys of woodlands by means of pairs of randomly selected trees. Ecology. 30:101–104.

Council on Environmental Quality and Federal Council for Science and Technology. 1974. The role of ecology in the federal government. Report of the committee on ecological research. December 1974, 78pp.

Cowles, H. C. 1904. The work of the year 1903 in ecology. Science 19:879–85.

Curtis, J. T. 1959. The vegetation of Wisconsin. Univ. of Wisconsin Press, Madison, 657pp.

Darnell, R. M. 1970. The new ecology. Bioscience. 20:746–48.

Daubenmire, R. 1947. Plants and environment. J. Wiley, New York, 424pp.

Deevey, 1964. General and historical ecology. Bioscience 14:33–35.

Del Moral, R., and C. H. Muller. 1970. The allelopathic effects of *Eucalyptus camaldulensis*. Am. Mid. Nat. 83:254–82.

Dethier, V. 1970. Chemical interactions between plants and insects. pp. 82–102. In E. Sondheimer and J. B. Simeone (Eds.) Chemical Ecology. Academic Press, New York.

Ecology Study Committee, 1965. Summary report with recommendations for the future of ecology and the Ecological Society of America [15 April 1965].

Edmondson, Y. H. 1971. Some components of the Hutchinson legend. Limnol. Oceanog. 16:157–61.

Egerton, F. N. 1962. The scientific contributions of Francois Alphonse Forel, the founder of limnology. Schweizerische Zeitschrift Fur Hydrologie. 214:181–99.

Egerton, F. N. 1976. Ecological studies and observations in America before 1900. In B. J. Taylor and T. J. White (eds.). Evolution of issues and ideas in America, 1776–1976. Univ. of Oklahoma Press, Norman, 1976.

Egler, F. E. 1942. Vegetations as an object of study. Phil. of Sci. 9:245–60.

Egler, F. E. 1951. A commentary on American plant ecology based on the textbooks of 1947–1949. Ecology. 32:673–95.

Elton, C. 1927. Animal ecology. Macmillan, New York, London, 209pp.

Elton, C. 1933. The ecology of animals. John Wiley and Sons Inc., New York, 97pp.

Fleming, D. 1972. Roots of the new conservation movement: Perspectives. Am. Hist. 6:7–94.

Forbes, S. A. 1887. The lake as a microcosm. Bull. Ill. Nat. Hist. Survey. 15:537–50.

Forbes, S. A. 1907. On the local distribution of certain Illinois fishes: An essay in statistical ecology. Bull. Ill. Lab. Nat. Hist. 7:1–19.

Forbes, S. A. 1909. An ornithological cross-section of Illinois in autumn. Bull. Ill. State Lab. Nat. Hist. 7:305–35.

Fretwell, S. D. 1975. The impact of Robert MacArthur on ecology. Ann. Rev. Ecol. Syst. 6:1–13.

Frey, D. (ed.) 1963. Limnology in North America. Univ. of Wisconsin Press, Madison, 734pp.

Ganong, W. F. 1904. The cardinal principles of ecology. Science. 19:493–98.

Gates, D. M. 1968. Toward understanding ecosystems. Adv. Ecol. Res. 5:1–36.

Gates, D. M., and R. B. Schmerl. 1975. Perspectives of biophysical ecology. Springer-Verlag, New York, 609pp.

Gauch, H. G., Jr., G. B. Chase, and R. H. Whittaker. 1974. Ordination of vegetation sample of Gaussian species distributions. Ecology. 55:1382–90.

Gleason, H. A. 1917. The structure and development of the plant association. Bull. Torrey Bot. Club. 44:463–81.

Gleason, H. A. 1926. The individualistic concept of the plant association. Bull. Torrey Bot. Club. 53:1–20.

Gleason, H. A. 1936. Twenty-five years of ecology, 1910–1935. Mem. Brooklyn Bot. Gard. 4:41–49.

Goldman, C. R. 1968. Aquatic primary production. Am. Zool. 8:31–42.

Goodall, D. W. 1952. Quantitative aspects of plant distribution. Biol. Rev. 27:194–245.

Green, R. H. 1974. Multivariate niche analysis with temporally varying environmental factors. Ecology. 55:73–83.

Greig-Smith, P. 1957. Quantitative plant ecology. Butterworth's Sci. Publ., London, 198p.

Grinnell, Jr. 1917. The niche relationships of the California thrasher. Auk. 34:427–33.

Haeckel, E. 1891. Planktonic studies: a comparative investigation of the importance and constitution of the pelagic fauna and flor. (Trans.) Rept. U.S. Fish Comm. 1889–91. pp. 565–641.

Haeckel, E. 1898. The evolution of man. 2 vols. Kegan Paul, London.

Hairston, N. G. 1969. On the relative abundance of species. Ecology. 50:1091–94.

Hariston, N. G., F. Smith, and L. B. Slobodkin. 1960. Community structure, population control and competition. Am. Nat. 94:421–25.

Hardin, G. 1962. The competitive exclusion principle. Science. 131:1292–97.

Hardin, G. 1968. The tragedy of the commons. Science. 162:1243–48.

Harper, J. L. 1967. A Darwinian approach to plant ecology. J. Ecol. 55:247–70.

Harper, J. L. 1969. The role of predation in vegetation diversity. pp. 48–62. In Diversity and stability in ecological systems. Brookhaven National Laboratory, New York, No. 22.

Harper, J. L. 1972. Review of H. S. Horn. The adaptive geometry of trees. Science. 176:660–61.

Harshberger, J. 1911. Phytogeographic survey of North America. G. E. Stechert, New York, 790pp.

Haskell, E. F. 1940. Mathematical systematization of "environment," "organism" and "habitat." Ecology. 21:1–16.

Hesse, R. 1924. Tiergeographie auf oekologischer grundlage. Fischer, Jena, 613pp.

Hesse, R., W. C. Allee, and K. P. Schmidt. 1937. Ecological animal geography. H. Wiley and Sons, New York, 597pp.

Hutchinson, G. E. 1940. Review of Clements F. E. and V. E. Shelford. 1939. Bio-ecology. Ecology. 21:267–68.

Hutchinson, G. E. 1957a. Concluding remarks, Cold Spring Harbor Symp. Quant. Biol. 22:415–27.

Hutchinson, G. E. 1957b. A treatise on limnology. Vol. I. J. Wiley and Sons Inc. New York. 1015pp. Vol. II. 1967. 1115pp.

Huxley, L. 1903. Life and letters of T. H. Huxley. Macmillan Company, New York, 3 vols.

Jeffers, J. N. R. 1972. Mathematical models in ecology. Blackwell Sci. Publ., Oxford, 398pp.

Johnson, H. A. 1970. Information theory in biology after 18 years. Science. 168:1545–50.

Juday, C. 1940. The annual energy budget of an inland lake. Ecology. 21:438–50.

Just, T. (ed.). 1939. Plant and animal communities. Am. Midl. Nat. 21:1–255.

Kadlec, J. A. 1971. A partial annotated bibliography of mathematical models in ecology. Analysis of Ecosystems I.B.P. Univ. of Michigan, Ann Arbor. Unnumbered.

Kuhn. 1962. The structure of scientific revolutions. Univ. of Chicago Press, Chicago, 210pp.

Kittredge, J. 1948. Forest influences. McGraw-Hill Inc., New York, 394pp.

Kohn, A. 1971. Phylogeny and biogeography of Hutchinsonia: G. E. Hutchinson's influence through his doctoral students. Limnol. Oceanog. 16:173–76.

Kolata, G. G. 1972. Theoretical ecology: beginnings of a predictive science. Science 183:400–401 (cont. p. 450).

Lane, P. A., G. H. Lauff, and R. Levins. 1974. The feasibility of using a holistic approach in ecosystem analysis. pp. 111–11 in S. A. Levins (ed.). Ecosystem analysis and prediction. Proc. SIAM-SIMS Conference, Alta, Utah, July 1–5, 1974.

Leopold, A. 1933. Game management. Scribners, New York, 481pp.

Leopold, A. 1949. A Sand County almanac. Oxford Univ. Press, New York, 226pp.

Levandowsky, M. 1972. Ecological niches of sympatric phytoplankton species. Am. Nat. 106:71–78.

Levins, R. 1968. Evolution in changing environments. Princeton Univ. Press, Princeton, N.J., 120pp.

Lewontin, R. C. (ed.). 1968. Population biology and evolution. Syracuse, Univ. Press Syracuse, N.Y., 206pp.

Lewontin, R. C. 1970. The meaning of stability. pp. 13–23 in Diversity and stability in ecological systems. Brookhaven Nat'l. Lab. Symp. Biol. 22.

Lindeman, R. 1941. Seasonal food-cycle dynamics in a senescent lake. Am. Mid. Nat. 26:636–73.

Lindeman, R. L. 1942. The trophic-dynamic aspect of ecology. Ecology. 23:399–418.

Lussenhop, J. 1974. Victor Hensen and the development of sampling methods in ecology. J. Hist. Biol. 7:319–37.

Lutz, H. J., and R. F. Chandler. 1946. Forest soils. John Wiley and Sons Inc., New York, 514pp.

MacArthur, R. H. 1957. On the relative abundance of bird species. Proc. Nat. Acad. Sci. U.S. 43:293–95.

MacArthur, R. H. 1966. Note on Mrs. Pielou's comments. Ecology. 47:1074.

MacArthur, R. H. 1972. Geographical ecology. Harper and Row, New York, 269pp.

Maguire, B. Jr. 1973. Niche response structure and the analytical potentials of its relationship to the habitat. Am. Nat. 107:213–46.

Margaleff, R. 1958. Information theory in ecology. Gen. Syst. 3:36–71.

McDougall, W. B. 1927. Plant ecology. Lea and Febiger, Philadelphia, 326pp.

McIntosh, R. P. 1962. Raunkiaer's "law of frequency." Ecology. 43:533–35.

McIntosh, R. P. 1967. The continuum concept of vegetation. Bot. Rev. 33:130–87.

McIntosh, R. P. 1974a. Commentary—an object lesson for the new ecology. Ecology. 55:1179.

McIntosh, R. P. 1974b. Plant ecology 1947–1972. Ann. Miss. Bot. Gard. 61:132–65.

McIntosh, R. P. 1975. H. A. Gleason—"individualistic ecologist," 1882–1975. Bull. Torrey Bot. Club. 102:253–73.

Merriam, C. H. 1894. Laws of temperature control of the geographic distribution of terrestrial animals and plants. Nat. Geogr. Mag. 6:229–38.

Mitchell, R. 1974. Scaling in ecology. Science. 183:1131.

Moore, B. 1920. The scope of ecology. Ecology. 1:3–5.

Mortimer, C. H. 1956. E. A. Birge: an explorer of lakes. pp. 163–211. In Sellery, G. C. 1956. E. A. Birge: a memoir. Univ. Wisconsin Press, Madison.

Muller, C. H. 1970. The role of allelopathy in the evolution of vegetation. pp. 13–31. In P. R. Ehrlich (ed.). Biochemical coevolution. Oregon State Univ. Press, Corvallis.

Nice, M. M. 1937. Studies in the life history of the song sparrow I. Trans. Linn. Soc. 4:1–247.

Nichols, G. E. 1928. Plant ecology. Ecology. 9:267–70.

Odum, E. P. 1953. 1st ed. Fundamentals of ecology. 1st ed. W. B. Saunders Co., Philadelphia, 546pp. 2d. ed., 1958. 3d ed., 1971.

Odum, E. P. 1964. The new ecology. Bioscience. 14:14–16.

Odum, E. P. 1968. Energy flow in ecosystems. A historical review. Am. Zool. 8:11–18.

Odum, H. T. 1971. Environment, power and society. Wiley-Interscience, New York, 331pp.

Oosting, H. J. 1948. The study of plant communities. W. H. Freemand and Co., San Francisco, 389pp.

Osburn, F. 1948. Our plundered planet. Little, Brown and Co., Boston, 217pp.

Paine, R. T. 1971. The measurement and application of the calorie to ecological problems. Adv. Ecol. Syst. 2:145–62.

Passmore, J. 1974. Man's responsibility for nature. Charles Scribner's Sons, New York, 213pp.

Patil, G. P., E. C. Pielou, and W. E. Waters. 1971. Statistical

Ecology. 3 vols. Pennsylvania State Univ. Press, University Park.
Patten, B. C. 1959. An introduction to the cybernetics of the ecosystem: the trophic dynamic aspect. Ecology. 40:221–31.
Patten, B. C. 1971. Systems analysis and simulation in ecology. Vol. I. Academic Press Inc., New York, 610pp.
Pearl, R. 1925. The biology of population growth. Knopf, New York, 260pp.
Pearl, R., and L. J. Reed. 1920. On the rate of growth of the population of the United States since 1790 and its mathematical representation. Proc. Nat. Acad. Sci. 6:275–88.
Pearse, A. S. 1926. Animal ecology. McGraw-Hill Book Co., Inc., New York, 417pp.
Pielou, E. C. 1966. Comment on a report by J. H. Vandermeer and R. H. MacArthur concerning the broken stick model of species abundance. Ecology. 47:1073–74.
Pielou, E. C. 1969. An introduction to mathematical ecology. Wiley Interscience, New York, 286pp.
Pielou, E. C. 1972. On kinds of models. Science. 171:981–82.
Reichle, D. E., and S. I. Auerbach. 1972. Analysis of ecosystems. pp. 260–80. In J. A. Behnke (ed.). Challenging biological problems. Oxford Univ. Press, New York.
Rice, E. L. 1964. Inhibition of nitrogen-fixing and nitrifying bacteria by seed plants. Ecology. 45:824–837.
Riley, G. A. 1971. Introduction. Limnol. Oceanog. 16:177–79.
Rolston, H. Jr. 1975. Is there an ecological ethic? Ethics 85:93–109.
Rosen, R. 1972. Review of: Trends in General Systems Theory. Klir (ed.) Wiley-Interscience, New York. Science. 177:508–509.
Schimper, A. F. W. 1898. Pflanzengeographie auf physiologischer Grundlage. Jena., Fischer, 870pp. Trans. by Wm. A. Fischer. 1903. Plant geography upon a physiological basis. Clarendon Press, Oxford, 839pp.
Schoener, T. W. 1972. Mathematical ecology and its place among the sciences. Science. 178:389–91.
Sears, P. B. 1935. Deserts on the march. Univ. Oklahoma Press, Norman, 231pp.
Sears, P. B. 1944. The future of the naturalist. Am. Nat. 78:43–53.
Sears, P. B. 1949. Integration at the community level. Am. Sci. 37:235–42.
Sears, P. B. 1956. Some notes on the ecology of ecologists. Sci. Monthly. 83:22–27.
Shelford, V. E. 1913. Animal communities in temperate America. Univ. Chicago Press, Chicago, 368pp.
Shelford, V. E. 1929. Laboratory and field ecology. Williams and Wilkins Co., Baltimore, 608pp.
Shelford, V. E. 1938. The organization of the ecological society of America. Ecology. 19:164–66.
Shelford, V. E. 1963. The ecology of North America. Univ. Illinois Press, Urbana, 610pp.
Slobodkin, L. B. 1965. On the present incompleteness of mathematical ecology. Am. Sci. 53:347–57.
Slobodkin, L. B. 1968. Aspects of the future of ecology. BioScience. 18:16–23.
Slobodkin, L. B. 1969. Pathfinding in ecology. Science. 164:817.
Slobodkin, L. B. 1974. Comments from a biologist to a mathematician. pp. 318–29. In S. A. Levin (ed.). Proc. SIAM-SIMS Conference, Alta, Utah, July 1–5, 1974.
Smith, J. M. 1974. Models in ecology. Cambridge Univ. Press, New York, 146pp.
Spurr, S. M. 1964. Forest ecology. Ronald Press Co., New York, 352pp.
Stauffer, R. C. 1957. Haeckel, Darwin and ecology. Quart. Rev. Biol. 32:138–44.
Stoddard, H. L. 1932. The bobwhite quail. Scribners, New York, 559pp.
Taylor, W. P. 1936. What is ecology and what good is it? Ecology. 17:333–46.
Terborgh, J. 1970. Distribution on environmental gradients: theory and a preliminary interpretation of distributional patterns in the avifauna of Cordillera Vilcabama, Peru. Ecology. 52:24–40.
Transeau, E. N. 1926. The accumulation of energy by plants. Ohio J. Sci. 26:1–10.
U.S. National Committee for the International Biological Program. 1974. U.S. participation in the International Biological Program. Report No. 6. National Academy of Sciences, Washington, D.C., 166pp.
Vandermeer, J. H. 1972. Niche theory. Ann. Rev. Ecol. Syst. 3:107–32.
VanDyne, G. M. 1969a. Implementing the ecosystem concept in training in the natural resource sciences. pp. 327–367. In G. M. Van Dyne (ed.). The ecosystem concept in natural resource management. Academic Press, New York.
Van Dyne, G. M. (ed.). 1969b. The ecosystem concept in natural resource management. Academic Press, New York, 383pp.
VanDyne, G. M. 1972. Organization and management of an integrated ecological research program. pp. 111–172. In I.N.R. Jeffers (ed.). Mathematical models in ecology. Blackwell's Scientific Publications, Oxford.
Van Valen, L., and F. Pitelka. 1974. Commentary: Intellectual censorship in ecology. Ecology. 55:925–26.
Verhulst, P. F. 1838. Notice sur la loi que la population suit dans son accroissement. Corresp. Math. et Phys. 10:113–21.
Vogt, W. 1948. Road to survival. William Sloane Assoc., Inc., New York, 335pp.
Warming, E. 1895. Plantesamfund-grundtrak of den okologiska plantegeogrefi. Copenhagen. Trans. 1909 as Oecologie of plants. Oxford, Clarendon Press, 422pp.
Watt, K. E. F. 1962. Use of mathematics in population ecology. Annu. Rev. Entomol. 7:243–60.
Watt, K. E. F. ed. 1966. Systems analysis in ecology. Academic Press, New York, 276pp.
Watt, K. E. F. 1968. Ecology and resource management: A quantitative approach. McGraw-Hill Book Co., Inc., New York, 450pp.
Watt, K. E. F. 1973. Principles of environmental science. McGraw-Hill Book Co., Inc., New York, 319pp.
Weaver, J. E., and F. E. Clements. 1929. Plant ecology. McGraw-Hill Book Co., Inc., New York, 601pp.
Welch, R. S. 1935. Limnology. McGraw-Hill Book Co., Inc., New York, 471pp.
Wheeler, W. M. 1902. Natural history, "Oecology" or "Ethology"? Science. 15:971–76.
White, L., Jr. 1968. Dynamo and virgin reconsidered. Massachusetts Inst. Tech. Press, Cambridge, Mass., 186pp.
Whittaker, R. H. 1962. Classification of communities. Bot. Rev. 28:1–239.
Whittaker, R. H. 1967. Gradient analysis of vegetation. Biol. Rev. 42:207–64.
Whittaker, R. H., and P. P. Feeney. 1971. Allelochemics: chemical interactions between species. Science. 171:757–70.
Whittaker, R. H., S. A. Levin, and R. B. Root. 1973. Niche, habitat and ecotope. Am. Nat. 107:321–38.
Wilson, E. O. 1969. The new population biology. Science. 163:1184–85.
Woodbury, A. M. 1954. Principles of general ecology. The Blakiston Co. Inc., Toronto, 503pp.
Young, O. R. 1964. The impact of general systems theory on political science. General Systems. 9:239–53.

TRENDS IN FISHERY RESEARCH

J. L. McHugh

Trends in Fishery Research

J. L. McHugh

Office for the International Decade of Ocean Exploration
National Science Foundation
Washington, D. C.

Introduction

As soon as I started to plan this chapter and to delve into the literature I began to regret that I had been so quick to accept the task. The subject is so vast and the literature so scattered that no individual could do justice to it in a short time. When that individual is an administrator, years removed from active participation in fishery research, the task is almost impossible. The assignment would have been more appropriate for a university professor and active research worker.

However, there were compensating circumstances. Adequate treatment of the subject would call for a good-sized book. The assignment called for a single chapter of moderate size. Therefore, it became mandatory to limit the scope of the essay. I have done this by adopting a simplified definition of fishery research and its objectives, and by tracing only the major trends. I have been helped immeasurably by a number of my colleagues to whom I turned for advice. They are not to be blamed for the deficiencies of this chapter, but I must acknowledge that it would have been even more deficient without their help.

In the early stages of preparation I ran into a problem that has plagued all historians—the past is much easier to evaluate than the present. The period prior to World War II can be seen clearly in perspective. The major trends and accomplishments have been identified by frequent review and discussion over the past quarter-century. Some of the trends in the postwar period also stand out clearly. The strong emphasis on population dynamics is an outstanding example. But through the 1950's and into the 1960's has come a growing realization that the simple prewar concepts of scientific fishery management are not very useful in practice and that successful fishery management must be based on scientific understanding of the resource as it interacts with all the physical and biological variables in its environment. Because man is forcing the natural dynamic equilibrium out of balance in an increasing variety of ways and with increasing momentum, economics and social-political forces have assumed greater importance. It now is becoming abundantly clear even to many laymen that man's relations with his environment are exceedingly complex, and that control is not likely to be successful until all parts of the puzzle are linked. Thus, fishery ecology, difficult though it is, will not be understood very clearly, and fishery management will not be very effective, until we know how to relate them to balanced multiple use of the environment.

The observer who tries to detect contemporary trends in fishery research is like the fly on the oil painting—too close to see the picture. My former teacher (Hubbs, 1964), in his review of the history of ichthyology in the United States after 1850, remarked that he treated the history of the period after 1925 "with disproportionate and increasing brevity, largely because the activities and publications of this period are already best known to present-day workers. The period since 1950—virtually current events—is scarcely mentioned." He virtually closed his account at about 1935, explaining that "to continue would infringe on the domain of current events. Another reason is that to do so would call for a treatment of at least as many persons as I have already mentioned, or for an arbitrary and likely unjust omission of many who have contributed notably to ichthyology in America over the past quarter century...." To be equally cautious in writing a history of fishery research would be to ignore many of the most interesting and exciting developments. Thus, I have been forced to delve into the recent literature, often with a frustration bordering on panic. At best, I have been able to follow trends imperfectly, and often, I am sure, without finding the most important references.

The recent literature proved to be something of a problem. The major large works are well known, and the pre-World War II trends can be followed fairly well by consulting relatively few publications. But some important advances have been published in small papers or in obscure journals, and I undoubtedly have missed some for these reasons. After 1945, the task becomes increasingly difficult, not only because the numbers of printed pages are increasing at an accelerated rate, but also because the diversity of subject matter has increased. Papers from fishery laboratories are appearing in an increasingly wide assortment of journals, many of which one would not be likely to consult for literature on fisheries. Indeed, the young fishery scientist today is likely to have a much more specialized background than his senior colleagues (McHugh, 1968). Most trends can be seen only dimly, if at all, in the literature of the past 10 or 15 years. Some future historian will have to evaluate the last 25 years of the first century of fishery science.

Hubbs (1964) illustrated graphically the publication explosion in ichthyology by illustrating the acceleration in numbers of pages in annual volumes of the journal *Copeia*. Similar trends can be demonstrated for the fishery literature by examining trends in numbers of pages in the *Transactions of the American Fisheries Society* or the *Journal of the Fisheries Research Board of Canada*. The growing diversity of subject matter and variety of journals can be seen by analyzing the total output of the Division of Biological Research of the Bureau of Commercial Fisheries.

For example, in 1966 scientists of the Division of Biological Research of the Bureau of Commercial Fisheries published 271 scientific papers, reports, and articles. These appeared in 105 different journals or books, 8 of which were federal government serials, and 97 of which were state, private, foreign, or international publications. Of all these titles published in 1966 only 25 percent appeared in federal series. Of the non-federal series 68 contained only one title each. This illustrates vividly the increasing difficulty of tracing the fishery literature. With respect to the work of federal fishery scientists in the United States it has led many people to underestimate the breadth and the quality of their accomplishments.

The Beginnings of Fishery Research

The century which began in 1871 encompasses virtually the entire history of truly scientific fishery research, and the literature is voluminous and very widely scattered. Ten states had established fishery commissions prior to 1871, beginning with Massachusetts in 1865 (Smiley, 1883) but serious scientific inquiry into fishery matters began in 1871 when Spencer F. Baird arrived at Woods Hole, Massachusetts, to begin his duties as the first United States Commissioner of Fisheries. Woods Hole was the original seat of American fishery research. Thus, 1971 marks the centennial of another important American fishery institution, the United States Bureau of Fisheries, now the Fish and Wildlife Service. Galtsoff (1962) has provided a detailed history of the development of marine science at Woods Hole from the beginning. The text by Rounsefell and Everhart (1953) reviews accomplishments in fishery science through its early development.

Although the general trend has been similar, freshwater fishery research has developed at a different pace, and in somewhat different directions than marine fishery research. Fresh waters are relatively small in area and volume as compared with the sea, and for this and other reasons most bodies of fresh water can be treated essentially as closed systems. The number of species in fresh waters is fewer, and their interrelationships are less complex. For these reasons the ecology of fresh waters is easier to understand and to manipulate, and fishery science has been applied most successfully in fresh water fishery management. Several excellent chapters in this volume deal in various ways with freshwater fishery science. For this reason I have chosen to emphasize marine fishery research, although freshwater research is not ignored entirely. The elements of freshwater fishery science are given by Lagler (1956), and Carlander (1969) provides a valuable index to the literature and summarizes available data on life histories. Another useful recent volume is *Inland Fisheries Management* (Calhoun, 1966). The history of fishery research in the Great Lakes has been reviewed by Hile (1952, 1954). A valuable recent text is Ricker (1968).

Americans have played an important part in the development of fishery science, but many major contributions have been by scientists in the British Commonwealth countries, in Europe, and in Asia. This volume marks the centennial of a New World organization, and for this reason I have stressed the work of the United States and Canadian scientists. It was neither possible nor logical to ignore the work of foreign scientists entirely, and their work is cited as necessary.

Fishery research had its origin in man's reactions to a universal natural phenomenon, the phenomenon of fluctuation in abundance. From the very early days of fishing it was found that catches were variable. We now know that these fluctuations were caused by annual changes in the success of spawning, or by changes in environmental conditions which altered patterns of migration and geographic distribution. The natural inclination then, as today, was to blame decreased catches upon man's activities. When fishing was responsible for the decline, the nature of the effect was misunderstood, as it often is today. It was not clear then, as it is not clear to many people today, that in a fully developed, but still healthy fishery, the catch per unit of effort is much less than it was when the virgin stock was first fished, the average size of fish is considerably less, and the stock and its biology are altered in various other ways. Fishery research began because people wanted to know the reasons why their catches fluctuated and how to correct the situation.

It is easy, armed with hindsight, to be critical of the past. One way to begin a history of fishery research would be to observe that development of fishery science in America was inhibited by the philosophy of Spencer F. Baird and his immediate successors, who placed great importance on hatchery culture as a solution to the problems of marine and freshwater fisheries. As a consequence, the federal government embarked upon a vigorous, and apparently completely futile, program of fish culture which persisted for more than sixty years. It may come as a surprise to many that the Federal Laboratory at Woods Hole had virtually no year-round scientific program until after the end of World War II. Yet a full-time program of fish propagation which started in 1873 continued until it was terminated by the war in 1941. In the first twelve years of operation the total budget for scientific inquiry was $47,000, an average of less than $4,000 a year, while total expenditures on fish propagation were more than $800,000 or an average of about $70,000 a year (Smiley, 1883). From 1871 until 1946, Woods Hole had no permanent full-time scientific laboratory director (Galtsoff, 1962). The scientific program was almost entirely a summer operation and most of the scientists were independent investigators.

The American Fisheries Society had a hand in this great emphasis on fish culture at Woods Hole. In February 1872, the American Fish Culturists' Association (the Society's predecessor) passed a resolution urging the United States government to take measures for introduction and artificial propagation of shad, salmon, and other valuable food fishes throughout the country, especially in waters common to several states. Congress responded almost immediately be appropriating $15,000 for this purpose in 1872 (Smiley, 1883). This was the genesis of the marine hatchery at Woods Hole.

A quick review of the fish cultural operations of the United States government shows that at least 73 species were reared in hatcheries at one time or another. These included at least 47 freshwater species, 13 anadromous, and 12 marine. Marine species made up an increasing proportion of the total as time went on, and by 1940 more than 98 percent of the eggs and 75 percent of the fry released were marine, principally flounders, cod, and pollock. Releases of anadromous species were important during the first half of this period, but fell off in the second decade of the present century.

On the positive side, there is little reason to doubt that Spencer Baird was the father of fishery science in the United States. He was a distinguished scientist, who understood very clearly for his day the complicated nature of the relationships between organisms and their environment. He recognized that fluctuations in catches could be caused by natural phenomena as well as by man. His original program outline demonstrated this concept very clearly, for it contemplated oceanographic and meteorological investigations, biology, ecology, parasitology, and even a primitive approach toward an understanding of population dynamics. He believed firmly in the value of providing research space for distinguished scholars who spent one or more summers at Woods Hole. Dr. Baird also firmly believed that research and teaching should go together, but he was never able to accomplish this ambition. His interest in very broad studies of the ocean was reflected in the cruises of the *Albatross*, launched in 1882 and decommissioned in 1921. The cruises of the *Albatross* are well known for their important contributions to deep sea exploration in many parts of the world. Dr. Baird also recognized the merits of creating an informed public.

Meager budgets and the uncertainty of appropriations until late in the year made it very difficult for Spencer Baird to maintain his policy of accommodating visiting scientists. This lack of support undoubtedly encouraged a group of university professors to establish the now famous Marine Biological Laboratory. Development of MBL was certainly encouraged by the policies of Commissioner Marshall McDonald, who from 1888 to 1895 placed major emphasis on fish culture and routine administration rather than scientific research. Nevertheless, from 1871 through the 1920's a number of federal government and independent scientists made the Woods Hole fishery laboratory a temporary base from which many important contributions to marine biology and fishery science were made.

The second research laboratory of the Commission of Fisheries was founded at Beaufort, N. C., in 1901. The first in Canada was the Atlantic Biological Station of the Fisheries Research Board of Canada in 1898. Several federal and university marine laboratories were established prior to 1920, but the state laboratories came later. The California State Fisheries Laboratory was first, founded in 1917 by Dr. W. F. Thompson. It was followed in 1920 by the Institute for Fisheries Research at the University of Michigan and the Ontario Fisheries Research Laboratory, the first by Dr. Carl L. Hubbs, the second by Professor B. A. Bensley and Dr. W. A. Clemens. Highlights of the history of world fishery research are contained in Hiatt (1963).

The Woods Hole fishery laboratory fell on hard times during the depression, as did all fishery research in North America. In the middle 1930's, there was talk of closing Woods Hole entirely. With the advent of World War II closure became a necessity, and the only investigation that remained was the oyster research program of Paul S. Galtsoff. The decade prior to the war, difficult though it was, was marked by a radical change in the scientific program from laboratory research and faunal studies to an emphasis on oceanic fisheries. This change paralleled a general transformation taking place in fishery science everywhere, from relatively undirected research to serious attempts to measure variations in the stocks of fish and to explain their causes. During this period, emphasis in marine fisheries was being placed on analysis of catches, abundance of year classes, rates of growth and death, migrations and movements, and measures of fishing effort. The trend was distinctly toward quantitative biology. The science of fishery population dynamics was being born. This evolution proceeded vigorously after World War II, culminating in a growing interest in mathematical models and a recognition that if scientific fishery management is to succeed it must understand the dynamic relationships of total communities, rather than of individual species or stocks of fish. Another fairly recent development has been the growing realization that economics and the other social sciences cannot be ignored.

Herrington (1954), reviewing progress in the first half of the 20th century, painted a somewhat gloomy picture. He recognized that some progress had been made, notably in developing a scientific understanding of the relation between fishing intensity and sustainable yields; e.g., Pacific salmon and halibut, Northwest Atlantic haddock, North Sea trawl fisheries, and certain fisheries in the East China Sea. But he concluded that progress had been hampered by the lack of adequate scientific information and by public views and policies that placed a premium on opinions rather than on facts. Although public attitudes now are changing, largely because man's fearsome capacity to ruin his environment is being recognized widely, Herrington's conclusions are still appropriate today.

Harkness *et al* (1954) presented a brief philosophical report on the status of fishery research at mid-century. Needler (1958) reviewed the work of the Fisheries Research Board of Canada over the previous 50 years. The reminiscences of Clemens (1958, 1968) are particularly unusual. Galtsoff (1958) reviewed recent developments in fishery research in the Gulf of Mexico, a region once described by a fishery scientist as a biological desert, but now a major fish and shellfish producing area. Walford's (1958) book is of especial interest because it emphasizes the gaps in knowledge at that time. Most of these gaps remain to be filled. Kesteven (1966) reviewed the problems of organization and administration of fishery research, and Lucas (1964) has written a particularly thoughtful paper on marine fishery research.

The Descriptive Phase

Fishery science began with the study of ichthyology, and in that sense it is very old, going back to the days of Linnaeus or even earlier. The early American ichthyologists found a wealth of material to be described and identified, and they also made many observations on life histories. Undoubtedly, some of those early naturalists worried about the future of the rich fishery resources of the new world. But organized measures to preserve the abundance of fishery resources were not possible in a pioneer land with very loose governmental structures. It was not until after the end of the Civil War in the United States (1865), and until Confederation in Canada (1867), that the need for conservation was clearly recognized or that arrangements for fishery research and management were possible. It is interesting that such action developed quickly after these important events, and evidence that the two young countries were taking their responsibilities seriously. Indeed, the bison and the passenger pigeon were still abundant in many parts of North America in 1865-1867, and their extinction in the wild did not come until about 1900.

The history of ichthyology in the new world has been covered masterfully by Dymond (1964), Myers (1964), and Hubbs (1964) on the occasion of the 50th anniversary of the American Society of Ichthyology and Herpetology. The major events and the leading naturalists and scientists of the period from 1492 to about the beginning of World War II are well covered in these three papers. It is sufficient to say that Dymond identified the beginning of scientific fish studies in Canada with the Franklin expeditions in search of a Northwest Passage, which produced Richardson's *Fauna Boreali–Americana* (1836). Myers says that North American Ichthyology began about the time the Philadelphia Academy was formed (1812), and that it is marked by publication of Mitchill's *Fisheries of New York* (1814). Hubbs, describing the period after 1850, tells how the ichthyologists led the way in establishing fishery science as we know it today, which is concerned with fluctuations in abundance of fishery resources, the role of man in producing such fluctuations, and measures which can be taken to achieve and maintain optimum yields from these resources.

The following people had a major influence in giving birth to this ecological offshoot of ichthyology: Spencer F. Baird, Theodore N. Gill, George B. Goode, David Starr Jordan, Barton W. Evermann, Charles H. Gilbert, Hugh M. Smith, Lewis Radcliffe, Norman B. Scofield, William F. Thompson, Carl L. Hubbs, Henry B. Bigelow, Willis Rich, Samuel F. Hildebrand, and William C. Schroeder in the United States, and E. E. Prince and Andrew Halkett in Canada. The important contributions of some of these scientists have been recognized by giving their names to federal or state fishery research vessels.

Invertebrate fishery science, especially in molluscan research, developed in quite different ways and a different group of scientists took part. Some of the major early contributors were Addison E. Verrill, John A. Ryder, Richard Rathbun, W. K. Brooks, Wesley R. Coe, Edwin G. Conklin, Bashford Dean, Francis H. Herrick,

Paul S. Galtsoff, Victor L. Loosanoff, A. E. Hopkins, Herbert F. Prytherch, Trevor Kincaid, Julius Nelson, and Thurlow C. Nelson in the United States, and Joseph Stafford, A. W. H. Needler, C. Roy Elsey, and J. C. Medcof in Canada. It is worth noting that at least four of the leaders of North American fishery science today obtained some of their early experience, and published, on molluscan biology: Wilbert M. Chapman, A. W. H. Needler, Donald L. McKernan, and Milner B. Schaefer.

The Exploratory Phase

Despite early and intense concern with artificial propagation as a solution to fishery problems, scientific research was not neglected. At the beginning this took the form of exploratory surveys. In the first summer at Woods Hole the scientists collected large numbers of fishes and studied their spawning, growth, distribution, and food. They investigated also the plankton, mussel beds, sea stars and other bottom invertebrates, and made temperature observations. Baird and Gill also completed a general plan for study of natural history of fishes and for observing the effects of fishing on the fish populations. According to Galtsoff (1962) this work "laid down the foundation of a new branch of science." Subsequently, the surveys shifted to the Bay of Fundy, with headquarters at Eastport, Maine; then to bases at Casco Bay, Maine, and Norwalk, Conn.; then back to Woods Hole in 1875 to compare results with the 1871 surveys. These surveys continued in various places and in various forms until the 1920's.

In addition to the voluminous reports of the Commissioner of Fish and Fisheries, published annually beginning in 1872, there came out of these exploratory surveys some notable scientific publications still used extensively today. Among the most important were the monumental *The Fisheries and Fishery Industries of the United States* by George Brown Goode (1884, 1887), and *Oceanic Ichthyology* by Goode and Bean (1895). In 1912, an important survey of the Gulf of Maine began. The results were slow in coming out in print, but they were worth waiting for: *Fishes of the Gulf of Maine* by Bigelow and Welsh (1925), *Plankton of the Offshore Waters of the Gulf of Maine* (Bigelow, 1926), and *Physical Oceanography of the Gulf of Maine* (Bigelow, 1927). In the meantime, work was underway at Stanford University on another important ichthyological contribution: *Fishes of North and Middle America* by Jordan and Evermann (1896-1900).

The years prior to World War II produced other important faunal surveys in the marine waters of the West and East coasts of North America. The results appeared in papers by Evermann and Goldsborough (1907), Walford (1931, 1937), Dymond (1932), Barnhart (1936), Clemens and Wilby (1946), Hildebrand and Schroeder (1928), Hay and Shore (1918), and many others. Actually, such explorations are an essential and continuing part of fishery research, for it is important not only to know what fauna and flora have been taken in particular regions, but also their variations in space and time. Thus, taxonomy and ecology or population dynamics must continue hand in hand. A few of the valuable recent works are Galtsoff (1954); Lagler *et al* (1962); Alverson *et al* (1964); Leim and Scott (1966); Williams (1965); Peréz Farfante (1969); and *Fishes of the Western North Atlantic*, a massive series still in preparation.

Life Histories

One of the first steps in a program of research on a fishery resource is to study the life history of the species. This includes its spawning habits, as to place, time and special features; identification and development of eggs and larvae; age and growth; food and feeding; migrations and movements; and fluctuations in abundance. Such questions have occupied much of the attention of fishery scientists over the past 100 years. An example of the wealth of information that has been gathered on a single species, the sockeye salmon (*Oncorhynchus nerka*), is provided by Foerster (1968).

Spawning and Early Development

Studies of fish eggs are valuable for a variety of purposes. The eggs of fishes are nonmotile, and those of most commercial species are planktonic, buried in the bottom, attached to vegetation or other substrates, or carried by the parent until they hatch. Because they cannot move unless they are carried by their mothers, they can be collected in proportion to their absolute abundance in the natural environment, which is not true for the free-swimming stages. Thus, properly planned and conducted quantitative sampling of eggs in the plankton or on the bottom can be used to make accurate estimates of the magnitude of the spawning population, provided the fecundity of females is known. An early, relatively large-scale application of this method is the work of Sette and Ahlstrom (1948) on the Pacific sardine *(Sardinops caerulea)*. This work has been carried forward by Ahlstrom (1967) showing how the abundance of Pacific sardine *(Sardinops caerulea)* and Pacific northern anchovy *(Engraulis mordax)* have changed in the past 20 years.

Even under sampling conditions which do not allow absolute estimates of abundance of eggs and larvae, relative abundance can be compared in time and space. Such techniques were used in Atlantic haddock *(Melanogrammus aeglefinus)* and mackerel *(Scomber scombrus)* research by Walford (1938) and Sette (1943a) respectively, to study distribution and survival of eggs and larvae.

Estimates of abundance of eggs and larvae have been used with varying success for prediction of adult abundance. Johann Hjort (1914) believed that "infant" mortality was the principal cause of fluctuations in abundance, and this hypothesis has influenced the thinking of fishery scientists ever since. Sette's work (1943) showed that this might not be true for Atlantic mackerel, but the difficulty of sampling larval and young stages of fishes, especially since their ability to escape the sampling gear increases with size and age, makes it difficult to prove the point one way or the other for most species (Marr, 1956). It is clear from studies of oysters and other sessile invertebrates that setting of pelagic larvae varies widely (Loosanoff, 1966), but we know also that these animals suffer substantial and sometimes catastrophic mortalities from disease (Hewatt and Andrews, 1954), freshets and other biological and physical phenomena. Thus, forecasting abundance of commercial-sized fish from abundance of young is risky.

An understanding of spawning and early development also is obviously useful in artificial rearing of fish and shellfish. As long as reasonable precautions are taken (such as to provide adequate aeration and to control temperature and other environmental factors) eggs of most fishes and many invertebrates are remarkably easy to fertilize and to incubate to hatching. This is one of the reasons why artificial propagation was so popular. It was carried on by laymen, many of whom became remarkably skilled at the mechanics of fertilizing and holding eggs, but who knew nothing of the scientific aspects of their craft. It is now recognized that fish culture as practiced until recently was largely a sheer waste of energy and funds.

Another very important application of egg and larval studies is the opportunity afforded to study quantitatively the entire fish fauna of broad ecological zones in the ocean. An outstanding example is the work of Ahlstrom (1965) in the California Current System. This work began in 1948 with the expanded research program of the California Cooperative Fishery Investigations, stimulated by the decline of the Pacific sardine *(Sardinops caerulea)* fishery. Ahlstrom has been successful in working out the early developmental stages of several hundred fish species. His material (also Ahlstrom, 1966) has been used to demonstrate the dynamic relationships between sardine and anchovy *(Engraulis mordax)* by Murphy (1966), and provided basic information that demonstrated the wide geographic distribution and apparently great abundance of the Pacific hake *(Merluccius productus)* and that led in turn to establishment of a new fishery (Alverson and Larkins, 1969).

Age and Growth

The importance of age and growth studies became apparent very early in fishery research. Fluctuations in abundance of the resources were often accompanied by changes in size of fish, and it was obviously useful to try to find ways of tracing fish of a particular age through the fishery year by year. Petersen (1894) noticed that length frequency distributions of samples from fish catches often showed several modes and he deduced that these modes represented year classes. He demonstrated that these modes progress toward the larger sizes with the seasons, and that a year later the first mode will reach the point occupied by the second mode a year earlier. The weakness of this method is that growth in length slows with increasing age and the spread of sizes within a year class increases, so that the overlap is successively greater between successive ages. Thus, the method works best with the younger fish in the stock. Until a more direct method of age determination was found, the Petersen length-frequency method, with all its weaknesses, was the only one available. The numbers of fish of each age were estimated by assuming that the length frequency distribution curve for each age was a normal distribution, and partitioning the sample on that basis.

For some marine animals, especially invertebrates, length is still the best criterion of age. Crustacea have no permanent hard parts on which a record of age can be preserved, but they increase substantially in size with each molt. Thus, age can be estimated if the growth rate is known (Berry, 1967).

According to Parrish (1956), the scale method of age determination was first published by Hoffbauer (1898), who also showed that growth of scale and growth of fish are closely related. There followed a series of studies by various workers to prove the scale method. Particularly ingenious was Hjort's (1914) method of following a dominant age group in samples from the fishery and showing that the number of annuli increased by one each year. Later, Lea (1929) used an equally ingenious method to follow particular year classes through a fishery. He found that scale abnormalities, such as an unusually poor growth zone, could be followed in successive years. In North America, Huntsman (1918, 1919) studied the growth of scales and the scale method of age determination, and Van Oosten (1928) used scales to deduce certain features of the life history of lake herring *(Leucichthys artedi)* and wrote a critique of the scale method. In Britain, Graham (1929) published a detailed review of the literature on fish age determination. Another classic American study is the work of Hile (1936) on the cisco *(L. artedi)*.

Einar Lea (1911) was the first fishery scientist to use scales to study fish growth. Making use of the fish-length:scale-length relationship he was able to calculate the length of each fish at the time each annulus was formed. Later, it was discovered (Lee, 1912) that the body:scale relationship was not quite as simple as it had been assumed. "Lee's phenomenon" describes the tendency for growth rates back-calculated from the scales of larger and older fish to be progressively lower. The phenomenon has not yet been completely explained.

Various other hard parts of the body have been used with varying success to determine age. Otoliths have been used by many workers (Graham, 1929), sometimes more successfuly than scales. Similar annuli can be found on cross sections of vertebrae (tunas), opercular bones, and spiny fin rays (Palmen, 1956) (catfishes).

Food and Feeding

For anyone familiar with the flora and fauna of an area, studies of the food of fishes are relatively easy. The simplest type of food study is to examine stomach contents. Many studies of this type have been made by fishery scientists all over the world, but only a few have an important bearing on fishery problems. Too often, such studies are undertaken because the material is available from investigations planned for other purposes. Few fish food studies have been based on adequate sampling throughout the year, in all parts of the geographic range of the species, and over a sufficient period of time to demonstrate long-term variations in composition of the food. Although most of them are selective in their feeding, fishes are better collectors than man. Properly conducted food studies should provide valuable information on aquatic flora and fauna.

Some of the best studies of fish nutrition and feeding efficiency have come from hatchery work (Tunison, *et al,* 1939; Phillips, *et al,* 1939; Hublow, *et al,* 1959; Halver, 1963). For example, the tables of rations for trout feeding developed by Deuel *et al* (1944) and others are still used by hatchery operators.

Scientists in several countries have begun to use sophisticated methods to study food and growth and energy flow in ecosystems. An introduction to the literature on this subject can be obtained by reference to such authors as Petersen (1918), Ivlev (1955, 1961), Winberg (1956, 1960, 1962), Fry (1957), Parker and Larkin (1959), and Larkin and Ayyangar (1961). Recently, Paloheimo and Dickie (1965, 1966a, b) investigated thoroughly the food and growth of fishes and the influence of physiological and environmental factors.

Migration and Movements

Most fishes and commercially important crustaceans are migratory, and these migrations have been of great interest to fishery scientists. Fishermen have always known of the migratory habits of their prey, by observing the comings and goings of salmon and other anadromous species, by seasonal changes in their catches, or by noting the differential times of arrival on and departure from fishing grounds at different latitudes. Schmidt's classic studies of eel migrations (1932) were based on the sizes of larvae taken in different parts of the Atlantic Ocean.

A powerful tool for studying fish migrations and movements (useful also for a variety of other important purposes) was provided when the first successful tag was developed. Rounsefell and Kask (1945) made a thorough review of fish tagging and tagging methods; this review was brought up to date by Rounsefell (1963). According to their study, the first successful tagging experiment was done by the U.S. Fish Commission in 1873 on Atlantic salmon in Maine (Atkins, 1876). A small aluminum rectangle was attached to the dorsal fin with platinum wire. Credit for inventing the first tag is often given to the early Danish fishery scientist C. G. J. Petersen (Petersen, 1894) probably because his became one of the most widely used and successful of all fish tags. This well-known tag consists of two discs attached one on each side of the body by a metal pin or wire through the muscles.

Another interesting tag is the body cavity tag developed by Nesbit (1933) to mark squeteague *(Cynoscion regalis)*. This was a brightly colored celluloid strip, inserted into the body cavity through a small incision in the body wall. Rounsefell and Dahlgren (1933) developed a modification of this tag to be used on industrial fish, in this case, Pacific herring *(Clupea pallasii)*. Their modification was a small, numbered nickel strip which could be recovered by an electromagnet in the meal chute of a reduction plant. Dahlgren (1936) developed an electronic detector to recover the tags with the fish as they were being unloaded from the boat. This technique allowed precise identification of the locality of recapture. The search for a satisfactory way to mark active, fast swimming fishes like tunas was fruitless for a long time. The problem was finally solved when the "spaghetti" or streamer tag was perfected in 1952 (Wilson, 1953; Clemens, 1961) following preliminary experiments by Alverson and Chenoweth (1951).

Crustacea are particularly difficult to mark because tags tend to be shed at molting. Van Engel developed a method for blue crabs *(Callinectes sapidus)* by attaching the tag with surgical steel wire to the musculature through the suture between abdomen and thorax (McHugh and Ladd, 1953). Simpson and Shippen (1968) inserted plastic spaghetti tags through the arthral muscle at the junction of cephalothorax and abdomen. Shrimp (Neal, 1969) have been tagged also with Petersen disk tags, with tiny plastic tags inserted in the musculature of the first abdominal segment, and with color-coded magnetic tags of fine wire developed initially for marking young salmon (Jefferts, Bergman, and Fiscus, 1963).

For detailed studies of movements or migrations, modern electronics has permitted development of relatively small tags that emit a signal. The first models were used on salmon in the Columbia River (Trefethen, 1956). Smaller models which the fish can be induced to swallow have been developed for salmon and tuna (Trefethen, personal communication).

Various kinds of mutilation have been used to mark fish and shellfish. Among the earliest extensively used methods was the marking of salmon by removal of fins (Foerster, 1936). Lobsters *(Homarus americanus)* have been marked by punching holes in the telson (Appellof, 1909; Wilder, 1953), a method which survives molting. Hot branding of young salmon has been tried (Groves and Novotny, 1965) and cold branding apparently is equally successful (Mighell, 1969). Young salmon also have been marked by starvation (Major and Craddock, 1962) or by injection of certain antibiotics (Weber and Ridgway, 1967). Both methods leave recognizable marks in the scales and bones.

Staining by injection or immersion is a useful technique for shrimp (Neal, 1969). The injection method, apparently developed by Menzel (1955), appears to be the most successful. By this technique it has been demonstrated that young pink shrimp migrate from the bays of the Florida peninsula to the

Tortugas fishing grounds, a maximum distance of up to 100 miles.

Discussion of fish marking would not be complete without mention of natural tags. Scales are useful certificates of origin (Lea, 1929) whether unusual markings are induced by natural phenomena or artificial methods. Equally ingenious is the method of identifying continental origin of Pacific salmon caught at sea by the frequency of occurrence of parasites acquired in fresh water, a cestode from the river of Bristol Bay, Alaska, and a nematode from Kamchatka (Margolis *et al,* 1966).

Because marked fish and shellfish usually have higher mortality (Foerster, 1936) and slower growth than unmarked animals, much attention has been given to these problems. Anaesthetics have been found to reduce mortality from handling (Zein-Eldin, 1963; Toth, 1966). Adjustment for differential mortality may be made from experimental evidence (Berry, 1967).

I have discussed marking techniques in considerable detail because they are an important tool in fishery research. Marking has demonstrated the extensive and complicated migrations of Pacific salmon (INPFC Annual Reports and Bulletins) and many other species of fish and shellfish. This demonstration, and the sense of personal involvement associated with recovery of a marked fish by him, appeals strongly to the layman, and has been the reason for some ill-conceived and poorly conducted tagging programs. A well-planned and carefully conducted and analyzed tagging program offers much more information to the fishery scientist, providing data on rates of growth and death, subpopulations, population size, and other important data. In fact, marking is so important a tool that it is often raised to the status of a program in scientific planning, ranking equally with such matters as survival, growth, behavior, and other biological phenomena.

Recent comprehensive works on fish migrations and movements are Hoar (1959), Hasler's (1966) *Underwater Guideposts,* and Harden Jones' (1968) *Fish Migration.*

Diseases and Parasites

Diseases and parasites have been of interest to fishery scientists since fishery research began. Disease became an important matter as artificial culture developed, and the literature on fish diseases is extensive. The two fish disease laboratories of the Bureau of Sport Fisheries and Wildlife have made many important contributions to scientific knowledge of disease in freshwater fishes (Rucker *et al,* 1954; Snieszko *et al,* 1965; Snieszko, 1970). Much less is known about diseases of marine fishes (Sindermann, 1970), but the literature is extensive, nevertheless.

Mass mortalities of shellfish, especially oysters, have been attributed to disease. The "Malpeque disease" dealt a hard blow to the oyster industry of Prince Edward Island, Canada, in the 1920's (Logie, 1956). The fungus *Dermocystidium marinum* at times causes heavy mortality of oysters in the Gulf of Mexico and Chesapeake Bay (Mackin, 1961; Hewatt and Andrews, 1954). During the past 13 years, the oyster industry of Delaware Bay and the lower part of the Chesapeake Bay has been virtually wiped out by an epizootic traced to a haplosporidian parasite, *Minchinia nelsoni* (Sindermann, 1968). No direct methods of control of oyster diseases are known, but infections often can be avoided by taking advantage of ecological knowledge and growing oysters in places where they can survive but their diseases and parasites cannot.

Predators

Predator and coarse fish control has long been a popular solution to fishery problems. Such control is not without merit under some circumstances, as demonstrated by the work of Foerster and Ricker (1942) and Ricker and Gottschalk (1941). The best known example of effective large-scale control of predation is the sea lamprey control program in the Great Lakes (Ann. Repts. Gt. Lakes Fish. Comm.), in which the ammocete larvae are selectively poisoned with chemicals. The success of this program depended upon research, to understand the life history and ecology of the lamprey and to develop control techniques.

Shellfish growers are constantly plagued by predation. The most destructive predators are sea stars, boring snails, crabs, and certain fish species, Control is possible by physical means, either by fencing the grounds or by growing oysters off the bottom, but these methods are expensive in capital outlay and maintenance costs. The shellfish laboratory at Milford, Conn.

has developed methods for chemical control of sea stars and oyster drills (Loosanoff, 1961).

Subpopulations

From the early days of fishery science great interest has been shown in methods of identifying subpopulations of fish species. This interest probably is a result of the early dominance of fishery science by taxonomists. Subpopulation identification is considered important because management for maximum sustainable yield requires that each stock or race be managed separately. It is clear that if two or more stocks of fish are harvested by the same fishery the total yield probably will be less than the combined yield of the individual stocks because it is almost impossible to harvest each at its optimum level. If, for example, for practical reasons a resource made up of two or more subpopulations is being managed by imposing a single total catch quota it is inevitable that some stocks will be overfished and some underfished and the maximum sustainable catch will be less than the total maximum sustainable yield of the individual subpopulations. An impressive example of the damage that can be done under such circumstances is provided by the Antarctic whaling industry, in which the differences between species were ignored by setting a single quota based on the blue whale unit, an arbitrary management unit based on oil yield. All other things being equal, it takes about the same amount of effort to catch one whale, whatever the species. Thus, the blue whale was certain to be the preferred species, and the quota gave it no protection. The four important species in the Antarctic were reduced one by one, more or less in decreasing order of value of a single animal (Chapman, 1964).

The original techniques for distinguishing subpopulations of fishes were based on morphometric studies—measurements of body proportions and counts of meristic characters. Usually, the credit for discovering that a single species may be subdivided into races or subpopulations is given to Heincke (1898) for his monumental study of variations in body form and meristic characters of Atlantic herring *(Clupea harengus)*. Counts of vertebrae and fin rays, differences in body proportions and differences in structure of the scales were used by European scientists to distinguish stocks of cod, haddock and hake (Parrish, 1956), and plaice (Graham, 1956), among other species. In the United States and Canada, the earliest studies were those of Thompson (1917b) on Pacific herring and Hubbs (1925) on Pacific herring, sardine, and anchovy. Thompson concluded that there were differences between samples from British Columbia waters and San Francisco Bay, and Hubbs confirmed this conclusion. The first large scale and comprehensive subpopulation study in North America, on Pacific herring, was made by Rounsefell (1930). Of the three meristic and one morphometric characters studied, he found that vertebral number was the most important and he reported vertebral counts from a large series of samples caught throughout the range of the species, from the Aleutian Islands to San Diego Bay. He concluded that the Pacific herring is divided into a number of separate and distinct populations. Later, Rounsefell and Dahlgren (1935) used the analysis of variance, a new and powerful method of statistical analysis recently developed by R. A. Fisher, to delineate races of herring in southeastern Alaska. Tester (1937, 1948) made similar studies in British Columbia and found that herring in those waters also did not intermingle freely. Clark (1947) and Hart (1933) used vertebral counts to identify populations of Pacific sardine, and McHugh (1951) separated stocks of Pacific northern anchovy from counts of vertebrae, fin rays, and gill rakers. Unusual features of the last-named study were that some populations which could not be distinguished from vertebral counts differed in mean numbers of fin rays, and that there is a sexual difference in fin ray counts.

Morphometric studies were not confined to herring-like fishes. Townsend (1936) studied variations in meristic characters of 19 flounder species from the North Pacific and constructed a key based on vertebral and caudal and dorsal fin ray counts. Perlmutter (1947) found differences in mean fin-ray counts of blackback flounder *(Pseudopleuronectes americanus)* between localities. Scott (1954) found differences not only in dorsal and anal fin-ray counts in yellowtail flounder from three Atlantic coast fishery areas, but also in body proportions, otoliths, scales, growth rates, age at maturity, and time of spawning. Vernon (1957) found marked differences between three races of landlocked salmon *(Oncorhynchus nerka)* in a single lake in a number of meristic, morphometric, and biological characters. Royce (1964) concluded that more than one subpopulation of yellowfin tuna *(Thunnus albacares)* inhabits the tropical Pacific Ocean.

Recently, a number of ingenious new techniques have been developed. Vrooman (1964) used serological methods to conclude that there are three genetically distinct populations of Pacific sardine. Sindermann and Mairs (1969) used serological techniques to investigate races of Atlantic herring. Serology and electrophoresis have been used by various workers to detect races of tuna (Cushing, 1964) and a variety of fish species. Ridgway and Klontz (1961) discussed blood types of Pacific salmon. Margolis *et al* (1966) reported on racial identification of sockeye salmon by tagging, parasite infestation, scale characteristics, serology, and morphology. The morphological studies were planned and analyzed to use discriminant and distance functions to achieve maximum separation of stocks. These studies were used to determine the continent of origin of salmon caught on the high seas, far from their home streams, in the region where salmon of Asian and North American origin intermingle.

Generally speaking, the newer and more direct and sophisticated techniques of subpopulation identification have confirmed the earlier indirect methods. Serological, immunological, chromatographic and electrophoretic methods are particularly promising because they can detect differences of genetic origin (Farris, 1957). The 1961 Symposium of the Tenth Pacific Science Congress[1] is a particularly useful reference, as is de Ligny (1969).

Physiology and Behavior

The internal and external reactions of fishery organisms to stimuli are important for a variety of reasons. Anadromous fishes, for example, undergo great physiological changes as they migrate to the sea (Brett, 1964), as they prepare at sea for upstream migration, and as they proceed to their spawning grounds. A knowledge of these changes becomes especially important as the environment becomes altered by hydroelectric and other developments in the rivers (Trefethen, 1968). Tagging of any species requires that the animal be in good condition, and it is important to understand the effects of catching and handling on their chances of survival. It is important also to know how fish react to physical and chemical gradients in the sea and to seasonal changes in hydrographic conditions. Knowledge of fish behavior can be applied to reducing the cost of locating and catching fish.

Although these subjects are relatively new in fishery research, the literature is already extensive. One of the first quantitative studies, on reactions of fishes to gradients of dissolved gases, was by Shelford and Allee (1913). For clupeid fishes the extensive review by Blaxter and Holliday (1963) is valuable. In 1967, FAO held a conference on fish behavior in relation to fishing techniques and tactics.[2] The background papers and proceedings of this conference are a useful guide to the literature. The list of experts includes 332 names. Some other useful references are Fleming and Laevastu (1956), Kesteven (1960), Magnuson (1963), Woodhead (1966), Cahn (1967), Ingle (1968), and Marler and Hamilton (1966).

The first broad review of fish physiology was the two-volume work edited by Margaret Brown (1957). This is being extended and brought up to date by Hoar and Randall in a three-volume work, of which the first volume was published in 1969.

International Fishery Research

The oldest permanent international organization concerned with the ocean and its living resources is the International Council for the Exploration of the Sea, founded in 1902 with headquarters in Copenhagen. The great interest of the Council in fisheries is clear from the early issues of its Rapports et Procés Verbaux and by the first issue of the Journal du Conseil, its scientific journal, in which the two major papers are fishery papers (Hjort, 1926; Russell and Edser, 1926). The Council also has been a pioneer in relating oceanography to fisheries.

[1] A symposium on immunogenetic concepts in marine population research. Amer. Nat. 96(889): 193-246.

[2] FAO Fisheries Repts. 62, 3 vols., FRm/R62.2 (Tri), Rome 1969.

The roles of the international fishery commissions in fishery research have been reviewed by Herrington (1960), McKernan (1961), and in a series of recent publications by FAO (e.g., Carroz, 1965).

Canada and the United States jointly also have been pioneers in international fishery research. The first bilateral Commission was the International Fisheries Commission (now the International Pacific Halibut Commission) founded by treaty between the two countries in 1924. But before this event, the United States and Canada had joined with Japan and the Soviet Union to negotiate the International Fur Seal Convention in 1911. These two commissions, Fur Seal and Halibut, are generally recognized as having come closer to achieving their objectives than any other such international organization. They were followed by the International Pacific Salmon Commission (1930), the International Whaling Commission (1937), the Inter-American Tropical Tuna Commission and the International Commission for the Northwest Atlantic Fisheries (1950), the International North Pacific Fisheries Commission (1953), and the International Great Lakes Fisheries Commission (1955). In 1959, the Tortugas Shrimp Convention (Cuba and United States) was ratified, but the Commission met only once and is now inactive. In 1969, the International Convention for the Conservation of Atlantic Tunas was ratified, and the new Commission held its first meeting in December 1969. Another important international body, with quite different functions, is the Department of Fisheries of the Food and Agriculture Organization of the United Nations (FAO). Canada and the United States participate actively in the work of the first eight commissions, and also in the affairs of FAO.

The international fishery commissions fall into two general types as far as research is concerned. The Halibut, Salmon, and Pacific Tuna Commissions employ scientists and do their own research, each under the leadership of a well-known fishery scientist. The others rely upon the scientists of the individual countries to do the necessary research under the guidance of the respective commission. The Shrimp Commission was not active long enough to establish a research program. The Atlantic Tuna Commission has not yet established a research program, but it will undoubtedly be of the second kind.

There is much argument about which arrangement produces the best research and best serves the needs of the convention concerned. Proponents of the first arrangement maintain that a centralized international staff, under competent leadership, can best maintain the scientific objectivity necessary to meet the scientific goals of the commission. They argue that scientists employed by member nations, when they come together to compare and collate their research, cannot maintain this objectivity. Experience has shown that there is merit in this argument. So far, there is no doubt that the three centralized commissions have been remarkably successful in meeting their responsibilities. Conduct and analysis of their research are much simplified when they come under one director, and these commissions have been wise and fortunate in choosing outstanding fishery scientists as directors of investigations.

Proponents of the second type argue that the research of an international scientific staff is more difficult to evaluate and control, and that the staff is likely to become hidebound in its approach over a period of time. They consider that the critical scientific scrutiny to which the research of the individual member nations is necessarily subjected keeps the national scientific teams alert and fosters good research.

Each point of view has its merits. The commissions with centralized staff, difficult though their problems are, have somewhat simpler terms of reference and more well-defined objectives than the others, and to a degree this explains their generally successful performance. The others have also produced many excellent and imaginative pieces of research and have made excellent progress toward their goals. Time will tell which approach is better, for the commissions with decentralized programs, as a group, are much younger. In large degree, the type of organization is dictated by political considerations.

All active international fishery commissions publish scientific papers and reports, which together constitute an important library of fishery research literature. These journals, available in most comprehensive fishery libraries, are too well known to require listing.

The central objective of the FAO fishery program is to assist the developing nations to improve their utilization of the resources of the sea. There is no continuing research program of the types sponsored by the independent fishery commissions, but the FAO staff has some distinguished scientists and it is able to call upon the world scientific community for assistance. FAO sponsors regional fishery councils, such as the Indo-Pacific Fisheries Council, and it has been instrumental in establishing some of the independent fishery commissions, such as the new Atlantic Tuna Commission. FAO established the whale stock assessment group which helped the Whaling Commission to resolve its problems in the Antarctic. FAO also publishes compendia of scientific information, such as its *Species Synopses*, and convenes international expert meetings on fishery science and publishes the proceedings. These publications also are important elements in the fishery research literature.

FAO maintains close liaison with the Intergovernmental Oceanographic Commission of the United Nations Educational, Scientific, and Cultural Organization (UNESCO). A staff member of the Department of Fisheries is assigned to IOC headquarters in Paris as liaison officer. Lucas (1969) has published an excellent review of international fishery oceanographic trends.

It is appropriate here to recognize the important contributions of three men in the United States who have a major influence on international fishery research and development. These are the three fishery scientists who have held the position of Special Assistant for Fisheries and Wildlife to the Secretary of State: Wilbert M. Chapman, the first incumbent (1948-1951), William C. Herrington (1951-1966), and Donald L. McKernan (1966 to date). Special mention should also be made of the scientists who established the research programs of the Halibut, Salmon, and Pacific Tuna Commissions: W.F. Thompson, halibut and salmon, and Milner B. Schaefer, tuna.

Population Dynamics

In the North American concept, fishery research, as contrasted with research on fishes, means research directed toward management for sustained yields. The theory of fishing states that for each species or stock of fish there is a rate of removal which produces the maximum sustainable biological yield. The purpose of fishery research, according to this concept, is to find out how fishing affects the stocks, so that the rate of removal can be regulated accordingly (Schaefer, 1968). In the early days of fishery research, it must have been recognized intuitively that a fishery resource is capable of sustainable surplus production that can be taken by man. Otherwise, the scientists would have believed that a stock could not remain in equilibrium under human predation, and that fishing would drive the resource to extinction. But is is obvious that this was not recognized clearly by most early students of fisheries, for a recorded interest in quantitative population dynamics did not develop until almost the end of the second decade of the 20th century. The slow start is not surprising, because the living resources of the inland lakes and rivers, and especially the ocean, were so vast and so little known that a great deal of exploration was necessary, as it still is. But from the earliest days fishermen and scientists were concerned with fluctuations in abundance (Hjort, 1914, 1926).

In America the trend toward fishery population dynamics began with W. F. Thompson and his studies of Pacific halibut *(Hippoglossus stenolepis).* Even before the first International Pacific Halibut Convention entered into force in 1924, Thompson had made a detailed study of the Pacific halibut for the Commissioner of Fisheries of the Province of British Columbia (Thompson, 1916, 1917a). Meanwhile in Russia the biological basis of fishery management also was being developed (Baranov, 1918). This work did not come to the attention of North American fishery scientists until several years later. Thompson and Bell published the first American application of population dynamics in 1934. Since that time this branch of fishery science has been developed to relatively sophisticated levels by W. E. Ricker and others in Canada, by R. J. H. Beverton and S. J. Holt and others in the United Kingdom, and by Milner B. Schaefer and others in the United States.

Although many fishery scientists now believe that population dynamics is the heart of fishery research, others are not so certain, at least in the sense that population dynamics is understood today. Management of fisheries is a complex and difficult art, and it is not so easy to apply scientific knowledge successfully to management. Many kinds of information, only remotely related to the dynamic relationships of fish stocks to fishing and other sources of mortality, are needed for rational harvesting of common property aquatic resources. These include, but are not restricted to: biological and technical studies such as fish behavior, exploratory fishing, and gear development; economic research relating to harvesting, as well as to the processing-distributing segments of the industry; and social-political research and development to find out how to create a receptive constituency.

Some of the world's most successful fishing nations find it difficult to understand the North American and Western European preoccupation with population dynamics and fishery management. An interesting point of view was expressed by Kasahara (1961) in the introduction to his lecture series at the University of British Columbia:

> It appears that fishery biology should not restrict itself to the question of conservation, which is primarily concerned with the means to protect what is now being exploited. More attention should be paid to the development of fisheries and the exploitation of new resources. Fishery biology should not be a conservative science. As a result of the emphasis on the question of conservation, the role of population dynamics also seems to have been overemphasized. Even for the question of conservation, there are so many cases in which the present methods of population dynamics are not useful, either because basic ecological characteristics of the species involved are not sufficiently known, or because the abundance of the population changes tremendously by unknown environmental factors.

Japan, the Soviet Union, and other nations which have developed self-sufficient distant-water mothership fishery fleets, able to operate anywhere in the world ocean, can afford to set aside the question of conservation as long as there remain unfished or underfished resources anywhere in the world. Once the full potential of the sea to yield economically harvestable protein is reached, these nations also may be forced to adopt the North American-western European concept of conservation, or turn to other sources for their supply of protein food.

The idea of an equilibrium between births, growth, natural deaths, and rate of fishing may be traced back to Petersen (1894, 1903), who combined these quantities into a definition of "overfishing." But the quantitative dynamic concept now commonly referred to as the theory of fishing was developed more or less similtaneously by Baranov (1918, 1925) and Thompson (1916, 1917, 1919, 1936, 1937; Thompson and Bell, 1934). The theory since has been elaborated by American, Canadian, and European scientists (Russell, 1931; Buckmann, 1932; Hjort, Jahn and Ottestad, 1933; Graham, 1935 a,b; Ricker 1940, 1954, 1958; Beverton and Holt, 1957; Schaefer, 1943, 1954, 1957a; Chapman, 1961, 1964; Gulland, 1964). The literature is much more extensive than this listing, but the literature cited in these works will lead to most of the important historic papers on the subject.

The essence of the theory of fishing is that each stock of fish is resilient to fishing mortality so that, up to a certain limit, the stock adjusts by increased survival and growth, reaching a new level of equilibrium with its environment. From the biological point of view it is important to recognize that equilibrium is possible over a wide range of fishing intensity, from zero to levels considerably greater than is necessary to take the maximum sustainable yield. Thus, as a fishery develops, the equilibrium catch rises to a maximum and then falls, until (theoretically, at least) the stock can no longer remain in equilibrium. The principal job of the mathematician-biologist is to identify the level of fishing intensity that will produce the maximum sustainable biological yield. Obviously, this end is difficult to accomplish because the information required (principally catch, fishing effort, recruitment, growth, and mortality from fishing and from natural causes) is subject to uncontrollable errors of estimating, and each quantity varies seasonally and annually. The theoretical basis of fishery population dynamics can be developed more easily than it can be applied to fishery management.

Various special studies have been made to obtain estimates of the parameters needed for quantitative fishery science. These studies have dealt with such questions as how to obtain accurate records of catch and effort, how to measure recruitment, growth, and mortality, and how to partition natural and fishing mortality. Sette (1943b) examined these problems in detail and outlined a program by which the necessary information for a study of the Pacific sardine fishery could be obtained with acceptable accuracy. Clark and Marr (1955) examined these questions in depth in the light of sardine research accomplished up to that time. The two authors were not able to agree on all points (a reflection on the inadequacy of the data), but they agreed that it was not possible at that stage of research results to explain why sardine year classes fluctuate widely in abundance, or to predict the catch, or to explain adequately what effect, if any, the fishery had upon future catches.

Many ingenious ways have been developed to estimate the quantities needed to measure the effect of fishing on the stocks. For example, Baranov (1918) demonstrated the effect of fishing in reducing the mean age of fish in the stock. Huntsman (1918) independently derived the same conclusion. Baranov (1918) and Heincke (1922) developed the "catch curve" method of estimating mortality, which is based on the principle that, once fish are fully recruited into the fishable stock, the decline in frequency of successive ages is a measure of mortality. If the stock is virgin or nearly so, this method can give an estimate of natural mortality. If the stock is being fished, it estimates natural plus fishing mortality. Petersen (1897) was apparently the first to use marked fish to estimate rates of fishing. Silliman (1943), by selecting two periods with different but relatively constant rates of fishing, was able to distinguish fishing mortality of Pacific sardines. Beverton and Holt (1957) developed a formula for partitioning fishing and natural mortality when fishing mortality was changing. In 1945, Silliman demonstrated a method of computing mortality from length frequency distributions, for use when age determination are not available. He also (Silliman and Gutsell, 1958) demonstrated experimentally the validity of the theory of fishing under conditions much simpler than are found in nature. Marr (1960) discussed in considerable detail the causes of variations in the catch of Pacific sardines.

The dynamic relationships between fish stocks, their environment, and the fisheries can be understood if the size of the stock is known only in relative terms. Many studies have used indices of abundance rather than absolute measures of stock size. But absolute estimates are desirable, and much effort has been devoted to such estimates. Mark and recapture experiments were among the early techniques. A simple method, using only catch and fishing effort, was developed by DeLury (1947). The various methods were reviewed thoroughly by Ricker (1958). A recent approach using catch and effort is that of Murphy (1965).

Much attention has been given recently to estimates of the total potential yield of living resources of the sea. Estimates have ranged all the way from about 200 million to 2 billion metric tons. From considerations of basic production Ryther (1969) has reviewed previous estimates. He has concluded that it is unlikely that the potential sustainable yield of fish to man is appreciably greater than 100 million tons, and that much of the potential expansion must come from resources for which no harvesting technology or markets yet exist. Alverson *et al* (1970) disagreed substantially with Ryther's conclusions.

Meanwhile, the realization has been growing that it is not so easy to manage a fishery, stock by stock, according to the narrower concepts of the theory of fishing. Individual stocks in the same environment react with each other in many ways. The model most commonly used to describe the effect of fishing on a resource shows that when the maximum sustainable yield is taken the standing crop is only 50 percent of its virgin magnitude. If the resource is fished even harder, as commonly happens, the stock can be reduced to considerably lower levels. As pointed out by Murphy (1966) the energy released can be captured successfully by another species. Such an energy transfer may explain the 20-fold increase in abundance of Pacific northern anchovy as the sardine stock fell to about 5 percent of its former abundance (Ahlstrom, 1967). Introducing interactions between two competing species into the calculations would appear to be a major step forward in sophisticated modeling of fisheries, but it is not likely that the problem is so simple.

Many examples of the need for balanced management can be found in the literature on fresh water fishery research, for example the work of Swingle (1950) on bass and bluegill populations. A striking example of the extent of interactions between fish species, their physical environment, and the fisheries in Lake Michigan is provided by Smith (1968). Watt (1968) devoted more than half of his book to the use of computers in working out the complexities of resource ecology. Patten (1969) has made a most interesting application of systems analysis to the marine community of the English Channel. Sette (1969) has speculated on the advantages of a balanced multi-species fishery. These trends will be followed with interest by fishery research scientists.

Hancock and Simpson (1962) have reviewed the status of knowledge of population dynamics of invertebrate populations. Biologists studying invertebrate fisheries have taken quite different research approaches, probably because their training and backgrounds are different. This approach has been especially true of research on mollusks, probably because most commercially valuable mollusks are sedentary animals, and their fisheries resemble farming to some degree. Hopkins and Menzel (1952) pointed out that the best time to harvest oyster crops should be judged on the basis of numbers of survivors as well as on the size of the oysters. McHugh and Andrews (1955) developed this concept, showing that Virginia oyster growers could obtain higher yields by harvesting more frequently. The computations are simple for planted oyster grounds because recruitment is controlled and the yield at any time can be determined simply by multiplying growth and death rates.

The dynamics of crustacean fishery stocks also have not been given the attention they deserve. Two problems have contributed to this neglect: these animals have no permanent hard parts from which age can be determined; and because they shed their outer skeleton periodically, they are difficult to tag. These problems have been resolved in various ingenious ways, already described. Thus, preliminary approaches have been made toward understanding the population dynamics of king crab resources (Internatl. N. Pacific Fish. Comm., Ann. Repts; Miyahara and Shippen, 1965), blue crabs (Fischler, 1965), and shrimp stocks (Kutkuhn, 1965; Berry, 1967).

Scientists have been slow also to apply the principles of population dynamics to marine mammal resources. The first studies of fur seal population dynamics apparently were those of Chapman (1961) and Nagasaki (1961) for the International Fur Seal Commission. It was not until the early 1960's that the International Whaling Commission began to consult quantitative biologists (Chapman, 1964).

It must be stressed that good fishery statistics are the basis of any successful study of population dynamics. The ideal biological statistical program includes not only good records of catch and effort by reasonably small geographic areas, but also biological data on size and age distribution, migrations, recruitment, growth, mortality, absolute abundance, and subpopulations, among other things. Some states have been collecting good catch and effort statistics for years. But few, if any, states yet have satisfactory routine systems of gathering biological statistics on the catch, and many do not record fishing effort in a form useful for catch and effort analyses. In most places, the necessary data on fishing effort, biology, and sometimes also catch, are still collected fishery by fishery, by biologists. Another serious problem is the lack of adequate data on catch and effort on marine sport fisheries. As Walford (1965) pointed out, the marine sport fisheries are too diffuse for routine and complete statistical gathering, hence the necessary data probably always must be obtained by a system of sampling. But as marine sport fisheries continue to grow the need for adequate information on catch and effort is increasing.

Fishery Economics

Space limitation will not allow a detailed discussion of this relatively new branch of fishery science, but it cannot be ignored in any account of fishery research. The first substantial publication on the subject appeared in 1943 as the record of a most penetrating and historic debate between William C. Herrington and Robert A. Nesbit on the topic "Fishery Management," during a forum discussion arranged by the Atlantic States Marine Fisheries Commission in 1942 to help the newly born Commission to develop policy. Essentially, the two points of view expressed were that scientific fishery management should have as its objective maximum biological yield (Herrington, 1943), and that the only logical objective is to seek maximum economic yield (Nesbit, 1943).

Herrington's approach assumed that maximum biological yield from a fishery is the important point, and it considered the fisherman's income and the total gain to the economy to be of less importance. Nesbit's view was that there is little point in managing a fishery which allows the fisherman only wages (and modest wages at that, for most of them) and no net gain to the economy. He proposed reducing the number of licenses until there remained just the number of fishermen and units of gear needed to take the surplus production. Nesbit was ahead of his time, and his proposal did not receive enthusiastic support from many colleagues. He was successful, however, in persuading the Maryland fishing industry and legislature to adopt a modified limited entry system, the Maryland Management Plan (Tiller 1944, 1945). This imaginative experiment has failed, largely for social-political reasons, but present-day fishery economists and most leading fishery biologists now agree that some form of limited entry, which essentially confers a kind of property right to the resource, is the only reasonable solution to rational management of wild fishery stocks.

The next recorded study of fishery economics was sponsored by the University of North Carolina (Taylor et al, 1961). Curiously, Taylor made no reference to Nesbit's work, and his recommendations give no indication that he was aware of the economic implications of limited entry. Taylor believed that economic factors exert strong control on fish catches, and he was not concerned about overfishing: "Heavy fishing may reduce a population, but the fishery arrests itself automatically as it becomes unprofitable and is discontinued or much diminished long before any species is totally 'fished out.' "

Taylor's recommendations did not indicate an understanding of the difference between total weight caught and the net economic gain from fishing. His eight principal recommendations included the following:

1. Increase the demand for fishery products;
2. Unite biology and economics to lay down general strategy;
3. Carry on various studies of price structure throughout the distribution chain, and conduct economic studies of specific segments of industry;
4. Improve techniques of finding and catching fish by studying fish behavior and improving fishing gear;
5. Resolve the basic conflict between the desire to increase the catch and the fear of overfishing.

Taylor's work was followed by more detailed studies of local fisheries, such as that by Quittmeyer (1957) in Virginia. The modern basis of fishery economics has been developed more recently by a relatively small group of economists in Canada and the United States (Gordon, 1954; Scott, 1962; Turvey, 1964; Sinclair, 1960; and Christy and Scott, 1966). The essence of the argument of the economists is the argument put forward by Nesbit nearly 30 years ago—that fishery management based on the biological principle of maximum sustainable yield makes no economic sense. Their solution is an optimal rate of fishing which produces the maximum net economic yield. The validity of this philosophy was challenged by Schaefer (1957b) and Frick (1957). The latest major economic study (Crutchfield and Pontecorvo, 1969) is of particular interest because it uses the Pacific salmon fisheries as a case history. The authors estimated that the net annual economic yield of the American and Canadian fisheries, if entry were limited to just that amount of fishing power necessary to take the surplus production, would be at least $50 million. From this basis, they concluded that the stakes are high enough to make it worth the effort to install a rational management program based on limited entry. This view may be too simplistic from the standpoint of anyone but a fisherman (McHugh, 1970).

The interest of fishery scientists in economic research was emphasized at the 98th annual meeting of the American Fisheries Society in a special session entitled "Economics and Fishery Resources."[3] Some fishery scientists and some economists believe that the theory of economic fishery management fails to take into account the realities of the social-political world, an argument certain to continue for some time. It will not be resolved until a satisfactory formula is found to describe the dynamic relationships between the living resources in their environment, the economics of man's use and misuse of the resources, and the social-political mecha-

[3] Trans. A.F.S. 98(2):347-378.

nisms by which rational management can be effected. A penetrating analysis of the broad problem was made by Hardin (1968).

Fish Farming

Fish farming, of one kind or another, has been going on since long before the time of Christ (Iversen, 1968; Yonge, 1960). Until very recently, culture techniques have been developed empirically. A milestone in the development of scientific fish culture in North America was the well known work of Foerster (1938) and Ricker in Canada, which demonstrated that artificial propagation of Pacific salmon as it was practiced in the 1920's and 30's gave no higher yields than natural reproduction. Subsequent work at Cultus Lake showed that predator control is a much more effective way of increasing survival of young salmon (Foerster and Ricker, 1942). Recent work in the Columbia River has shown that the contribution of individual hatcheries to silver and chinook salmon runs varies widely.[4] With adequate water supply, scientific feeding methods, and other scientific controls salmon can be produced at reasonable cost and historic runs can be restored.

Commercial trout production under close scientific control is a thriving business in some parts of the United States. Catfish farming is growing rapidly. It was estimated that in 1969 more than 12 million pounds of trout and about 40 million pounds of catfish were produced (Dillon, 1970). Freshwater fish culture is covered in several chapters of this book and the scientific as well as the farming aspects have been covered thoroughly in several recent studies. The science of fish farming is still too young for trends to be seen clearly. Interest of scientists in marine aquaculture is growing rapidly, as illustrated by the special session on Marine Aquaculture of the 98th annual meeting of A.F.S.[5] A rather complete review of artificial propagation may be found in Shelborne (1964).

Selective breeding is a continuing objective of fish culture. For warmwater fishes the subject was reviewed recently by Clemens (1968). The work of Donaldson (1963; Donaldson and Menasveta, 1961) on salmon and trout has received wide attention. Loosanoff and Davis (1963a) have discussed selective breeding of mollusks.

Scientific shellfish culture is not yet far advanced in North America, but the scientific basis for shellfish farming is being laid more and more rapidly (Loosanoff, 1965). For example, the shellfish research laboratory of the Bureau of Commercial Fisheries at Milford, Conn., founded in 1935, has developed techniques for artificial rearing of oysters and clams (Loosanoff and Davis, 1963b) and for predator control, and presently has in progress important research in genetics aimed at selective breeding of mollusks. Much of this work was stimulated by the studies of Dr. Imai and his colleagues in Japan (Imai *et al*, 1961; Ito and Imai, 1955). The Bureau of Commercial Fisheries laboratory at Galveston, Texas, the University of Miami Institute of Marine Science, and the Oceanographic Laboratory of Florida State University are perfecting techniques for shrimp culture (Anonymous, 1970).

Marine shellfish farming is more highly developed than marine fish farming, and both are more highly advanced in other parts of the world than North America. Two recent comprehensive works should be consulted: Iversen (1968) *Farming the Edge of the Sea* and Ryther and Bardach (1968) *The Status and Potential of Aquaculture.*

Some fishery experts have discounted the importance of aquaculture, arguing that the principal immediate objectives of world fisheries are to produce large quantities of fish at low cost to aid the developing countries. This is certainly a worthy objective as an instrument of foreign policy for the United States and the world, but it is hardly calculated to improve the economic position of the American primary producer unless he is also subsidized. From the standpoint of net economic gain to the national economy, which is clearly the goal of the present Administration, aquaculture is attractive, for it tends to eliminate the major problems associated with common property resources. For this reason a relatively small aquacultural production could be much more important to the economy than a large harvest from a wild resource. This fact should justify

[4] Report of Second Governors' Conference on Pacific Salmon, State Printing Plant, Olympia, Wash. 1963:166 p.

[5] Trans. A.F.S. 98(4):738-761.

relatively much larger investments in fundamental research for fish farmers.

Limnology and Oceanography

Fishery scientists and marine biologists have been active in the development of limnology and oceanography since the beginning. Limnology and freshwater fishery research and management have always been associated closely. I am setting this important subject aside on the assumption that it is reasonably well covered in the chapters dealing with freshwater fisheries. The International Council for the Exploration of the Sea, founded in 1902, perhaps marks the beginning of organized cooperative inquiry into fishery oceanography, but the birth of oceanography usually is attributed to the Challenger Expedition of 1872-1876, dates that coincide with the birth of fishery science in America.

Although some studies of physical and chemical oceanography were being made by fishery agencies in the United States and Canada before World War II, they were limited and spasmodic as compared with the rapid growth of oceanography during the past 25 years. The status of fishery hydrography in Western Europe prior to the war, with some attention to early post-war developments, was reviewed by Tait (1952). He emphasized the importance of temperature as a factor which determines the distribution and migrations of fish stocks and cited various bits and pieces of information which suggest that temperature, currents, and movements of water masses hold the answers to fluctuations in abundance and availability. Although we have advanced a long way in understanding oceanic ecology since Dr. Tait delivered the Buckland Lectures that were the origin of his book, many important questions are still unanswered. Hela and Laevastu (1961) published a more recent treatise on the subject, and Hela (1967) emphasized the complexity of such relationships and the need to understand air-sea interactions also.

Fishery scientists have played an important part in the postwar development of oceanography in the United States, as they have also in Canada. The genesis of the present United States national program in marine science came in the middle 1950's, when the Fish and Wildlife Service of the Department of the Interior joined with the Office of Naval Research and the Atomic Energy Commission in requesting the National Academy of Sciences to assist in developing a coordinated government-university seagoing program. This proposal led to establishment of the National Academy of Sciences Committee on Oceanography, which stimulated a whole series of developments, culminating in the Marine Resources and Engineering Development Act of 1966 (Public Law 89-454). These events helped to broaden the federal fishery research program, achieving substantially increased attention to "fundamental" research, a number of modern new laboratories and an impressive fleet of research vessels, and approximately a fourfold increase in the fishery research and exploratory budget in less than 10 years. Recent substantial federal budget reductions have reversed this trend and have led to curtailment or termination of laboratory research programs, closure of laboratories, and vessel tieups. Canadian fishery oceanographers are having similar difficulties. Scientists are concerned about the effects of these reductions upon a vigorous and relatively new scientific endeavor which was still maturing.

Much of the postwar development of fishery oceanography can be traced through the publications of federal and state fishery agencies, the international fishery commissions, and FAO. For Canada a most useful publication is the extensive index of papers published by Canadian fishery scientists or in Canadian fishery journals (Carter, 1968). No similar comprehensive index exists for the United States, although several limited catalogs have been published (e.g., Dees, 1963).

An understanding of the relationship between marine fish stocks and their environment leads to forecasting. Sometimes large-scale changes in the environment provide natural experiments, which show how marine animals react to environmental variations. Such dramatic incidents as the warming of western North Atlantic waters which took place in the first half of this century (Taylor *et al*, 1957) and the warm years (1957-1958) in the eastern Pacific (Sette and Isaacs, 1960; Radovich, 1961) were well monitored by fishery oceanographers. The reactions of animals and plants to such widespread changes are useful in interpreting the effects of shorter-term phenomena and in formulating hypotheses that can be tested by research.

Examples of recent accomplishments in fishery forecasting from oceanographic research are the successful predictions of time of arrival and abundance of skipjack tuna on fishing grounds in the vicinity of the Hawaiian Islands and of albacore along the Pacific coast of North America. The Bureau of Commercial Fisheries and the Fleet Numerical Weather Facility of Navy at Monterey, California, have cooperated in a successful two-way radio communications operation which takes ocean data from tuna fishermen and other sources and, after analysis, transmits information in usable form back to the fishing fleet. Blackburn (1965) made a broad but concise review of the relationships between oceanography and ecology of tunas, with an extensive review of upwelling and its relation to fish production.

Favorite (1969, 1970) has published a series of articles on fishery oceanography which show how ocean salinity, temperature, and circulation affect Pacific salmon distribution and food supply. The Japanese Society of Fisheries Oceanography in 1969 published a special number of its *Bulletin* entitled *Perspectives in Fisheries Oceanography* in honor of Professor Michitaka Uda. In addition to the paper by Dr. Lucas, cited elsewhere, this volume contains several other papers in English, or with abstracts in English, describing trends in fishery research in the United Kingdom, Germany, Australia, and other parts of the world and the relation between oceanography and fisheries.

Comments on Selected Fisheries

In reviewing the table of contents of this volume I noticed some curious omissions. Several chapters deal with commercial marine fishery resources, including many of the most important in order of value. But several valuable commercial resources are missing, and there is no chapter on marine sport fisheries. Of the groups of species represented by the chapter titles, salmon rank second in landed value in the United States, oysters and clams together third, tuna fourth, groundfish eighth, and halibut tenth. Sea herrings are relatively low in order of landed value, and Pacific sardine landings are insignificant. Omitted were shrimp, which as a group are by far the most valuable fishery resource in the United States, with a landed value of about $125 million in 1969; crabs, fifth in value; lobsters sixth; menhaden seventh, flounders ninth, and scallops eleventh. All these species together are valued at more than $400 million, or about 85 percent of the total landed value of United States domestic fisheries. The neglected groups have a landed value of over $200 million, which is more than 40 percent of the total landed value of all species.

Although no one has yet arrived at a satisfactory method of comparing commercial and sport fisheries in dollar value, there is no doubt that marine sport fisheries are important recreationally and economically. Without citing sources, de Sylva (1969) asserted that "at an overall expenditure of $103.19 per angler the annual contribution to the U.S. economy by the regular saltwater angler exceeds $850 million." He further says, "Sport fishing was worth $2.1 billion to the economy in 1955; it now exceeds $4 billion." I question his economics but cannot challenge his conclusion that sport fishing is big business.

The earliest substantial research on the American shrimp fishery was that of Lindner and Anderson (1956). As the shrimp fishery in the Gulf of Mexico developed after World War II, research on the resource shifted to the Gulf Coast states, universities, and federal laboratories. Reference has been made to shrimp research as appropriate in the preceding sections of this chapter. Berry (1967) has concluded that the Tortugas pink shrimp fishery has had no detrimental effects on the stock but that the total weight and value of the catch would increase if the fishery were regulated to catch the shrimp at a larger size. He found also that fluctuations in abundance of the brown shrimp resource in the northern and western Gulf of Mexico are caused by factors that apparently operate over a wide geographic range.

Menhaden, although seventh in total landed value, produce by far the greatest weight of fish landed by United States fishermen. The menhaden fishery now occupies the place formerly held by the Pacific sardine industry. Substantial research on the menhaden resource began only about 15 years ago. Scientific knowledge of the resource and the fishery was summarized by Reintjes (1969). McHugh (1969) compared the history of the sardine and menhaden fisheries.

It is surprising, in view of the well developed

state of fishery population dynamics, that quantitative methods were not applied long ago to research on fur seal and whale resources. The first studies of fur seal population dynamics apparently were those of Chapman (1961) and Nagasaki (1961). Perhaps the fur seal herd was responding so nicely to empirical management that the need for more precise methods was not recognized until, as the resource approached equilibrium, the effect of natural fluctuations in abundance began to trouble the managers.

In the Antarctic, the effect of modern whaling on the whale resources was clear to the practiced eye from inspection of the raw catch and effort data even before World War II put a temporary halt to pelagic whaling. The decline continued after the war with hardly a sign that there had been a holiday. It was not until the early 1960's that the International Whaling Commission turned the problem over to the scientists experienced in quantitative biology. The FAO stock assessment group, with considerable assistance from the Scientific Committee of the Commission, soon demonstrated that the stocks were in even worse condition than qualitative inspection of the data suggested (Chapman, 1964). This had a salutory effect on the actions of the International Whaling Commission in setting realistic quotas, and in fact prohibiting killing of blue and humpback whales. It is appropriate to point out here that marine mammal populations are much more sensitive and less resilient to fishing than most fishes and invertebrates because their fecundity is so much lower.

Epilogue

This chapter began with the thought that the intense preoccupation of a century ago with artificial culture inhibited the development of scientific fishery research. In a hundred years we have gone a long way in developing fishery research as a quantitative science. But we also are circling back to fish culture[6] as an important part of fishery science. Fish culture, like the other scientific fishery techniques, is important when it is based on sound scientific principles and its objectives are clearly understood. Fish culture can be a totally controlled process in which the fish are reared from egg to commercial size, as are trout in Idaho and catfish in Arkansas, or when it is desirable to stock lakes and streams for "put and take" sport fishing, as is done in many parts of the United States.

Fish culture does not need to include all life history stages to be successful. The promising preliminary results of the Columbia River salmon hatchery program (Ellis and Noble, 1959; Worlund *et al,* 1969) appear to demonstrate that healthy young fish can survive to maturity if released into the natural environment at a stage in which they can cope successfully with predatory and other environmental hazards. Some workers recently have had encouraging success in rearing small numbers of marine fishes from the eggs to advanced juvenile stages (Shelbourne, 1964; Bardach and Ryther, 1968). There is some question as to whether this is a reproducible scientific process yet, or whether the scientists who did these things had an inherent knack for fish rearing—a "scaly thumb" so to speak. There appears to be no reason why, with adequate facilities and techniques, we cannot learn to raise large numbers of marine fishes to any desired life-history stage. This prospect offers exiting possibilities for buffering naturally occurring fluctuations in abundance in herring-like fishes and other coastal species, thus improving substantially the economic stability of such fisheries. Cost will be the limiting factor.

Encouraging progress also has been made in the past two decades in shellfish research leading to hatchery techniques. Commercial shellfish hatcheries presently are operating on Long Island Sound and Chesapeake Bay (Hidu *et al,* 1969). A number of problems remain to be solved, but the possibilities for selective breeding for such desirable qualities as disease resistance, rapid growth, fatness, and flavor are exciting. Genetic investigations with these objectives are under way at the Laboratory of the Bureau of Commercial Fisheries in Milford, Connecticut.

The Japanese have been pioneers in marine fish culture. Much of the American work in artificial breeding of oysters and clams was stimulated by the work of Dr. Takeo Imai and his colleagues (Imai and Hatanaka, 1949; Imai, 1967; and numerous other papers). The success of Dr. Fujinaga (sometimes spelled Hudinaga)

[6] Marine Culture Session. Proc. 21st Ann. Sess., Gulf and Caribb. Fish. Inst. 1968(1969):136-185.

in rearing shrimp from egg to adult offers potential for shrimp culture in America (Fujinaga, 1963). It would be particularly interesting to examine the possibility of adapting his techniques to the blue crab or menhaden industry on the Atlantic coast. If young blue crabs or menhaden could be reared to a stage at which they would have a good chance of survival in the natural environment, this development might reduce the wide fluctuations in abundance (McHugh, 1969) which are so troublesome to the industry.

As fish culture is expensive, it is unlikely to be a solution to the problem of producing large quantities of animal protein at low cost to feed a hungry world. For this reason, the substantial recent advances in exploratory fishing and in quantitative fishery science have been perhaps the most encouraging development for immediate application to increasing fishery yields. It is being recognized more clearly, however, that the theory of fishing as applied to individual species or stocks of fish is not very useful in practice. Fishery scientists are now examining the interactions of stocks in the total community, recognizing that successful fishery management of wild stocks may require balanced harvesting of a variety of species. Freshwater fishery scientists have been aware for years of the importance of population balance in lakes or ponds, and they have developed considerable skill in maintaining desired balances (Swingle, 1950). The need for similar understanding of the marine environment is becoming clearer as the harvest of the sea increases.

The importance of understanding interactions between plant and animal species within an aquatic community and between them and their physical environment has long been recognized. The problem of resolving these complicated relationships was overwhelming until high speed computers became available. Aided by these new tools, some fishery scientists are engaged in model building to examine the nature of dynamic ecology. The subject of computer applications to fishery science was important enough to warrant a special session at the 98th annual meeting of the American Fisheries Society.[7] This is a worthy objective but also an uncertain one, because models are only as good as the basic data that go into them and the mental processes of their human builders. The technical sophistication of electronic computers is apt to convince people that their output is equally sophisticated. In dropping this word of warning, I cannot improve upon the words of one of my colleagues in a recent letter: "The model builders—the ultimate in fishery research. Computers do all the thinking and never fail to extract fully dependable conclusions by feeding scant, scurvy data through elegant analyses."

Watt (1968), discussing multiple linear regression methods, puts it this way:

> Blind application of multiple regression analysis leads to erroneous conclusions. A significant regression coefficient may not mean that the factor is important; rather, the factor may be highly correlated with some important factor that has not been included in the regression analysis. Furthermore, a factor may not appear to be important when in fact it is, because the wrong model has been postulated.
>
> In general, while empirical models relate causes to effects, they yield little insight into the mechanics of the casual pathway. Such models really tell us little about how to manage complex systems.
>
> Excessive dependence on linear multiple regression as a research tool can have an insidious effect on the development of knowledge in a biological field. This statistical tool was not developed because linear models describe the behavior of complex systems in a real world. Rather, a linear additive model is theoretically straightforward to deal with, and leads to a simple, if tedious, computational procedure. Trying to cram research findings into a model that is best from a statistical point of view—but poorest from a biological or management point of view—is a procedure that can only result in preventing the development of a suitable theoretical substructure for biology.

It is impossible to write about fishery research without referring also to fishery management, for that is the end objective. In attaining its ultimate objective, fishery science has not been notably successful in its 100-year history. The results of research have been applied with some success to sport fish management and to some commercial fish cultural operations. Generally speaking, the degree of success is related inversely to the size of the body of water, the complexity of its flora and fauna, and the degree to which the manager is able to exercise control. In bodies of water as large as the Great Lakes and the world ocean not many successful

[7] Trans. Am. Fish. Soc. 98(3):551-588.

examples of scientific fishery management can be cited.

In the public segment of the economy, the case for fishery research and management usually has been made on emotional arguments, in which for the most part the problems are misidentified or improperly arranged in order of importance, their causes are misunderstood, and the solutions are the wrong solutions. In the United States federal government, adoption of a new program planning and budgeting system (first used by former Secretary McNamara in the Department of Defense) a few years ago has forced fishery executives to examine much more closely than before the goals and objectives of fishery research and management, the alternative ways of achieving these ends, and the economic benefits that might be achieved per unit cost. The end results have not been particularly promising, for these analyses have led the Administration to doubt the value of spending public funds on fishery research and development, and this conclusion has led in turn to a substantial reduction in the budget of the Bureau of Commercial Fisheries. The future of federal fishery research in the United States will depend upon careful re-examination of goals, objectives, and benefits to be attained.

Korringa (1969), reviewing the triumphs and frustrations of fishery scientists, suggested that they may have failed to explain clearly the consequences of controlled fishing, and that they may have worded their conclusions too cautiously. This alleged failure gets at one of the basic problems of natural resource utilization. The good scientist is trained to weigh all possible interpretations. Knowing better than anyone else how incomplete his data are, he naturally qualifies his conclusions. The layman, especially when he has a particular point of view to express, is not bound by scientific objectivity. Korringa saw four possible lines of development for world fisheries:

1. Development of unrestricted "tramp fisheries" in which the fleets would go wherever fish are available in commercial quantities. This trend is already under way.

2. Development of managed fisheries, the most attractive alternative to the scientist. But Korringa saw the diffiulties of reaching international agreement and recognized that this course may continue to lead to broader national jurisdictions over sea fisheries.

3. Fish farming presents many problems but also offers attractive advantages over fisheries based on open access resources. The principal advantages are maximum yield per unit area of water surface and property rights over the resource.

4. Sport fisheries may be judged more important. This trend is already well developed in the United States, where marine sport fishing is important and is growing steadily and where pressures are being exerted to restrict many kinds of commercial fishing in coastal waters.

With the exception of the first alternative, adequate scientific knowledge is a prerequisite to successful management. It behooves fishery scientists to be more vigorous in making this point clear.

Acknowledgments

The history of any technical subject is a history of people, and the number of people who have made important contributions to a branch of science even as small as fishery research is legion. Some people have stood out as important actors throughout their professional careers. Others have made important contributions behind the wings. To a considerable degree the personality of the individual determines whether he comes forth as a universally recognized leader in his field, whose name will live in the records of his field of specialization but whose influence as an administrator or teacher is less obvious. There is no single criterion of success, but perhaps the most important is the frequency with which a scientist's publications have been cited in the scientific literature. The scientists who attain greatest stature usually do so through their publications and through active participation in public and professional affairs. Others get their satisfaction vicariously by guiding the work of their colleagues, or carry on quietly and ably in the laboratory and in the field, revealing their competence only to a few close associates.

For these reasons, accepting an assignment to write a chapter on a subject as broad and important as the history of fishery research is a sure way to make enemies. My original plan called for a series of brief sketches of the leaders in American fishery science. The list quickly grew to such a length as to make it impossible for a single chapter, and probably even for a book. For the same reasons I rejected the idea of listing names. On the other hand, it was equally impossible to avoid giving names at all, because among the major merits of any scientific dissertation are the literature citations which give the reader a broader view of the subject if he wishes to explore it. Inevitably, if the author's literature research has been reasonably thorough, the names of most of the leaders will be included in the citations. This will not necessarily identify all of the giants of fishery research nor eliminate all of lesser rank, but it will include most of the leaders.

As additional insurance against slighting the leading fishery scientists I asked a small, but broadly representative group of friends to help me. I requested that they provide, from their personal knowledge and experience, a list of what they considered to be the major milestones in fishery research in the past 100 years, and the names of the leaders. To avoid bias as far as possible I selected correspondents at random from a list of those in the United States and Canada whom I considered to be leaders. Not all responded, which was not surprising, because they are all busy men. Several correspondents went beyond what I had asked them to do, and far beyond anything I would have had the temerity to ask or to expect. This cooperation saved me many hours of library research for which I am most grateful. Through this correspondence I received invaluable assistance from Elbert H. Ahlstrom, Ralph Hile, Victor L. Loosanoff, Stanford H. Smith, Ralph P. Silliman, Albert L. Tester, Donald L. McKernan, O. E. Sette, and William C. Herrington. Many others helped me by answering questions or by providing literature. To list them all would be impossible.

Literature Cited

Anonymous. 1970. Food fish facts–Shrimp. U.S. Dept. Interior, Bureau of Comm. Fish., Index for 1969, Vol. 31:40-41.

Appellof, A. 1909. Untersuchungen uber den Hummer. Bergens Mus. Skr., NyRaekke 1(1):1-79.

Ahlstrom, Elbert H. 1965. Kinds and abundance of fishes in the California Current region based on egg and larval surveys. Calif. Mar. Res. Comm., Calif. Coop. Ocean. Fish. Inv., Repts. 10:30-52.

———. 1966. Distribution and abundance of sardine and anchovy larvae in the California Current region off California, 1951-64; A summary. U.S. Fish and Wildl. Serv., Spec. Sci. Rept. -Fish. 534:71 p.

———. 1967. Co-occurrences of sardine and anchovy larvae in the California Current region off California and Baja California. Calif. Mar. Res. Comm., Calif. Coop. Oceanic Fish. Inv., Repts. 11:117-135.

Alverson, Dayton L., and Harry H. Chenoweth. 1951. Experimental testing of fish tags on albacore in a water tunnel. U.S. Fish and Wildl. Serv., Comm. Fish. Rev. 13(8):1-7.

———, and Herbert A. Larkins. 1969. Status of knowledge of the Pacific hake resource. Calif. Mar. Res. Comm., Calif. Coop. Ocean. Fish. Inv., Repts. 13:24-31.

———, A.T. Pruter, and L.L. Ronholt. 1964. A study of demersal fishes and fisheries of the Northeastern Pacific Ocean. Inst. Fisheries, Univ. Brit. Columbia, H.R. MacMillan Lectures in Fisheries: 190 p.

———, A. R. Langhurst, and J. A. Gulland. 1970. How much food in the sea? Science 168:503-505.

Atkins, Charles G. 1876. Atlantic salmon, Rept. U.S. Comm. Fish. for 1873-74 and 1874-75: xxx-xxxii.

Baranov, Th. I. 1918. On the question of the biological basis of fisheries. Inst. for Sci. Ichthyol. Invest., Proc. 1(1): 81-128. Reports from the Division of Fish Management and Scientific study of the fishing industry 1(1), Moscow.

———. 1925. On the question of the dynamics of the fishing industry. Narod. kom. snabzh., Bull. rybn. khoz. 8:7-11.

Bardach, John E., and John H. Ryther. 1968. The status and potential of aquaculture. Vol. II. Particularly fish culture. U.S. Dept. Commerce, Clearinghouse for Fed. Sci. & Tech. Information, PB 177, 768:vi + 225 p.

Barnhart, Percy S. 1936. Marine fishes of Southern California. Univ. Calif. Press, Berkeley:209 p.

Berry, Richard James. 1967. Dynamics of the Tortugas (Florida) pink shrimp population. A thesis submitted in partial fulfillment of the requirements for the degree of Doctor of Philosophy in Oceanography, Univ. of Rhode Island:xviii + 160 p.

Beverton, R.J.H., and S.J. Holt. 1957. On the dynamics of exploited fish populations. Min. Ag. Fish. Food U.K., Fish. Inv., Ser. 2, 19:533 p.

Bigelow, Henry B. 1926. Plankton of the offshore waters of the Gulf of Maine. Bull. U.S. Bur. Fish. 40, Pt. 2:1-509.

———. 1927. Physical oceanography of the Gulf of Maine. Bull. U.S. Bur. Fish. 40, Pt. 2:511-1027.

———, and William Welsh. 1925. Fishes of the Gulf of Maine. Bull. U.S. Bur. Fish. 40, Part 1:567 p. (Revised in 1953 as Bigelow, Henry B. and William C. Schroeder, U.S. Fish and Wildl. Serv., Fishery Bull. 74:viii + 577 p.).

Blackburn, M. 1965. Oceanography and the ecology of tunas. *In:* Oceanogr. Mar. Biol. Ann. Rev. 3, H. Barnes (ed.), Geo. Allen and Unwin Ltd., London: 299-322.

Blaxter, J.H.S. and F.G.T. Holliday. 1963. The behavior and physiology of herring and other clupeids. *In:* Advances in Marine Biology, Vol. 1, F.S. Russell (ed.):261-393.

Brett, J.R. 1964. The respiratory metabolism and swimming performance of young sockeye salmon. J. Fish. Res. Bd. Canada 21(5):1183-1226.

Brown, Margaret E. (ed.). 1957. The physiology of fishes. Academic Press Inc., N.Y. Vol. 1. Metabolism:xiii + 447 p., Vol. 2. Behavior:xi + 526 p.

Buckman, Adolf. 1932. Die frage nach der zweckmassigkeit des schutzes untermassiger fische und die voraussetzungen fur ihre beantwortung. Cons. Int. Expl. Mer. Rapp. Proc.-Verb. 80(7):16 p.

Cahn, Phyllis H. (ed.). 1967. Lateral line detectors. Indiana Univ. Press, Bloomington:xiv + 496 p.

Calhoun, Alex (ed.). 1966. Inland Fisheries Management. Calif. Dept. Fish and Game, Sacramento:vi + 546 p.

Carlander, Kenneth D. 1969. Handbook of Freshwater Fishery Biology. Iowa State Univ. Press, Ames, Iowa. 720 p.

Carroz, J.E. (ed.). 1965. Inter-American Tropical Tuna Commission (IATTC). *In:* Establishment, structure, functions and activities of international fisheries bodies. FAO Fish. Tech. Pap. 58:ii + 30 p.

Carter, Neal M. 1968. Index and list of titles, Fisheries Research Board of Canada and associated Publications 1900-1964. Fisheries Res. Bd. Canada, Bull. 164:xviii + 649 p.

Chapman, Douglas G. 1961. Population dynamics of the Alaska fur seal herd. Trans. 26th N. Amer. Wildl. and Nat. Res. Conf:356-369.

———. 1964. Final report of the Committee of Three Scientists on the special scientific investigation of the Antarctic whale stocks. Internatl. Whaling Comm., 14th Ann. Rept.:40-92.

Christy, Francis T., and Anthony Scott. 1966. The Common Wealth in Ocean Fisheries. The Johns Hopkins Press, Baltimore, Md:xiii + 281 p.

Clark, Frances N. 1947. Analysis of populations of the Pacific sardine on the basis of vertebral counts. Calif. Dept. Fish and Game, Fish. Bull. 65:26 p.

———, and John C. Marr. 1955. Population dynamics of the Pacific sardine. Calif. Marine Res. Comm., Calif. Coop. Oceanic Fish Inv., Repts. 1 July 1953-31 March 1955:12-52.

Clemens, Harold B. 1961. The migration, age and growth of Pacific albacore *(Thunnus germo)*, 1951-1958. Calif. Dept. Fish and Game, Fish. Bull. 115:128 p.

Clemens, Howard P. 1968. A review of selection and breeding in the culture of warmwater food fishes in North America. Proc. World Symposium on Warm-Water Pond Fish Culture. FAO Fish. Rept. 44(4):67-80.

Clemens, W.A. 1958. Reminiscences of a Director. Jour. Fish. Res. Bd. Canada 15(5):779-796.

———. 1968. Education and fish. An autobiography by Wilbert Amie Clemens. Fish. Res. Bd. Canada, Ms. Rept. 974:iv + 102 p.

———, and G.V. Wilby. 1946. Fishes of the Pacific coast of Canada. Bull. Fish. Res. Bd. Canada 68:368 p. (2nd. ed. published 1961:443 p.).

Crowe, Beryl. 1969. The tragedy of the commons revisited. Science 166:1103-1107.

Crutchfield, James A., and Giulio Pontecorvo. 1969. The Pacific salmon fisheries. A study of irrational conservation. Resources for the Future, Inc. (The Johns Hopkins Press), Washington, D.C.:xii + 220 p.

Cushing, D.H. 1969. Upwelling and fish production. FAO Fish. Tech. Pap. 84:iii + 40 p.

Cushing, J.E. 1964. The blood groups of marine animals. Adv. Marine Biol. 2:85-131.

Dahlgren, Edwin H. 1936. Further developments in the tagging of the Pacific herring, *Clupea pallasii.* J. Cons. 11(2):229-247.

Dees, Lola T. 1963. Index of fishery biological papers by U.S. Fish and Wildlife Service authors appearing in non-governmental publications 1940-56. U.S. Dept. Interior, Bureau Comm. Fish., Circ. 151:138 p.

de Ligny, Wilhelmina. 1969. Serological and biochemical studies on fish populations. *In:* Oceanogr. Mar. Biol. Ann. Rev. 7, H. Barnes (ed.), Geo. Allen and Unwin Ltd., London:411-513.

DeLury, Daniel B. 1947. On the estimation of biological populations. Biometrics 3(4):145-167.

deSylva, Donald P. 1969. Trends in marine sport fisheries research. Trans. Am. Fish. Soc. 98(1):151-169.

Deuel, C.R., D.C. Haskell, and A.V. Tunison. 1944. Feeding tables for trout. N.Y. Conservation Dept., Albany, N.Y.

Dillon, Olan W. 1970. The growing importance of commercial fish farming in U.S. The Amer. Fish Farmer 1(2):20.

Donaldson, Lauren R. 1963. Can the stocks of anadromous fish be improved in quantity and quality by selective breeding? Rept. 2nd Governors' Conf. on Pacific Salmon, Olympia, Wash: 102-104.

———, and Deb Menasveta. 1961. Selective breeding of Chinook salmon. Trans. Am. Fish. Soc. 90(2):160-164.

Dymond, J.R. 1932. The trout and other game fishes of British Columbia. Bull. Biol. Bd. Canada 32:51 p.

———. 1964. A history of ichthyology in Canada. Copeia 1964(1):2-33.

Ellis, C. H., and R.E. Noble. 1959. Calculated minimum contributions of Washington's hatchery releases to the catch of salmon on the Pacific coast and the costs assessable to hatchery operations. Wash. State Dept. Fish., Fish. Res. Papers 2(2):88-99.

Evermann, B.W., and E.L. Goldsborough. 1907. The fishes of Alaska. Bull. U.S. Bur. Fish. 26:219-360.

Farris, D.A. 1957. A review of paper chromatography as used in systematics. U.S. Fish and Wildl. Serv., Spec. Sci. Rep. Fisheries 208:35-38.

Favorite, Felix. 1969-70. Fishery Oceanography, Parts I-VI. Commercial Fisheries Review 31 (7-12 incl.), 32(1):32-34, 36-40, 34-40, 29-32, 35-39, 45-50.

Fischler, Kenneth J. 1965. The use of catch-effort, catch-sampling, and tagging data to estimate a population of blue crabs. Trans. Am. Fish. Soc. 94(4):287-310.

Fleming, Richard H. and Taivo Laevastu. 1956. The influence of hydrographic conditions on the behavior of fish. FAO Fish. Bull. 9(4):181-196.

Foerster, R.E. 1936. The return from the sea of sockeye salmon *(Oncorhynchus nerka)* with special reference to percentage survival, sex proportions and progress of migration. J. Biol. Bd. Canada 3:26-42.

———. 1938. An investigation of the relative efficiencies of natural and artificial propagation of sockeye salmon *(Oncorhynchus nerka)* at Cultus Lake, British Columbia. J. Fish. Res. Bd. Canada 4(3):151-161.

———. 1968. The sockeye salmon. Bull. Fish. Res. Bd. Canada 162:xv + 422 p.

———, and W.E. Ricker. 1942. The effect of reduction of predaceous fish on survival of young sockeye salmon at Cultus Lake. J. Fish. Res. Bd. Canada 5(4):315-336.

Frick, Harold C. 1957. The optimum level of fisheries exploitation. J. Fish. Res. Bd. Canada 14(5):683-686.

Fry, F.E.J. 1957. The aquatic respiration of fish, Vol. 1, Metabolism, Margaret E. Brown (ed.), Academic Press:1-63.

Fujinaga, M. 1963. Culture of Kuruma-Shrimp *(Penaeus japonicus)*. Indo-Pac. Fish. Council, Current Affairs Bull. 36:10-11.

Galtsoff, Paul S. (ed.). 1954. Gulf of Mexico. Its origin, waters, and marine life. U.S. Dept. Interior, Fish and Wildl. Serv., Fish. Bull. 55:xiv + 604 p.

———. 1958. A decade of progress in fishery biology of the Gulf and Caribbean area. Gulf and Caribb. Fish. Inst., Proc. 10th Ann. Sess:16-21.

———. 1962. The Story of the Bureau of Commercial Fisheries Biological Laboratory, Woods Hole, Massachusetts. U.S. Dept. Interior, Fish and Wildl. Serv., Bureau of Comm. Fish., Circ. 415:iii + 121 p.

Goode, George Brown, and a staff of associates. 1884-1887. The fisheries and fishery industries of the United States. Govt. Printing Off., Washington, D.C.:Sec. I:xxxiv + 895 p; II:ix + 787 p; III:xviii + 239 p; IV:148 p; V:xx + 881 p.

———, and Tarleton H. Bean. 1895. Oceanic ichthyology. Smithson. Inst., Washington, D.C.:xxxv + 553 p. and atlas xxiii + 26 p., cxxiii pl.

Gordon, H. Scott. 1054. The economic theory of a common property resource: The fishery. J. Pol. Econ. 62:124-142.

Graham, Michael. 1929. Studies of age determination in fish. Pt. II. A survey of the literature. Min. Agric. Fish., Fish. Inv. Ser II, xi(3):50 p.

———. 1935 a. Modern theory of exploiting a fishery, and application to North Sea trawling. J. Cons. 10(3):264-274.

———. 1935 b. Review of Thompson and Bell, Biological Statistics of the Pacific halibut fishery (2), Rep. Int. Fish. Comm. 8:49 pp., 1934. J. Cons. 10(2):210.

———. 1956. Plaice. *In:* Sea Fisheries. Michael Graham, ed., Edward Arnold Ltd., London: 332-371.

Groves, Alan B., and Anthony J. Novotny. 1965. A thermal marking technique for juvenile salmonids. Trans. Am. Fish. Soc. 94(4):386-389.

Gulland, J.A. 1964. Manual of methods of fish population analysis. FAO Fish. Tech. Pop. 40:62 p.

Halver, John E. 1963. Status of salmon nutrition research. Rept. 2nd. Governors' Conf. on Pacific Salmon, Olympia, Wash.:94-94.

Hancock, D.A., and A.C. Simpson. 1962. Parameters of marine invertebrate populations. *In:* The exploitation of natural animal populations, E.D. LeCren and M.W. Holdgate, ed., John Wiley and Sons. N.Y.:29-50.

Harden Jones, F.R. 1968. Fish Migration. Edward Arnold (Publishers) Ltd., London:viii + 325 p.

Harkness, W.J.K., Justin W. Leonard, and Paul R. Needham. 1954. Fishery research at mid-century. Trans. Am. Fish. Soc. 83:212-216.

Hardin, Garrett 1968. The tragedy of the commons. Science 162(3859):1243-1248.

Hart, John Lawson. 1933. A report on the investigation of the life-history of the British Columbia pilchard. Rept. Brit. Col. Commr. Fish. 1933:H60-H70.

Hasler, A.D. 1966. Underwater guideposts. Univ. Wisconsin Press:155 p.

Hay, W.P., and C.A. Shore. 1918. The decapod crustaceans of Beaufort, N.C., and the surrounding region. Bull. U.S. Bur. Fish. 35:369-475. (Revised as: Williams, Austin B. 1965. Marine decapod crustaceans of the Carolinas. U.S.F.W.S. Fishery Bull. 65(1):xi + 298 p.).

Heincke, F. 1898. Naturgeschichte des Herings. I. Die Lokalformen und die Wanderungen des Heringes in den europaischen Meeren. Abh. Deutsch. Seefish. Ver. 2(1):cxxxvi + 128 p.

———. 1922. The over-fishing of the North Sea and the action of the close time of the war on its plaice population. Part I. Der Fischerbote 14:365-389. (Translation No. 117 in Dept. of the Interior Library, Washington, D.C.).

Hela, I., and T. Laevastu. 1961. Fisheries hydrography. Fishing News (Books) Ltd., London:137 p.

Hela, Ilmo. 1967. Utilization of physical oceanography in the service of marine fisheries. Proc. Finnish Acad. Sci. Lett. 1965:157-187.

Herrington, William C. 1943. Some methods of fishery management and their usefulness in a management program. U.S. Fish and Wildl. Ser., Spec. Sci. Rep. 18:3-22, 55-59.

———. 1954. 50 years of progress in solving fishery problems. Proc. Gulf and Caribbean Fish. Inst., 6th Ann. Sess:81-96.

———. 1960. How international fishery commissions operate to promote conservation of high seas resources. Proc. Gulf and Caribb. Fish. Inst. 1959:18-21.

Hewatt, Willis G., and Jay D. Andrews. 1954. Oyster mortality studies in Virginia. I. Mortalities of oysters in trays at Gloucester Point, York River. Texas Jour. Sci. 1954(2):121-133.

Hiatt, Robert Worth. 1963. World directory of hydrobiological and fisheries institutions. Amer. Inst. Biol. Sci., Washington, D.C.:vii + 320 p.

Hidu, Herbert, Claus G. Drobeck, Elgin A. Dunnington, Jr., Willem H. Roosenburg, and Robert L. Beckett. 1969. Oyster hatcheries for the Chesapeake Bay region. Nat. Resources Inst., Univ. Md., Spec. Rep. 2:ii + 18 p.

Hildebrand, Samuel F., and William Schroeder. 1928. Fishes of Chesapeake Bay. Bull. U.S. Bur. Fish. 43, Pt. 1:366 p.

Hile, Ralph. 1936. Age and growth of the cisco *Leucichthys artedi* (Le Seuer) in the lakes of the Northeastern Highlands, Wisconsin. U.S. Fish and Wildl. Serv., Bur. Fish., Fishery Bull. 19:211-317.

———. 1952. 25 years of federal fishery research on the Great Lakes. U.S. Fish and Wildl. Serv., Spec. Sci. Rept. -Fish. 85:48 p.

———. 1954. Changing concepts in fishery research on the Great Lakes. Gulf and Caribb. Fish. Inst., Proc. 6th Ann. Sess:64-70.

Hjort, Johan. 1914. Fluctuations in the great fisheries of northern Europe viewed in the light of biological research. Cons. Perm. Int. Explor. Mer, Rapp. Proc-Verb. 20:228 p.

———. 1926. Fluctuations in the year classes of important food fishes. J. Cons. 1(1):5-38.

Hjort, Johan, Gunnar Jahn, and Per Ottestad. 1933. The optimum catch. Norske videns. -akad. Oslo, Hval. skr. 7:92-127.

Hoar, W.S. 1959. The evolution of migratory behavior among juvenile salmon of the genus *Oncorhynchus*. J. Fish. Res. Bd. Canada 15:391-428.

———, and D.J. Randall (eds.). 1969. Fish physiology, Vol. 1. Excretion, ionic regulation, and metabolism. Academic Press Inc., N.Y.:465 p.

Hoffbauer, C. 1898. Die Altersbestimmung des Karpfen an seiner Schuppe. Allg. Fisch. Zeit., Jg. 23:341-343.

Holt, S.J. 1962. The application of comparative population studies to fisheries biology—An exploration. *In:* The exploitation of natural animal populations, E.D. LeCren and V.W. Holdgate, eds., John Wiley and Sons, N.Y.:51-71.

Hopkins, Sewell H., and R. Winston Menzel. 1952. How to decide best time to harvest oyster crops. Atlantic Fisherman 33(9).15, 36-37.

Hubbs, Carl L. 1925. Racial and seasonal variation in the Pacific herring, California sardine, and California anchovy. Calif. Fish and Game Comm., Fish bull. 8:23 p.

———. 1964. History of ichthyology in the United States after 1950. Copeia 1964(1):42-60.

Hublow, Wallace F., Joe Wallis, Thomas B. McKee, Duncan K. Law, Russell O. Sinnhuber, and T.C. Yu. 1959. Development of the Oregon pellet diet. Oregon Fish Comm., Res. Briefs 7(1):28-56.

Huntsman, A.G. 1918. The growth of scales in fishes. Trans. Roy. Canadian Inst. 12(1):63-101.

———. 1918. The Canadian plaice. Bull. Biol. Bd. Canada 1:1-32.

———. 1919. The scale method of calculating the rate of growth in fishes. Trans. Roy Soc. Canada, Ser. 3, 12(4):47-52.

Imai, Takeo. 1965. Mass production of molluscs by means of rearing the larvae in tanks. Venus, Japan. J. Malacol. 25(3-4):159-167.

———, and Masayoshi Hatanaka. 1949. On the artificial propagation of Japanese common oyster, *Ostrea gigas* Thunb., by non-colored naked flagellates. Bull. Inst. Agric. Res., Tohoku Univ., Sendai, Japan 1(1):39-46 (Japanese with English summary).

———, ———, R. Sato and S. Sakai. 1951. Ecology of Mangoku-Ura Inlet with special reference to the seed-oyster production. Sci. Rept. Res. Inst. Tohoku Univ. D. 1-2:137-156.

Ingle, David (ed.). 1968. The central nervous system and fish behavior. Univ. Chicago Press:272 p.

Ito, S., and T. Imai. 1955. Ecology of oyster bed. I. On the decline of productivity due to repeated cultures. Tohoku J. Agr. Res. 5:251-268.

Iversen, E.S. 1968. Farming the edge of the sea. Fishing News (Books) Ltd., London:301 p.

Ivlev, V.S. 1955. Experimental ecology on nutrition of fishes. Pishchepromizdat, Moscow. (Trans. D. Scott, Yale Univ. Press, New Haven, 1961).

———. 1961. On the utilization of food by plankton-eating fishes. Trudy Sevastopolskoi Biol. Sta. 14:188-201. (Fish. Res. Bd. Canada, Trans. Ser. 447).

Jefferts, K.B., P.K. Bergman, and H.F. Fiscus. 1963. A coded wire identification system for macro-organisms. Nature, London 198(4879):460-462.

Jordan, D.S., and B.W. Evermann. 1896-1900. The fishes of North and Middle America. Bull. U.S. Natl. Mus. 47(1) 1896, (2) 1898, (3) 1898, (4) 1900:ix + 1240, xxx, 1241-2183, xxiv, 2183a-3136, ci, 3137-3313, cccxcii pl.

Kasahara, Hiroshi. 1961. Fisheries resources of the North Pacific Ocean. Part I. H.R. MacMillan Lectures in Fisheries, Univ. of Brit. Columbia:v + 135 p.

Kesteven, G.L. (ed.). 1960. Symposium on fish behavior. Proc. 8th. Sess. Indo-Pac. Fish. Council., Sec. III. IPFC Secretariat, FAO Reg. Off., Bangkok:116 p.

———. 1966. Organization and administration of fisheries research. FAO Conf. on Fishery Admin. and Serv., FAS/BP/66/3:i + 18 p.

Korringa, Pieter. 1969. Triumphs and frustrations of the fishery biologist. FiskDir. Skr. Ser. HavUnders., 15(3):114-127.

Kutkuhn, Joseph H. 1965. Dynamics of a penaeid shrimp population and management implications. Cons. Perm. Int. Expl. Mer, Rapp. Proc.-Verb. 156:120-123.

Lagler, Karl F. 1956. Freshwater Fishery Biology, 2nd ed. Wm. C. Brown Co., Dubuque, Iowa:xii + 421 p.

———, John E. Bardach, and Robert R. Miller. 1962. Ichthyology. John Wiley and Sons, N.Y.:545 p.

Larkin, P.A., and K.V. Ayyangar. 1961. Applications of the Parker equation to growth of aquatic organisms. Proc. 11th Alaskan Sci. Conf.:103-124.

Lea, Einar. 1911. A study in the growth of herrings. Cons. Perm. Int. Expl. Mer, Pub. Circ. 61:35-57.

———. 1929. Investigations on the races of food fishes. III. The herring's scale as a certificate of origin; its applicability to race investigations. Cons. Internatl. Expl. Mer., Rapp. Proc.-Verb. 54(3):21.
Lee, Rosa M. 1912. An investigation into the methods of growth determination in fishes. Cons. Perm. Int. Expl. Mer, Pub. Circ. 63:35 p.
Leim, A.H., and W.B. Scott. 1966. Fishes of the Atlantic coast of Canada. Bull. Fish. Res. Bd. Canada 155:485 p.
Lindner, Milton J., and William W. Anderson. 1956. Growth, migrations, spawning and size distribution of shrimp *Penaeus setiferus*. U.S. Fish and Wildl. Ser., Fish. Bull. 56:555-645.
Logie, R.R. 1956. Oyster mortalities, old and new, in the Maritimes. Fish. Res. Bd. Canada, Prog. Rept. Atl. Coast Stas. 65:3-11.
Loosanoff, V.L. 1961. Recent advances in the control of shellfish predators and competitors. Gulf and Caribb. Fish. Inst., Proc. 13th Ann. Sess:113-127.
———. 1965. Mariculture. Its recent development and its future. J. Amer. Soc. Agric. Engineers 46:73, 93-94.
———. 1966. Time and intensity of setting of the oyster, *Crassostrea virginica*, in Long Island Sound. Biol. Bull. 130(2):211-227.
———, and Harry C. Davis. 1963a. Shellfish hatcheries and their future. U.S. Fish and Wildl. Serv., Comm. Fish. Rev. 25(1):1-11.
———, and ———. 1963b. Rearing of bivalve mollusks. *In:* Advances in Marine Biology (F.S. Russell, Ed.), Academic Press, London and New York, Vol. 1:1-136.
Lucas, C.E. 1964. Aspects of marine fisheries research. The Advancement of Science (U.K.) 21:1-11.
———. 1969. Present trends in the organization of world fisheries research. Bull. Japan. Soc. Fish. Oceanogr., Spec. No:23-29.
Mackin, J.G. 1961. Status of researches on oyster diseases in North America. Proc. Gulf and Caribb. Fish. Inst., 13th Ann. Sess. 1960:98-109.
Magnuson, John J. 1963. Tuna behavior and physiology, a review. Proc. World. Sci. Meeting on Biol. of Tunas and Related Species, FAO Fish. Rept. 6(3):1057-1066.
Major, Richard L., and Donovan R. Craddock. 1962. Marking salmon scales by short periods of starvation. U.S. Fish and Wildl. Serv., Spec. Sci. Rept.-Fish. 416:12 p.
Margolis, L., F.C. Cleaver, Y. Fukuda, and H. Godfrey. 1966. Salmon of the North Pacific Ocean—Part 6. Sockeye salmon in offshore waters. Bull. Internatl. N. Pac. Fish. Comm. 20:70 p.
Marler, P.R. and W.J. Hamilton III. 1966. Mechanisms of animal behavior. John Wiley and Sons, N.Y.:xi + 771 p.
Marr, John C. 1956. The "critical period" in the early life history of marine fishes. J. Cons. 21(2):160-170.
———. 1960. The causes of major variations in the catch of the Pacific sardine, *Sardinops caerulea* (Girard). Proc. World Sci. Meeting on biol. of sardines and related spp., FAO, Rome, 3:667-791.
McHugh, J.L. 1951. Meristic variations and populations of northern anchovy *(Engraulis mordax mordax)*. Bull Scripps Inst. Ocean. 6(3):123-160.
———. 1968. The biologist's place in the fishing industry. BioScience 18(10):935-939.
———. 1969. Comparison of Pacific sardine and Atlantic menhaden fisheries. Fisk Dir. Skr. Ser. HavUnders., 15:356-367.
———. 1970. Review of: The Pacific salmon fisheries, a study of irrational conservation; by Jones A. Crutchfield and Giulio Pontecorvo. Science 168: 737-739.
———, and Jay D. Andrews. 1955. Computation oyster yields in Virginia. Proc. Natl. Shellf. Assn. 45:217-239.
———, and E.C. Ladd. 1953. The unpredictable blue crab fishery. Natl. Fisheries Inst. Yearbook, 1953:127-129.
Menzel, R. Winston. 1955. Marking of shrimp. Science 121(3143):446.
Mighell, James L. 1969. Rapid cold branding of salmon and trout with liquid nitrogen. J. Fish. Res. Bd. Canada 26(10):2765-2769.
Mitchill, S.L. 1814. The fishes of New York described and arranged. Trans. Lit. Phil. Soc. N.Y. 1:355-492.
Miyahara, Takashi, and Herbert H. Shippen. 1965. Preliminary report of the effect of varying levels of fishing on eastern Bering Sea king crabs, *Paralithodes camtschatica* (Tilesius). Cons. Perm. Int. Expl. Mer, Rapp. Proc.—Verb. 156:51-58
Murphy, Garth I. 1965. A solution of the catch equation. J. Fish. Res. Bd. Canada 22(1):191-292.
———. 1966. Population biology of the Pacific sardine *(Sardinops caerulea)*. Proc. Calif. Acad. Sci. 34(1):1-84.
Myers, George S. 1964. A brief sketch of the history of ichthyology in America to the 1850. Copeia 1964(1):33-41.
Nagasaki, Fukuso. 1961. Population study on the fur seal herd. Tokai Reg. Fish. Res. Lab., Spec. Sub. 7:60 p.
Neal, R.A. 1969. Methods of marking shrimp. FAO Fish. Rep. 57(3):1149-1165.
Needler, A.W.H. 1958. Fisheries Research Board of Canada Biological Station, Nanaimo, B.C., 1908-1958. Jour. Fish. Res. Bd. Canada 15(5):759-777.
Nesbit, Robert A. 1933. A new method of marking fish by means of internal tags. Trans. Am. Fish. Soc. 63:306-307.
———. 1943. Biological and economic problems of fishery management. U.S. Fish and Wildl. Serv., Spec. Sci. Rep. 18:23-53, 61-68.
Palmen, Arthur T. 1956. A comparison of otoliths and interopercular bones as age indicators of English sole. Wash. State Dept. Fish., Fish. Res. Papers 1(4):5-20.
Paloheimo, J.E., and L.M. Dickie. 1965. Food and growth of fishes. I. A Growth curve derived from experimental data. J. Fish. Res. Bd. Canada 22(2):521-542.
———, and ———. 1966a. Food and growth of fishes. II. Effects of food and temperature on the

relations between body size and metabolism. J. Fish. Res. Bd. Canada 23(6):869-908.
———, and ———. 1966b. Food and growth of fishes. III. Relations among food, body size and growth efficiency. J. Fish. Res. Bd. Canada 23(8):1209-1248.
Parker, R.R., and P.A. Larkin. 1959. A concept of growth in fishes. J. Fish. Res. Bd. Canada 16(5):721-745.
Parrish, B.B. 1956. The cod, haddock, and hake. *In:* Sea Fisheries. Michael Graham (ed.)., Edward Arnold Ltd., London:251-331.
Patten, Bernard C. 1969. Ecological systems analysis and fisheries science. Trans. Am. Fish. Soc. 98(3):570-581.
Pérez Farfante, Isabel. 1969. Western Atlantic shrimps of the genus *Penaeus.* U.S. Fish and Wildl. Ser., Fish. Bull. 67(3):x + 461-591.
Perlmutter, Alfred. 1947. The blackback flounder and its fishery in New England and New York. Bull. Bingham Ocean. Coll. 11(2):92 p.
Petersen, C.G.J. 1894. On the biology of our flat-fishes and on the decrease of our flat-fish fisheries, App. IV. On the labelling of living plaice. Rept. Danish Biol. Sta. 4:140-143.
———. 1903. What is overfishing? J. Mar. Biol. Assn. U.K. (n.s.) 6:587-594.
———. 1918. The sea bottom and its production of fish-food. Rept. Danish Biol. Sta. 25:62 p.
Phillips, A.M., A.V. Tunison, C.M. McCoy, C.R. Mitchell, and E.O. Rodgers. 1939. The nutrition of trout. Cortland Hatchery. Rept. 8 for 1939.
Quittmeyer, Charles L. 1957. The seafood industry of the Chesapeake Bay States of Maryland and Virginia (A study in private management and public policy). Advisory Council on the Virginia Economy, Richmond:xx + 295 p.
Radovich, John. 1961. Relationships of some marine organisms of the northeast Pacific to water temperatures, particularly during 1957 through 1959. Calif. Dept. Fish and Game, Fish Bull. 112:62 p.
Reintjes, John W. 1969. Synopsis of biological data on the Atlantic menhaden, *Brevoortia tyrannus.* U.S. Fish and Wildl. Serv., Circ. 320:iv + 30 p. (FAO Species Synopsis No. 42).
Richardson, John. 1836. Fauna Boreali-Americana; or the zoology of the northern parts of British America containing descriptions of the objects of natural history collected on the late northern land expeditions under the command of Sir John Franklin R.N. Part 3, Fishes:1-327.
Ricker, W.E. 1940. Relation of "catch per unit effort" to abundance and rate of exploitation. J. Fish. Res. Bd. Canada 5(1):43-70.
———. (ed.). 1968. Methods for assessment of fish production in fresh waters. IBP Handbook 3, Blackwell Sci. Pubs., Oxford:xiii + 313.
———. 1954. Stock and recruitment. J. Fish. Res. Bd. Canada 11(5):559-623.
———. 1958. Handbook of computations for biological statistics of fish populations. Bull. Fish. Res. Bd. Canada 119:300 p.
———, and John Gottschalk. 1941. An experiment in removing coarse fish from a lake. Trans. Amer. Fish. Soc. 70:283-390.
Ridgway, G.J., and G.W. Klontz. 1961. Blood types in Pacific salmon. Bull. Internatl. N. Pac. Fish. Comm. 5:49-55.
Rounsefell, George A. 1930. Contribution to the biology of the Pacific herring, *Clupea pallasii,* and the condition of the fishery in Alaska. Bull. U.S. Bur. Fish. 45:227-320.
———. 1963. Marking fish and invertebrates. U.S. Fish and Wildl. Serv., Fishery Leaflet 549:12 p.
———, and Edwin H. Dahlgren. 1933. Tagging experiments on the Pacific herring, *Clupea pallasii.* J. Cons. 8(3):371-384.
———, and ———. 1935. Races of herring, *Clupea pallasii,* in southeastern Alaska. Bull. U.S. Bur. Fish. 48(17):119-141.
———, and W. Harry Everhart. 1953. Fishery Science, its Methods and Applications. John Wiley and Sons, Inc., New York:xii + 444 p.
———, and John L. Kask. 1945. How to mark fish. Trans. Am. Fish. Soc. 73:320-363.
Royce, William F. 1964. A morphometric study of yellowfin tuna *Thunnus albacares* (Bonnaterre). Bull. U.S. Fish and Wildl. Serv. 63(2):395-443.
Rucker, R.R., B.J. Earp, and E.J. Ordal. 1954. Infectious diseases of Pacific salmon. Trans. Am. Fish. Soc. 83:297-312.
Russell, E.S. 1931. Some theoretical considerations on the "overfishing problem." J. Cons. 6(1):3-20.
Ryther, John H. 1969. Photosynthesis and fish production in the sea. Science 166:72-76.
———, and John E. Bardach. 1968. The status and potential of aquaculture. Vol. I. Particularly invertebrate and algae culture. U.S. Dept. Commerce, Clearinghouse for Fed. Sci. and Tech. Information, PB 177, 767:vi + 261 p.
Schaefer, Milner B. 1943. The theoretical relationship between fishing effort and mortality. Copeia 1943(2):79-82.
———. 1954. Some aspects of the dynamics of populations important to the management of commercial marine fisheries. Bull. Inter-Amer. Trop. Tuna Comm. 1(2):27-56.
———. 1957a. A study of the dynamics of the fishery for yellowfin tuna in the eastern tropical Pacific Ocean. Bull. Inter-Amer. Trop. Tuna Comm. 2(6):245-285.
———. 1957b. Some considerations of population dynamics and economics in relation to the management of the commercial marine fisheries. J. Fish. Res. Bd. Canada 14(5):669-681.
———. 1968. Methods of estimating effects of fishing on fish populations. Trans. Amer. Fish. Soc. 97(3):231-241.
Schmidt, Johannes. 1920. Racial investigations. v. Experimental investigations with *Zoarces viviparus* L. Compt. Rend. Trav. Lab. Carsberg 14(9):14 p.
———. 1932. Danish eel investigations during 25 years 1905-1930. Carlsberg Found:16 p.
Scott, A. 1962. The economics of regulating fisheries. *In:* Economic Effects of Fishery Regulation, FAO Fish. Rep. 5, FLe/R5:21-96.
Scott, D.M. 1954. A comparative study of the yellowtail flounder from three Atlantic fishing areas. J. Fish. Res. Bd. Canada 11(3):171-197.

Sears Foundation for Marine Research. 1948–. Fishes of the Western North Atlantic. Mem. Sears Found. Mar. Res. 1(Pts. 1-5 published to date, 1948-1956).

Sette, O.E. 1943a. Biology of the Atlantic mackerel *(Scomber scombrus)* of North America. Part I: Early life history, including the growth, drift, and mortality of the egg and larval populations. U.S. Fish and Wildl. Serv., Fishery Bull 38:149-237.

———. 1943b. Studies on the Pacific pilchard or sardine *(Sardinops caerulea)*. I. Structure of a research program to determine how fishing affects the resource. U.S. Fish and Wildl. Serv., Spec. Sci. Rept. 19:27 p.

———. 1969. A perspective of a multi-species fishery. Calif. Mar. Res. Comm., Calif. Coop. Ocean. Fish. Inv. 13:81-87.

———, and E.H. Ahlstrom. 1948. Estimations of abundance of eggs of the Pacific pilchard *(Sardinops caerulea)* off Southern California during 1940 and 1941. J. Mar. Res. 7(3):511-542.

———, and John D. Isaacs (eds.). 1960. Symposium on "The Changing Pacific Ocean in 1957 and 1958." *In:* Calif. Coop. Oceanic Fish. Inv., Repts. 7, Pt. II:13-217.

Shelbourne, J.E. 1964. Artificial propagation of marine fish. *In:* Advances in Marine Biology, J.T. Russell, ed., Vol. 2, Academic Press, London and N.Y.:1-83.

Shelford, V.E. and W.C. Allee. 1913. The reactions of fishes to gradients of dissolved atmospheric gases. J. Exper. Zool. 14:207-266.

Silliman, Ralph P. 1943. Studies on the Pacific pilchard or sardine *(Sardinops caerulea)*. 5. A method of computing mortalities and replacements. U.S. Fish and Wildl. Serv., Spec Sci. Rept. 24:10 p.

———. 1945. Mortality rates from length frequencies of the pilchard. Copeia 1945(4):1910196.

———, and James S. Gutsell. 1958. Experimental exploitation of fish populations. U.S. Fish and Wildl. Serv., Bur. Comm. Fish., Bull. 58:215-252.

Simpson, Robert R., and Herbert H. Shippen. 1968. Movement and recovery of tagged King crabs in the Eastern Bering Sea 1955-63. Bull. Internatl. N. Pac. Fish. Comm. 24:111-123.

Sinclair, S. 1960. License limitation–British Columbia: A method of economic fisheries management. Dept. Fisheries of Canada, Ottawa:viii + 256 p.

Sindermann, Carl J. 1966. Diseases of marine fishes. Adv. in Mar. Biol., Academic Press, London and N.Y., Vol. 4:1-89.

———. 1968. Oyster mortalities, with particular reference to Chesapeake Bay and the Atlantic Coast of North America. U.S. Fish and Wildl. Serv., Spec. Sci. Rep.-Fish. 569:1-10.

———. 1970. Principal diseases of marine fish and shellfish. Academic Press, N. Y. and London: 369 p.

———, and D.F. Mairs. 1959. A major blood group system in Atlantic sea herring. Copeia 1959:228-232.

Smiley, Chas. W. 1883. Four tables showing the amount of public money appropriated for carrying on the United States Commission of Fish and Fisheries and the various State commissions from 1865 to 1882, inclusive. Bull. U.S. Fish Comm., Vol. III for 1883:149-152.

Smith, Stanford H. 1968. Species succession and fishery exploitation in the Great Lakes. J. Fish. Res. Bd. Canada 25(4):667-693.

Snieszko, Stanislas F. (ed.). 1970. Symposium on diseases of fishes and shellfishes. Am. Fish. Soc., Spec. Pub. (in press).

———, R.F. Nigrelli, and K.E. Wolf (eds). 1965. Viral diseases of poikilothermic vertebrates. Ann. N.Y. Acad. Sci. 126:1-680.

Swingle, H.S. 1950. Relationships and dynamics of balanced and unbalanced fish populations. Bull. Ala. Poly. Inst., Agric. Exp. Sta. 274:74 p.

Tait, John B. 1952. Hydrography in relation to fisheries. Edward Arnold and Co., London:xii + 106 p.

Taylor, Clyde C., Henry B. Bigelow, and Herbert W. Graham. 1957. Climatic trends and the distribution of marine animals in New England. U.S. Fish and Wildl. Serv., Fish. Bull. 57(115):293-345.

Taylor, Harden F. (and a staff of associates) 1951. Survey of Marine Fisheries of North Carolina. Univ. of N.C. Press, Chapel Hill:xii + 555 p.

Tester, Albert L. 1937. Populations of herring *(Clupea pallasii)* in the coastal waters of British Columbia. J. biol. Bd. Canada 3:108-144.

———. 1948. Populations of herring along the west coast of Vancouver Island on the basis of mean vertebral number, with a critique of the method. J. Fish. Res. Bd. Canada 7(7):403-420.

Thompson, W.F. 1916. Statistics of the halibut fishery in the Pacific. Rept. B.C. Commissioner of Fisheries for 1915 (1916):65-126.

———. 1917a. Regulation of the halibut fishery of the Pacific. Rept. B.C. Commissioner of Fisheries for 1916 (1917):23-34.

———. 1917b. A contribution to the life history of the Pacific herring: Its bearing on the condition and future of the fishery. Rept. B.C. Commissioner of Fisheries for 1916 (1917):S39-S87.

———. 1919. The scientific investigation of marine fisheries, as related to the work of the Fish and Game Commission in southern California. Calif. Fish and Game Comm., Fish Bull. 2:27 p.

———. 1936. Conservation of the Pacific halibut, an international experiment. Smithson. Inst., Rept. for 1935:361-382.

———. 1937. Theory of the effect of fishing on the stock of halibut. Rept. Int. Fish. Comm. 12:22 p.

———. and F. Heward Bell. 1934. Biological statistics of the Pacific halibut fishery (2) Effect of changes in intensity upon total yield and yield per unit of gear. Rept. Internatl. Fish. Comm. 8:49 p.

Tiller, R.E. 1944-1945. The Maryland fishery management plan. Md. Board of Natural Resources. Dept. of Research and Education, Educ. Ser. 1:6 p., 2:7 p., 5:6 p., 6:8 p.

Toth, Robert. 1966. Fish anesthetics. *In:* Inland Fisheries Management. Alex Calhoun, ed. Calif. Dept. Fish and Game, Sacramento:148-149.

Townsend, Lawrence D. 1936. Variations in the meristic characters of flounders from the northeastern Pacific. Rept. Internatl. Fish. Comm. 11:24 p.

Trefethen, P.S. 1956. Sonic equipment for tracking individual fish. U.S. Fish and Wildl. Serv., Spec. Sci. Rept.-Fish. 179:11 p.

———. 1968. Fish-passage research: Review of progress, 1961-66. U.S. Dept. Interior, Bu. Comm. Fish., Circ. 254:24 p.

Tunison, A.V., A.M. Phillips, C.M. McCay, C.R. Mitchell, and E.O. Rodgers. 1939. The nutrition of trout. Cortland Hatchery, Rept. 8 for 1939.

Turvey, Ralph. 1964. Optimization and suboptimization in fishery regulation. Amer. Econ. Review 54:64-76.

Van Oosten, John. 1928. Life history of the lake herring *(Leicichthys artedi)* (Le Seuer) of Lake Huron as revealed by its scales, with a critique of the scale method. Bull. U.S. Bur. Fish. 44:265-428.

Vernon, E.H. 1957. Morphometric comparison of three races of kokanee *(Oncorhynchus nerka)* within a large British Columbia lake. J. Fish. Res. Bd. Canada 14(4):573-598.

Vrooman, Andrew M. 1964. Serologically differentiated subpopulations of the Pacific sardine, *Sardinops caerulea*. J. Fish. Res. Bd. Canada 21(4): 691-701.

Walford, Lionel A. 1931. Handbook of common commercial and game fishes of California. Calif. Dept. Fish and Game, Fish and Game, Fish Bull. 28:138 p.

———. 1937. Marine game fishes of the Pacific coast from Alaska to the Equator. Univ. Calif. Press, Berkeley:205 p.

———. 1938. Effect of currents on distribution and survival of the eggs and larvae of the haddock *(Mellanogrammus aeglefinus)* on Georges Bank. U.S. Dept. Commerce, Bureau of Fisheries, Bull. 29:1-73.

———. 1958. Living resources of the sea. Opportunities for research and expansion. Ronald Press Co., N.Y.:xv + 321 p.

———. 1965. Research needs for saltwater sport fisheries: Experience in the United States. Canadian Fish. Rep., Spec. Issue on a Symposium on the Economic Aspects of Sport Fishing 4:81-90.

Watt, Kenneth E. F. 1968. Ecology and resource management. A quantitative approach. McGraw-Hill Book Co., New York:xii + 450 p.

Weber, Douglas, and George J. Ridgway. 1967. Marking Pacific salmon with tetracycline antibiotics. J. Fish. Res. Bd. Canada 24(4):849-865.

Wilder, D.G. 1953. The growth rate of the American lobster *(Homarus americanus)*. J. Fish. Res. Bd. Canada 10(7):371-412.

Williams, Austin B. 1965. Marine decapod crustaceans of the Carolinas. U.S. Fish and Wildl. Serv., Fish. Bull. 65:1-298.

Wilson, Robert C. 1953. Tuna marking, a progress report. Calif. Fish and Game 39(4):429-442.

Winberg, G.G. 1956. Rate of metabolism and food requirements of fishes. Nauchnye Trudy Belorusskogo Gosudarstvennogo Univ. imeni V.I. Lenina, Minsk:253 p. (Fish. Res. Bd. Canada, Trans. Ser. 194).

———. 1960. Primary productivity of bodies of water. Minsk. (English version in FRB Nanaimo library).

———. 1962. The energy principle in studying food associations and the productivity of ecological systems. Zool. Zhurn. 41(11):1618-1630. (Fish. Res. Bd. Canada, Trans. Ser. 433).

Woodhead, P.M.J. 1966. The behavior of fish in relation to light in the sea. *In:* Oceanography and Marine Biology, Vol. 4, H. Barnes (ed.), Hafner Pub. Co., N.Y:337-403.

Worlund, Donald D., Roy J. Wahle, and Paul D. Zimmer. 1969. Contribution of Columbia River hatcheries to harvest of fall chinook salmon. U.S. Fish and Wildl. Serv., Fish. Bull. 67(2):361-391.

Yonge, C.M. 1960. Oysters. Collins, London:xiv + 2-9 p.

Zein-Eldin, Z.P. 1963. Use of anesthetics in matabolism studies with penaeid shrimp. U.S. Fish and Wildl. Serv., Circ. 161:63 p.

WISCONSIN: THE BIRGE-JUDAY ERA

David G. Frey

1 David G. Frey

Wisconsin: The Birge-Juday Era

A remarkable chapter in the development of the science of limnology extends from 1875, when the young E. A. Birge became an instructor at the University of Wisconsin, to the early 1940's. Chancey Juday, who was Birge's close associate for more than four decades, retired in 1942 and died in 1944. Birge lived until 1950, just 15 months short of his 100th birthday, but although he was active during most of his later years, his last papers were published in 1941 (Juday and Birge, 1941; Juday, Birge, and Meloche, 1941). Juday and Hasler (1946) cite a paper by Birge as being in press in the 1945 volume of the *Transactions of the Wisconsin Academy of Sciences, Arts, and Letters,* but this was never published.

The accomplishments of these two men and their associates are outstanding. Since a mere listing of their more than 400 publications occupies 21 printed pages (Juday and Hasler, 1946), I cannot aspire in a single chapter to give a critical appraisal of this vast effort in terms of its overall impact on the development of limnology. This has already been done in part for Birge, and to a lesser extent for Juday, in a very lucid essay by Mortimer (1956), which may be read and reread with profit by limnologists young and old.

The studies of Birge and Juday, although they are largely what is known today as descriptive limnology, are of interest not merely for their limnological descriptions of Wisconsin lakes but also for their significant contribution to our understanding of limnological processes in general. "To summarize their impact on limnology in a few words is difficult; but I believe he [Birge] will be chiefly remembered because he laid bare the mechanics of stratification, and showed (with Juday) how the living processes of photosynthesis, respiration, and decay combine to produce a concurrent stratification of the dissolved gases. The Wisconsin partners will further be remembered for their chemical analyses and crop estimates of plankton; and for the extensive survey of water chemistry and plankton in northeastern Wisconsin" (Mortimer, 1956). To this should be added the pioneering studies of Birge and Juday and their associates on transmission of solar radiation by water.

Another important consideration is that Birge and Juday developed a program in limnology in which persons of many different primary interests participated—chemists, physicists, bacteriologists, algologists, plant physiologists, geologists, etc. Most of these persons were staff members and students from the University of Wisconsin, but during the operation of the Trout Lake Laboratory more and more persons from outside the state and even from outside the United States became associated with the program. Hence, the story of limnology in Wisconsin is not merely that of Birge and Juday, although they were the motivating force, but also that of their many associates. A chronological listing of the papers and reports arising from this total effort closely parallels the general development of the science of limnology, as reflected by changing rationale, methods of attack, and problems being investigated.

Contribution 719, Department of Zoology, Indiana University.

Fig. 1.1.—E. A. Birge. Portrait by Harold Hone.

Fig. 1.2.—Chancey Juday. Portrait by Harold Hone.

The men

Birge (Fig. 1.1) was born in 1851 in Troy, New York. He received his A.B. and A.M. degrees from nearby Williams College in Massachusetts, where he had already started working on Cladocera (Brooks *et al.*, 1951). "His early interest in the planktonic crustacea and the chance which brought him to the shores of Mendota combined to start him on an exploration of the world in which lake plankton live" (Mortimer, 1956).

Promotions were more rapid in those times. Birge became a professor at Wisconsin in 1879 after only four years as an instructor, including time off to complete his Ph.D. at Harvard in 1878. During 1880–81 he studied at Leipzig with Carl Ludwig, working on the nerve fibers and ganglion cells in the spinal cord of the frog. On his return to Wisconsin he constituted a one-man department of biology, teaching courses in zoology, botany, bacteriology, human anatomy, and physiology. Later when a separate Department of Zoology was organized, he served as its first chairman until 1906.

Birge became more and more involved in administrative work at the university. These facets of his life, as well as his impact on the university and community in general, are detailed in the book by Sellery (1956). Among other responsibilities he was appointed Dean of the College of Letters and Science in 1891, and he served as Acting President of the university from 1900–1903 and as President from 1918–25.

His early studies on the plankton Crustacea of Lake Mendota (Birge *et al.*, 1895; Birge, 1897) represent the first real beginning of limnology in Wisconsin and of Birge as a limnologist. His earlier studies (Birge, 1878, 1879, 1881, 1892, 1893) were primarily faunistic. The study on the seasonal distribution of the plankton in lakes led him directly into an investigation of thermal stratification and chemical changes in the hypolimnion.

Fortunately, through the establishment of the Wisconsin Geological and Natural History Survey in 1897, of which Birge served as Director until 1919, he was able to initiate a broad program of obtaining basic morphometric data on the lakes of southeastern Wisconsin, and he was also able to hire a full-time biologist to help direct and carry out the limnological activities of the survey. This biologist was Chancey Juday.

Juday (Fig. 1.2) was born in 1871 at Millersburg along the northern edge of the lake district in Indiana. Very likely as a boy he was stimulated by lakes and by the excitement of discovering the diversity of life they contain. At Indiana University, where he obtained the A.B. and A.M. degrees (in 1896 and 1897) and much later an honorary LL.D., he came into contact with Carl Eigenmann, who in 1895 had established a biological station on Turkey Lake (now known as Lake Wawasee) only a few miles from Juday's home (Frey, 1955). It was perhaps inevitable that Eigenmann and Juday should get together, and that Juday should participate in the summer research program at Turkey Lake. Juday's first papers (1896, 1897, 1903) are concerned with Turkey Lake and Winona Lake, to which the station was relocated in 1899, and with Lake Maxinkuckee (Juday, 1902, 1920*c*), where he spent some time in 1899 studying the amount of plankton in the water and the diel movements of the plankton Crustacea.

Juday was appointed Biologist of the Wisconsin Geological and Natural History Survey in 1900. His first assignment, appropriately, was to study the diel migration of zooplankton in Mendota and other lakes of southeastern Wisconsin (Juday, 1904*a*), but after only a year he had to withdraw because of health, and for the next few years he served on the biology or zoology staffs of the universities of Colorado and California. During these years he studied the fishes and fisheries of Colorado and Lake Tahoe (Juday, 1904*b*, 1905, 1906*a*, 1907*b*, 1907*d*, 1907*e*) and the marine Cladocera (Juday, 1907*a*) and ostracods (Juday, 1906*b*, 1907*c*) of the San Diego region.

In 1905 he rejoined the Wisconsin Geological Survey as Biologist, a position he held until 1931. In 1908 he was appointed Lecturer in Limnology in the Department of Zoology at the University of Wisconsin, and from this time until 1941 (serving as Professor of Zoology from 1931) he taught and directed the training of graduate students in limnology and fisheries.

The early efforts of Birge and Juday as a team were concentrated on the Madison lakes, especially Lake Mendota, and on other lakes of southeastern Wisconsin. These studies were either problem oriented or lake oriented. The volume on dissolved gases (Birge and Juday, 1911) Mortimer (1956) regards as "the most outstanding single contribution of the Wisconsin School."

Fig. 1.3.—Sketch of original three buildings at the Trout Lake Limnological Laboratory. From Juday and Birge, 1930.

This study led directly into quantitative studies of plankton standing crops (Birge and Juday, 1922) and still later to an investigation of the dissolved organic content of lake waters (Birge and Juday, 1926, 1927*a*, 1927*b*, 1934) as a means of studying the differences among lakes in their ability to produce organic matter.

After 1917 their effort shifted away from the Madison region. During the period 1921–24 they carried out an intensive chemical and biological investigation of Green Lake, the deepest (72 m) lake in the state and also the deepest lake in the United States (exclusive of the Great Lakes) between the Finger Lakes of New York and the mountain lakes in the West. Unfortunately the results were never completely analyzed and published (Juday, 1924*a*, 1924*b*).

The study on dissolved gases was based mainly on lakes in southeastern Wisconsin, although many lakes in the northeastern and northwestern lake districts were examined briefly. Birge and Juday believed it might be desirable to shift their base of activities from near Madison to the northern part of the state. Birge had previously spent part of the summer of 1892 in northern Wisconsin (Birge, 1893), and a preliminary survey in August 1924 (Juday and Birge, 1930) showed the lakes in the northeastern district to be diversified both in biology and in chemistry. Accordingly, in June 1925 a summer field station was established on Trout Lake (Fig. 1.3) with the close cooperation of the State Forestry Headquarters there. Juday served as the Director of this laboratory until his retirement in 1942. The approach here was not so much problem oriented or lake oriented, but rather it was concerned with surveying large numbers of lakes for various chemical and biological properties and studying the range of variation of these properties and their presumed controls, especially as related to drainage and seepage categories.

Many students, both undergraduate and graduate, were involved in these studies. Many senior investigators from the University of Wisconsin and from other states or nations were attracted to the Trout Lake Laboratory to conduct studies of interest. Some of the persons associated with this period of research are Manning, Pennak, Hasler, Twenhofel, Whitney, Woltereck, Kozminski, Wilson, Potzger, and others. Regardless of one's opinion concerning the value of survey-type programs, he must admit that a large volume of basic information concerning limnology derived from these efforts.

> If the aim of limnology is the better understanding of the environmental control of living processes, it is a debatable point whether, for a given effort, more knowledge is to be gained by concentrating on a problem selected for one lake or organism, or by the wider survey of the kind we are reviewing. Or, stated differently, did Birge [and Juday] advance more on the narrow front on Lake Mendota or in the wider campaigns in northeastern Wisconsin? This is a matter of opinion. . . . No doubt the future will show that both methods of attack, in their time and place, have value [Mortimer, 1956].

Although Birge and Juday did most of their research in Wisconsin, separately and together they carried on some short-term studies outside the state. From October 1907 to June 1908 Juday visited various limnologists and limnological laboratories in Europe (Juday, 1910), and in February 1910 he visited some lakes in Guatemala and Salvador. The resulting paper (Juday, 1915) represents one of the first studies in tropical limnology. Birge and Juday together investigated the Finger Lakes of New York (Birge and Juday, 1914, 1921) and likewise made a brief study of Lake Okoboji in Iowa (Birge and Juday, 1920). Other studies, in which the field work was carried out by their associates, concern lakes of the northwestern United States (Kemmerer *et al.*, 1923) and Karluk Lake, Alaska (Birge and Rich, 1927; Juday *et al.*, 1932).

Both men were active in national affairs, serving variously as president of the American Microscopical Society, American Fisheries Society, Ecological Society of America, and the Wisconsin Academy of Sciences, Arts, and Letters. Moreover, Juday was one of the persons instrumental in bringing about the birth of the Limnological Society of America, and he was elected president for its first two years. Juday was awarded the Leidy Medal by the Academy of Natural Sciences of Philadelphia in 1943, and Birge and Juday together were awarded the Einar Naumann Medal by the International Association of Limnology in 1950 in recognition of their important and numerous contributions to the field.

> They were not summer vacation limnologists; their approach was the opposite of dilettante. They were by no means averse to speculation; but first of all they assiduously collected the facts. The complexity of the questions [in the dissolved gases study] have "become more and more manifest as our experience has extended to numerous lakes and to many seasons. If this report had been written at the close of the first or second year's work it would have been much more definite in its conclusions and explanations than is now the case. The extension of our acquaintance with the lakes has been fatal to many interesting and at one time promising theories." Without such "extension of acquaintance" they might never have achieved that insight into the mechanisms of stratification, interplay of sun and wind, and the quantitative bonds between plankton activities and dissolved gases, which form the unique and really valuable core of their work [Mortimer, 1956].

These are good words to remember at a time, such as the present, when there is so much emphasis on speed of publication and length of personal bibliographies.

Further information on the biographies of Birge and Juday is given in the publications by Welch (1944), Noland (1945), Brooks *et al.* (1951), Sellery (1956), and Mortimer (1956).

The region

In terms of its effects on limnology and limnological processes, the climate of Wisconsin is continental, with cold winters and hot summers. "In 36 of the last 50 years a minimum of −40° F or lower has been recorded in some part of the State, while, on the other hand, there are very few summers during which the temperature does not reach 100° or higher in some localities" (Coleman, 1941). The temperature is considerably milder in the southern portion, with mean annual temperatures ranging from 47° F in the south to 39° F in the extreme north.

This has two consequences for limnology. In the first place, all the lakes become ice covered in winter. In the northeastern lake district the shallower lakes freeze over by mid-November, the deeper lakes not until early December. The ice generally disappears about the first of May (Juday and Birge, 1930). In southern Wisconsin, during the period 1852–1948 Lake Mendota on the average froze over on 14 December and thawed on 4 April, for an ice-cover duration of 112 days. The extremes in duration of ice cover ranged from 65 days in the winter of 1931–32 to 161 days in 1880–81. The earliest the lake has frozen over is 25 November 1857, and the latest it has thawed is 6 May 1957 (Ragotzkie, 1960).

Lake Wingra, which is nearby, is much smaller, and, as would be expected, it freezes earlier in winter than Mendota and thaws earlier in spring. The records, extending from 1877 to the present, are much less complete than for Mendota. They show, however, that on the average the lake froze over on 25 November and thawed on 29 March, for an average ice-bound period of 125 days (W. E. Noland, 1950).

Coupled with this long period of ice cover is a rather heavy snowfall. Some regions in the extreme northern part of the state average 55 to 60 in. per winter, whereas in the south the snowfall averages about 30 in. (Coleman, 1941). In those winters having a long duration of ice cover with a sufficient thickness of snow on top, at least the shallow lakes can experience great enough oxygen utilization beneath the ice to result in fish mortalities (Greenbank, 1945: this study was conducted in the neighboring state of Michigan, where conditions are roughly comparable to those of Wisconsin).

The second consequence of the continental climate for limnology is that as a result of the short springs and hot summers, thermal stratification in the lakes becomes established early, often with only an incomplete spring overturn in the smaller lakes, and within a relatively short time the temperature-depth curve exhibits the sharp sigmoid shape which is characteristic of our continental lakes throughout the summer but which is not attained by most western European lakes until late summer or even until the partial fall overturn. The three zones of a stratified holomictic lake—epilimnion, metalimnion (thermocline), and hypolimnion—are so sharply defined in the lakes of southern Wisconsin that Birge's definition of the thermocline as that stratum in which the temperature gradient equals or exceeds 1° C per meter of depth (Birge, 1904*b*; Birge and Juday, 1914) is quite realistic.

Also of importance to limnology is precipitation and its distribution. "The average annual precipitation in Wisconsin ranges from about 26 inches in some parts of the extreme north to about 34 inches along the extreme southern and southwestern borders. The wettest months are May to September, inclusive. . . . During this period the rainfall is fairly well distributed" (Coleman, 1941). The heaviest rainfall comes after the melting of the winter snows, which tends to reduce the vernal flood peaks in the rivers. Because of the abundant summer precipitation, all except the smallest streams flow the year around (Martin, 1932).

The bedrock geology of Wisconsin consists of pre-Cambrian and Paleozoic formations, covered by a mantle of glacial drift, except in the southwestern quarter of the state, which has long been known as the Driftless Area. There are no Mesozoic or Tertiary rocks in the state. The pre-Cambrian igneous and metamorphic rocks occur in the northern portion of the state (largely co-extensive with the Northern Highland and Lake Superior Lowland in Fig. 1.4), which is part of the Superior Upland (Fenneman, 1938) and is continuous with the vast pre-Cambrian shield of Canada. The region to the south is included in Fenneman's Central Lowland, although Martin (1932) prefers to subdivide this further into Central Plain, Western Upland, and Eastern Ridges and Lowlands provinces (Fig. 1.4).

The Central Plain is underlain mainly by a soft Upper Cambrian sandstone. The remainder of southern Wisconsin is underlain mainly by resistant Ordovician and Silurian dolomites, which, as a result of their arrangement on a broad anticlinal fold with a general north-south trend, produce a series of cuestas, with their escarpments facing westward in the Eastern Ridges and Lowlands province. Martin refers to southern Wisconsin as a belted plain, similar to those of eastern England and the Paris basin in France.

Elevations range from a low of 581 ft in the southeast along Lake Michigan to a high of 1,940 ft in the northern part of the state (Martin, 1932). In general the entire Northern Highland is a relatively high plateau with elevations ranging from 1,000 to 1,800 ft. The southwestern

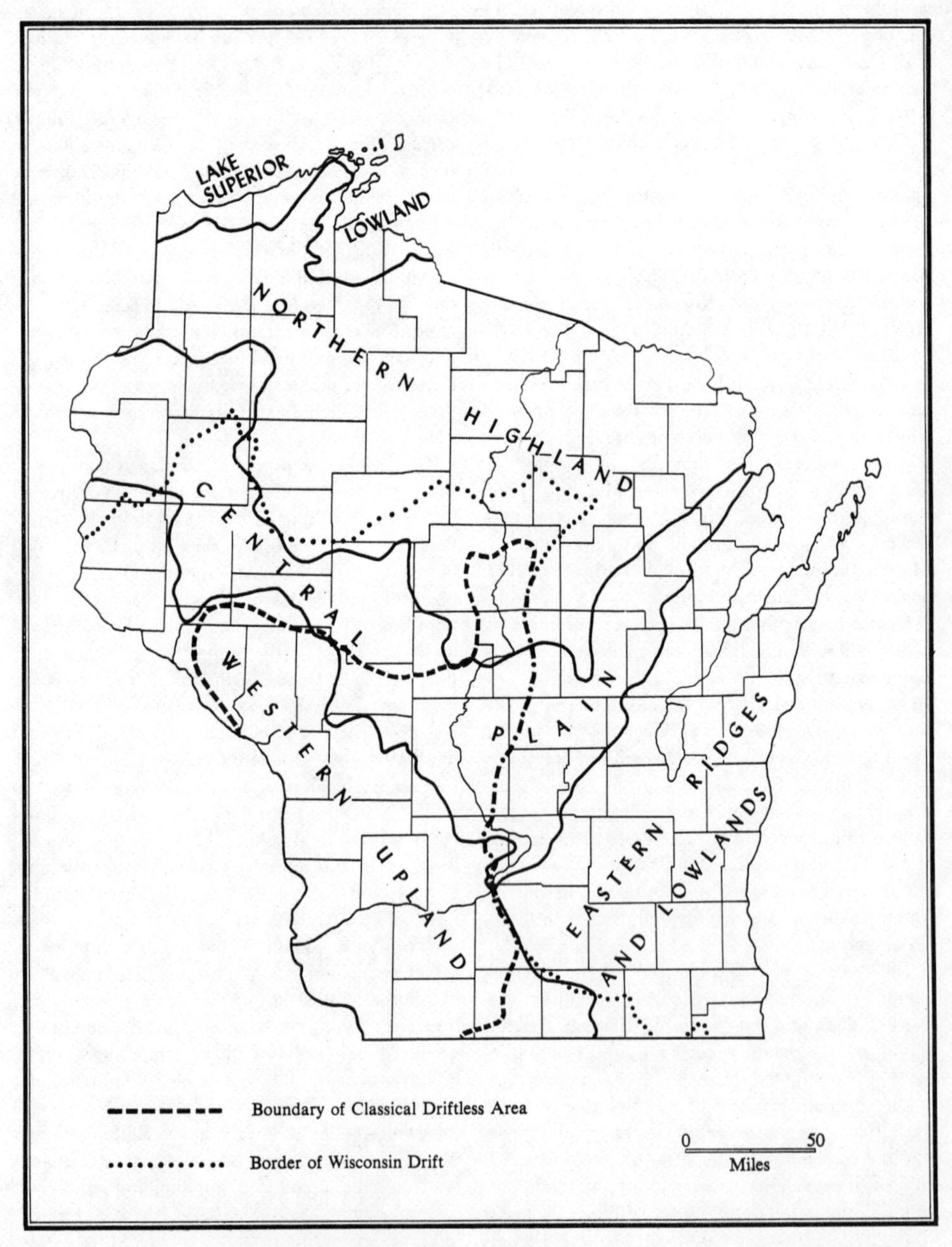

Fig. 1.4.—Physiographic provinces of Wisconsin. From Martin, 1932, and a map furnished by the Wisconsin Geological and Natural History Survey. Prepared by the University of Wisconsin Cartographic Laboratory.

quarter of the state in the Driftless Area, on the other hand, is a dissected region of steep valleys and well-developed dendritic drainages.

Three-quarters of the state definitely has been glaciated at one time or another. The most recent glacial deposits are from the Wisconsin (or Wisconsinan: see Frye and Willman, 1960) Stage, named by T. C. Chamberlin from its occurrence in Wisconsin (Chamberlin, 1894, 1895). The location of the Wisconsin Drift boundary is shown in Figure 1.4. It is well marked by an almost continuous system of end moraines. The Wisconsin Stage land surfaces are characterized by an abundance of such glacial features as drumlins, eskers, kames, and boulder trains and by several kettle moraines in the state, the one in southeastern Wisconsin, representing an interlobate moraine between the Green Bay and Michigan lobes, extending a linear distance of considerably more than 100 mi.

Between the Wisconsin Stage boundary and the boundary of the Driftless Area in the southwest are extensive regions of ground moraine, outwash, and glacial lake sediments. The Driftless Area was early noted because of the complete absence there of all igneous and metamorphic erratics, such as are common elsewhere in the state. More recently Black (1959, 1960) has claimed that "isolated deposits explainable only by glaciation are on crests of the highest ridges in all but LaCrosse County in the classical 'Driftless Area' of southwest Wisconsin. . . . Locally contorted bedrock, absence of thick residual soils and of weathering in the deposits, and absence of old loess further confirm glaciation of much, if not all, of the Driftless Area during the early Wisconsinan stage or all of it during a pre-Wisconsinan stage." The Driftless Area as originally mapped extends modestly into the neighboring states of Minnesota, Iowa, and Illinois. There are many small caves in the Driftless Area but almost none in the Eastern Ridges and Lowlands Province. Most of the karst features here either have been destroyed by glacial erosion or are buried beneath drift.

Largely because of the recency of its glaciation, Wisconsin is well provided with lakes. According to Voigt (1958) there are 8,830 lakes in the state, 4,136 of which have an area exceeding 10 acres. Included in this total are an unknown number of "flowage" lakes and reservoirs, although the vast majority of the lakes listed are natural. Every county has at least one lake, and, as would be expected, the counties in the Driftless Area generally have a smaller number and a lower density of lakes than do the counties in the glaciated region (Fig. 1.5). Vilas County leads with 968 lakes and a density of slightly more than one per mi^2, followed by Oneida with 830 lakes. It is little wonder that Birge and Juday established a laboratory at Trout Lake in Vilas County near the center of this northeastern lake district (Fig. 1.6).

For the state as a whole, lakes make up 2.6% of the total area, but for Vilas County they comprise 15.1% of the total land surface. Other counties have smaller percentages of their total areas in lakes. Only 1.6% of the area of Dane County is in lakes, although in this case the intensive studies on Mendota and the other lakes of the Yahara River chain (Fig. 1.7) more than compensate for any deficiency in number or acreage.

The largest natural lakes of the state are Winnebago, 558 km^2; Poygan, 44.5 km^2; Koshkonong, 40.0 km^2; Mendota, 39.4 km^2. Winnebago has a maximum depth of only 6.4 m, whereas nearby Green Lake is the deepest lake in the state, with a maximum depth of 72.2 m (Juday, 1914*a*).

There are three well-defined lake districts in Wisconsin (Juday and Birge, 1930; Martin, 1932): (1) eastern and southeastern Wisconsin, consisting of scattered, moderate-size lakes, (2) northeastern Wisconsin (Vilas, Oneida, Iron, Forest, Lincoln, and Langlade counties), containing 34% of all the lakes in 10% of the state's area, and (3) northwestern Wisconsin (chiefly Washburn, Burnett, Polk, Barron, and Chippewa counties), containing 21% of the lakes on 8% of the area. In addition there are Lake Pepin in the Mississippi River, which is backed up behind the delta of the Chippewa River, Lake St. Croix on the lower St. Croix River, and hundreds of small flood-plain lakes ("Altwässer") in the Mississippi bottomland. These latter have not been included in the state totals.

Most of the major rivers of the state arise in the Northern Highland. The master stream of the state is the Wisconsin River, which arises at Lac Vieux Desert on the Wisconsin-Michigan border and completely traverses the state, entering the Mississippi River near the southwestern corner (Fig. 1.8). The Black, Chippewa, and St. Croix rivers are other major drainages that enter

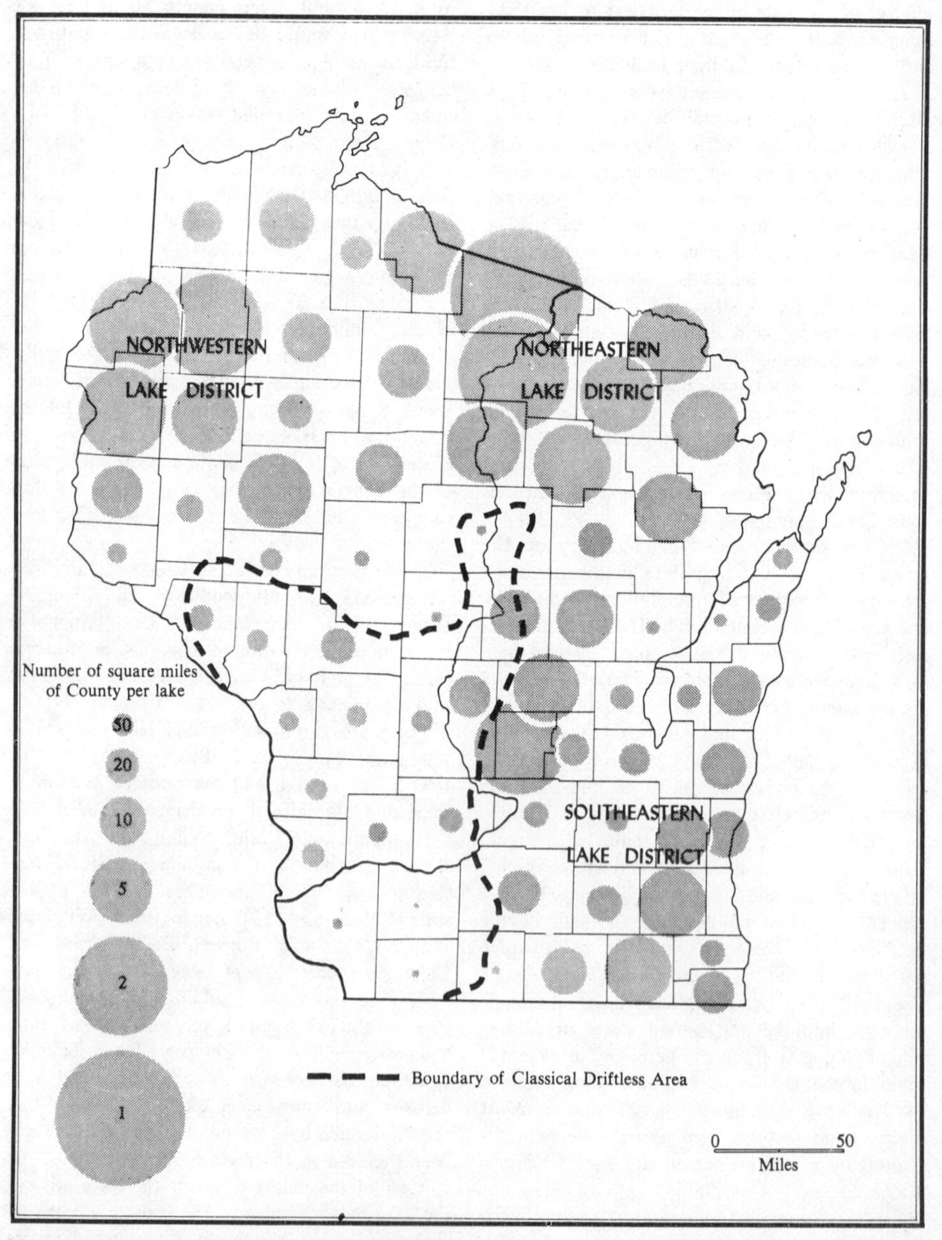

Fig. 1.5.—Density of lakes in Wisconsin by counties. The areas of the circles are proportional to the mean number of lakes per square mile. The scale at the left is the reciprocal of this, namely, the number of square miles of county for each lake. The dashed line represents the boundary of the Driftless Area, as originally described. Prepared by the University of Wisconsin Cartographic Laboratory.

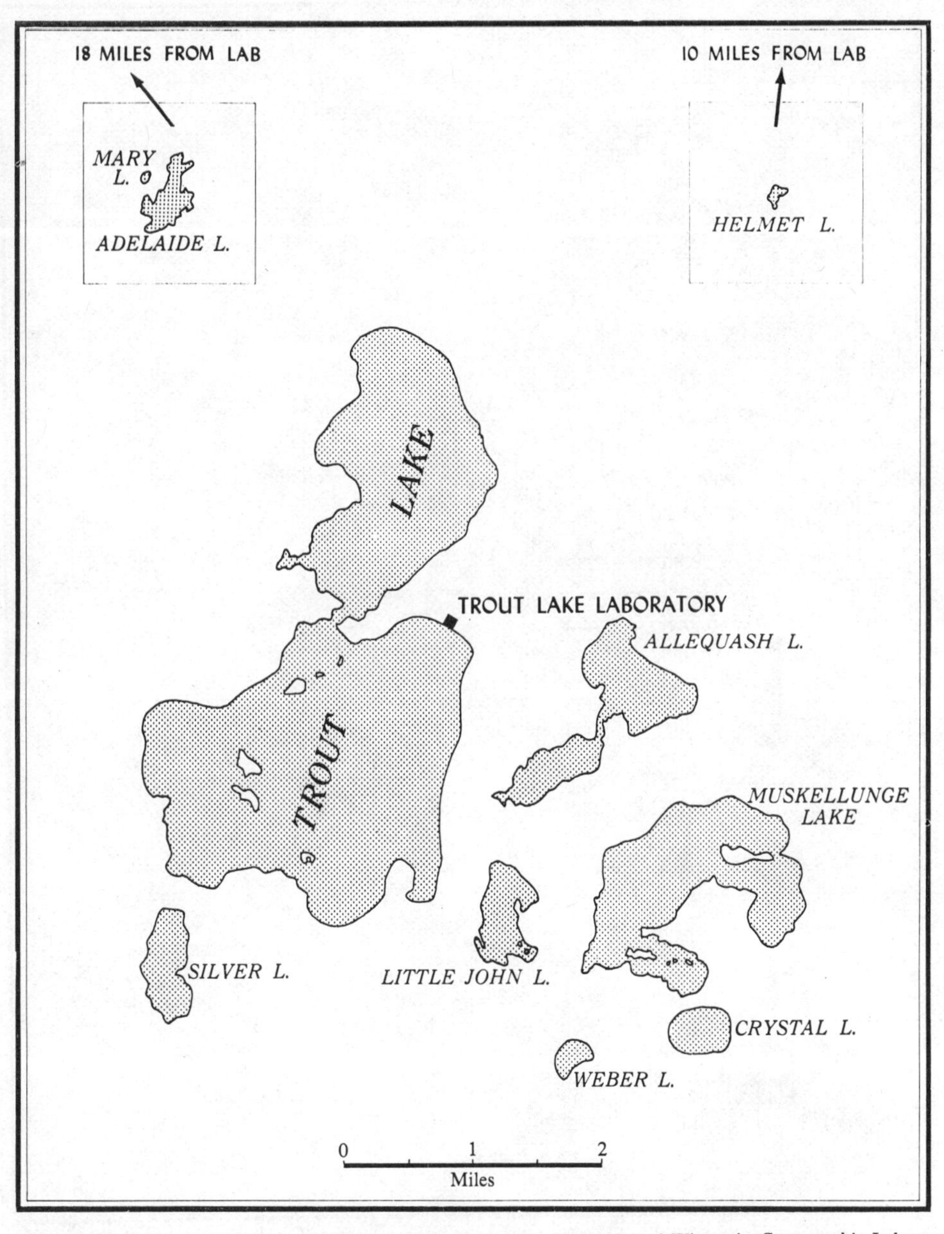

Fig. 1.6.—Location of the Trout Lake Limnological Laboratory and of some of the neighboring lakes that have been studied most intensively. Prepared by the University of Wisconsin Cartographic Laboratory.

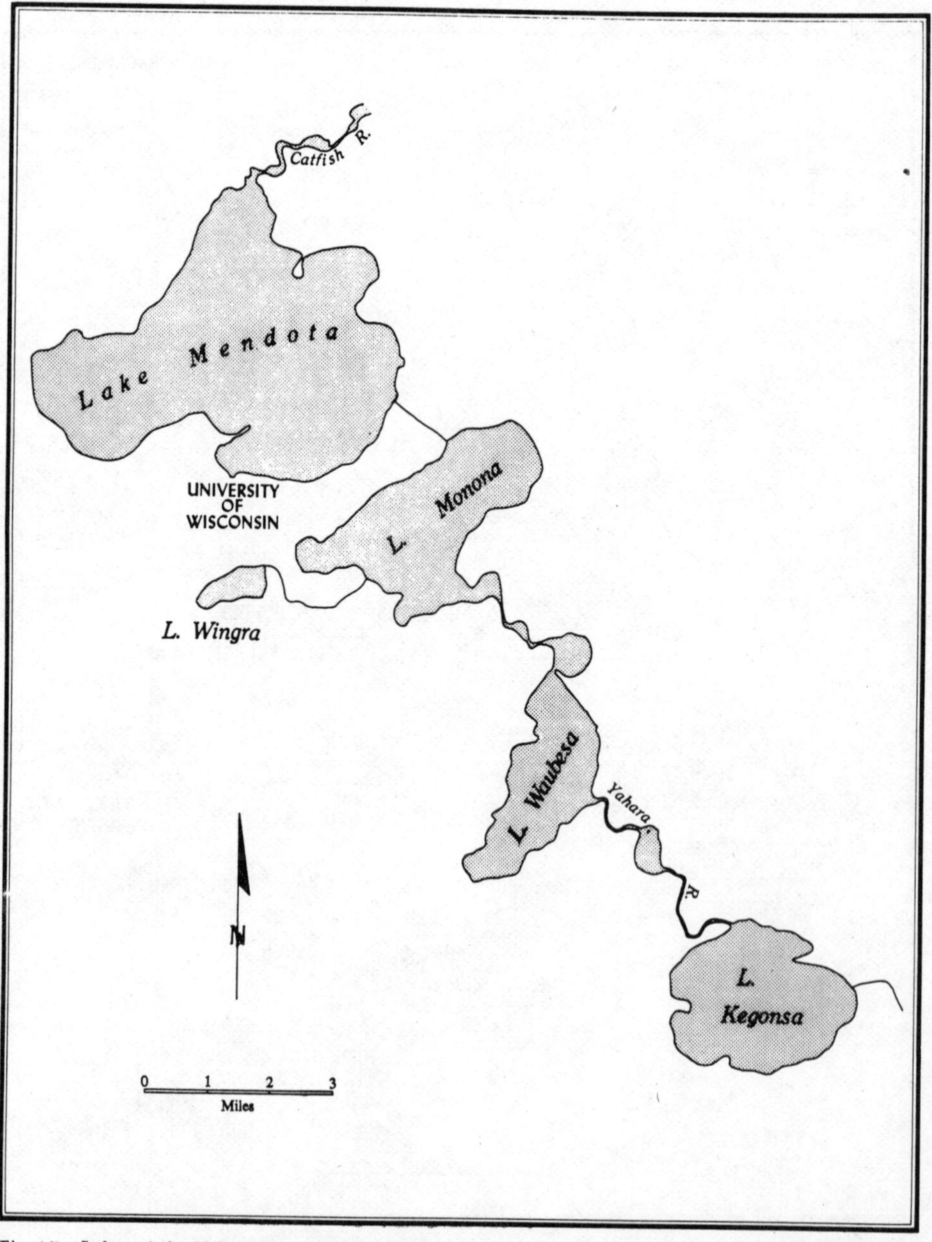

Fig. 1.7.—Lakes of the Yahara River chain at Madison, Wisconsin. The lakes are sometimes designated by numbers beginning at the south, rather than by the names shown in the figure. Thus, Kegonsa is alternately know as First Lake and Mendota as Fourth Lake. From Juday, 1914*a*, as prepared by the University of Wisconsin Cartographic Laboratory.

Fig. 1.8.—Major rivers of Wisconsin. The dashed line marks the approximate watershed divide between the Mississippi and Great Lakes drainages. From a map provided by the Wisconsin Geological and Natural History Survey, as prepared by the University of Wisconsin Cartographic Laboratory.

the Mississippi River in Wisconsin. A number of short and relatively minor streams drain into Lake Superior. The Menominee River along the northeast border and the Fox-Wolf River system drain into Green Bay of Lake Michigan. The Fox River in southeastern Wisconsin enters the Illinois River, which, along with the Rock River, enters the Mississippi in Illinois.

It is interesting that in early times the most important water route from the Great Lakes to the Mississippi was via the Fox River and the lower Wisconsin River. Where these rivers come close together, the present city of Portage marks the site of overland transport. The first known travelers over this route were Louis Jolliet and Father Jacques Marquette, who entered the Mississippi on 17 June 1673 and proceeded as far downstream as Arkansas (Martin, 1932). The next most important route was via the Bois Brule River to the St. Croix. This involved a short portage at Upper St. Croix Lake.

Through the efforts of various persons and agencies, basic information, including morphometry, has been obtained for quite a number of Wisconsin lakes. The Wisconsin Geological and Natural History Survey began a survey of the lakes of southeastern Wisconsin in the winter of 1897–98 and continued it more or less regularly for about five years. Hydrographic maps for some of these lakes and groups of lakes were issued separately by the survey. Juday (1914*a*) brought all this information together in a volume containing complete morphometric data, including maps, on 54 lakes and partial data on 185 others, including more than 120 in the northeastern and northwestern lake districts.

The book by Fenneman (1902) describes the physical geography of the southeastern lakes, especially their origin and history and the effects of lake waters on the shores. The formation of cusps, spits, bars, ice ramparts, etc. is very well described and illustrated by specific examples. One of the interesting points of the book limnologically is a discussion of the surficial sediments of Mendota, based on samples collected by Juday in the autumn of 1900 and the spring of 1901. Fenneman expected to find only mud at depths beyond the wave base, but in many places the change from sand to mud did not occur until depths of 40–50 ft. The slope was such that gravity alone could not be the explanation. Fenneman suggested that either the coarser materials had been actively transported down slope by currents or else the finer materials had been removed differentially. In these ideas he anticipated the recent conclusions by Lathbury and Bryson (1958) that turbidity currents have been active forces in shaping the bottom configuration of Mendota.

In a later paper Juday and Birge (1941) present hydrographic maps and morphometric data for 34 lakes in the northeastern lake district.

The bulletin by Voigt (1958) gives a complete listing of all lakes in the state with an area greater than 10 acres, including their general locations within the counties, their areas, whether or not contour maps are available, and the types of gamefish and panfish that are present. According to this bulletin, maps of approximately 1,000 lakes are available at a nominal charge. These maps show the depth contours, submerged bars, weedbeds, and principal access roads.

An earlier bulletin (Bordner, 1939) is restricted to the lakes of northern Wisconsin. It gives locations more precisely along with data on maximum depth, whether the lake is a seepage or drainage lake, and very general information on hardness, transparency, abundance of aquatic vegetation, and nature of surrounding land cover. The first portion of the book presents a general discussion of aquatic life, lake management, and general fishery biology.

The various lake districts differ greatly in their regional chemistry, but in spite of the many lakes studied in Wisconsin, this information has never been adequately summarized. The eastern and southeastern lakes are hard-water lakes by virtue of a highly calcareous drift and the underlying dolomites. The lakes of the northeastern district frequently have very soft water or at most are listed as medium-hard. The outwash here is non-calcareous, and the underlying bedrock consists of pre-Cambrian granites, quartzites, schists, etc. The mean calcium content of 20 seepage lakes in this region was only 1.07 mg/L, and of 13 drainage lakes, 6.4 mg/L (Juday and Birge, 1941). Associated limnological conditions are low conductivity, acid *p*H, low methyl orange alkalinity, and often a substantial dissolved color. The lakes in the extreme western portion of the northwestern lake district are mostly medium-hard-water lakes (Birge, 1910*e;* Birge and Juday, 1911), although Bordner (1939) considers them very hard-water lakes.

Limnology: 1879–1910

Cladocera

It is particularly appropriate that we begin the discussion of the limnology of Wisconsin with Cladocera, since this is how Birge began, and this is what led him into a study of diel migration and seasonal distribution of the planktonic entomostraca, and from there into the fundamental work with Juday on the stratification of lakes, its causes and consequences. Most of what we know about the systematics and distribution of Cladocera in North America is based upon Birge's careful studies prior to 1910. The treatment of Cladocera (Brooks, 1959) in the revision of *Fresh-Water Biology* is little changed from Birge's original treatment (Birge, 1918*a*), except for the genera *Daphnia* and *Bosmina* and the addition of a few miscellaneous species described since Birge's time.

Prior to the monograph in *Fresh-Water Biology* Birge had written just four major papers dealing with the systematics of Cladocera (1879, 1892, 1893, 1910*d*). His first paper, based on his Ph.D. thesis at Harvard (Birge, 1878), was presented orally before the Wisconsin Academy of Science when Birge was only 26 years old. It represented the efforts of a young and enthusiastic but relatively immature investigator. Nevertheless, it was the first major effort by anyone on the Cladocera of North America. Birge had only limited systematic literature available at this time, and since the realization that Cladocera are generally intercontinental in distribution had not yet been reached, Birge discovered many forms in his limited collecting in Massachusetts and Wisconsin that could not be readily identified. As a result, out of the 37 species he listed, 20 were described as new, including one new genus. Of these supposedly new species, only 5 survived to reach *Fresh-Water Biology,* and one of these has since been reduced to a synonym. Unfortunately, although Birge himself soon realized that many of these names he proposed were synonyms of previously described species, at no place in his later publications does he list the synonymy of the species recognized from North America, and some of his early descriptions are quite inadequate to enable an outsider to do this.

With the exception of a brief and relatively unimportant note (Birge, 1881), Birge's next paper on Cladocera was a list of the species occurring in the Madison lakes (Birge, 1892). A total of 65 species and varieties (not all of which are currently recognized) was listed, of which 60 were collected in Lake Wingra and 38 in Lake Mendota. Birge suggested that larger lakes are poorer in species, especially those occurring in littoral situations. Most important, however, was his growing realization that species of Cladocera are widely distributed. Only 9 of the forms listed for the Madison lakes were considered peculiar to the United States, and only 4 of these had no close morphological equivalents known from other continents. "If the species of Cladocera have so wide a range as appears from Sars' observations on Australian Cladocera, and from my work here, it is not probable that many species are strictly local. . . . No doubt some species are strictly local, confined to a small area, or the product of life conditions existing there and not elsewhere. But the chance that this is true in any given case is small, and all well marked species should be looked for in every suitable locality. We should expect also that a locality especially favorable to the development of the Cladocera would contain a very large fraction of the fauna of the region."

In the summer of 1892 Birge spent several days in northeastern Wisconsin (Vilas County) for the express purpose of collecting Cladocera. Including some collections from a few other places, Birge recovered 69 species and varieties, bringing the total list for the state to 84. He noted that 64 of these are common to both North America and Europe, providing further evidence for the intercontinental distribution of the Cladocera. A high-school teacher (Merrill, 1893), working under Birge's direction, described the detailed morphology of Birge's new genus (*Bunops*) and made detailed comparisons between the species involved and a closely related species described by Daday from Hungary. Birge's genus is still valid, although the species has since been synonymized with Daday's species.

Birge's interest in the systematics of the Cladocera continued. One gets the impression from his correspondence that he perhaps hoped eventually to tackle the Cladocera of the world. In 1903 he purchased a very extensive and world-wide collection of Cladocera from Jules Richard, who earlier had abandoned his work on Cladocera when he became the Director of the Oceanographic Institute at Monaco. Birge also purchased Richard's reprint collection on the Cladocera. Realizing that his knowledge of the Cladocera in the southern United States was very deficient, he undertook a brief collecting trip to New Orleans and Texas in 1903, and he profited greatly from the subsequent collecting by and correspondence with Mr. E. Foster of the New Orleans *Daily Picayune.*

Birge's chapter on Cladocera for *Fresh-Water Biology* was completed by 1910 (Birge, 1910*d*), even though the volume was not published until 1918. His paper of 1910 detailed the taxonomic changes that were to appear in his monograph and presented some of his general views regarding relationships among the Cladocera. These last two efforts on the Cladocera (1910*d*, 1918*a*) were those of a mature, thoroughly competent investigator. But, whatever Birge's goals may have been originally, he abandoned further work on the systematics of Cladocera after 1910. Moreover, no one else has subsequently undertaken so detailed a study of Cladocera in general. His monograph in *Fresh-Water Biology* remains to this day the definitive treatment of our Cladocera, except as it has been modified by Brooks (1959) in the revised edition.

In the monograph (1918*a*) Birge listed 113 species for the United States, of which he described 14. He noted that:

> A majority of the species found in this country are found also in Europe. Where a species is peculiar to this region it is often but slightly different from the European form. The student of Cladocera should presume that any species is probably intercontinental, although it may prove to be more restricted in its range. The study of our forms has not gone far enough to enable us to speak of the local distribution of each species within the general area which it covers, but it is known that rare species are very irregularly distributed. On the whole, the fauna of the various regions of the country is strikingly similar, but with some forms peculiar to each region. The southern states contain numerous species which are common to them and to South America, but are not found in the northern states.

A word of caution should be inserted, however. Birge himself recognized that the monograph was incomplete. For the small species of *Alona,* for example, he noted that "there are species and numerous varieties besides those listed." Our intensive work on the chydorids and their exuviae in lake sediments has revealed quite a substantial number of species even in the Midwest that are not included in Birge's monograph. The gap is greatest in the South, where quite a number of new species and new records of species, previously described from South America and elsewhere in the Tropics, occur. A person using the monograph should not assume that it is complete. Specimens not agreeing with the descriptions or figures may, in fact, be undescribed.

Both Birge and Juday at various times examined collections made by other persons and reported on the species recovered. These papers are of limited value, since they contain very little information besides names of species and places and dates of collecting. These comprise reports by Birge on Lake St. Clair and some nearby lakes in Michigan (Birge, 1894), Turkey Lake in Indiana (Birge, 1895*a*), some lakes of the Sierras and Rocky Mountains in which he described *Macrothrix montana* (Birge, 1904*a*), and Lakes Amatitlan and Atitlan in Guatemala (Birge, 1908*b*). Except for the recording of the marine cladoceran, *Evadne tergestina,* near San Diego (Juday, 1907*a*) and the listing of Cladocera in Turner's Lake, Maine (Juday, 1923*b*), all the rest of Juday's papers involving the identification of Cladocera are concerned with Canada and Alaska (Juday, 1920*a*, 1926*a*, 1927; Juday and Muttkowski, 1915).

Zooplankton

Although in his early years at Wisconsin Birge was concerned chiefly with the systematics of the Cladocera, he was not unaware of their ecology. The shift in Birge's dominant interest from the systematics of the Cladocera to their ecology and finally to limnology in its broadest sense was occasioned by a brief and incompletely documented paper by Francé (1894) on diel migration, which demonstrated that in Lake Balaton the zooplankters come to the surface at night and do not descend to greater depths until about dawn, where they remain until early afternoon. Cladocera migrated first, followed by copepods. The pattern was modified somewhat by weather conditions.

Since Mendota is about twice as deep as Balaton, although much smaller in area, Birge was interested in ascertaining how extensive the migration might be in a deeper lake and also at what rate it occurred. Accordingly, he assigned this problem to two undergraduate students—Olson and Harder—for their undergraduate theses. They did most of the collecting and all the counting. Birge designed a vertical tow net that could be opened at any desired depth by means of a messenger and then closed again by a second messenger after the net had been pulled through a desired thickness of water, generally 3 m. By this means the 18-m water column at a station marked by a buoy was sampled by 3-m strata at 3-hour intervals day and night during four periods from 5 July to 4 August 1894.

The results were surprising and unanticipated (Birge *et al.*, 1895). During July only the uppermost 12 m of the lake were tenanted by copepods and Cladocera. More than 90% occurred in the uppermost 9 m, and about 50% occurred in the top 3 m. (In a subsequent paper Birge [1898] stated that 95% or more of the zooplankton occurred above the thermocline.) Of the 6 dominant species, the only variation from this pattern of greatest abundance towards the top and decrease with depth was *Daphnia pulicaria* (*D. schödleri*), which was largely confined between 6 and 15 m depth in its distribution. These patterns of distribution were constant day and night. There was no evidence for any diel migration, unless it was confined within the uppermost 3 m.

These findings so surprised and stimulated Birge that he continued sampling into the autumn. At this time he also began studying the thermal structure of the lake. He noted that in the autumn the distribution of the Cladocera changed gradually over a two-week period from the surface concentration characteristic of summer to an approximately uniform distribution from surface to bottom, which coincided with the cooling phase of the lake and the development of homothermy. Concerning the summer distribution, Birge noted that "the crustacea apparently stopped [in their vertical distribution] rather abruptly either somewhat above or somewhat below the 10 m level." This would mark the approximate upper limit of the thermocline, but there is no evidence that Birge at this time was aware of the summer thermal structure of lakes.

What started as the assignment of a routine research problem to undergraduates became Birge's introduction to limnology. The questions raised by the vertical distribution of the planktonic Crustacea stimulated him to continue these studies for two more years (Birge, 1898). He obtained 333 series of six successive 3-m samples each over the 2-year period, and of equal importance he measured the temperature distribution in the water column on each sampling date, using the simplest of equipment—a Meyer flask and a laboratory thermometer graduated to 0.2° C. When a thermophone was finally obtained in July 1896, Birge checked the accuracy of the Meyer flask method and decided that, for the purposes of his present study at least, the temperature readings were sufficiently accurate. Whereas in the previous paper all results of zooplankton distribution had been expressed as percentages, in the present study they were expressed as numbers of organisms per m³ or per m² of lake surface. Birge calculated an efficiency factor for the net by comparing the net catches with those of a metal tube 3 m long and 10 cm in diameter.

The seasonal and vertical distributions of the planktonic crustaceans were described in detail for the two-year period, and these were then related to the annual temperature cycle of the lake. Birge was surprised that the changing seasonal abundance of the limnetic Crustacea was determined primarily by fluctuations in abundance of the seven perennial species. The periodic species contributed relatively little to the total. He noted in addition that "most of the littoral forms of crustacea also appear occasionally in the plankton, especially after storms, as also do Hydrachnids and Ostracoda." These, however, comprised less than 1% of the total plankton. The course of the cycles of abundance was remarkably similar in the various years, although the magnitude of the peaks varied. Birge believed the location of the peaks was determined primarily by temperature, as affecting the cycles of growth and reproduction of the various species, and the magnitude of the peaks by food supply. He made a special point of the occurrence in the plankton of *Chydorus sphaericus*, which is generally considered to be a littoral form, noting that it occurs almost exclusively when blue-green algae are abundant.

The pattern of vertical distribution noted in the previous paper was confirmed and extended: the entomostraca occur throughout the water column during the cold months of the year and are confined mainly above the thermocline in summer. Through attempts to determine the exact lower limit of the Crustacea, Birge concluded that "the crustacean population usually passes into the thermocline and often towards its lower part, but that here it ends often with great abruptness." Furthermore, "the number of algae also declines very rapidly at the thermocline and those which are obtained below this level are dead or dying." Birge concluded that:

> The crustacea are not excluded from the deeper water of the lake by the low temperature of the water, as is proved by the occurrence of the same species in the far colder water of other lakes in the same district. The exclusion is due to the accumulation of the products of decomposition in the lower water, which remains entirely stagnant after the thermocline has been formed, and is never exposed to the action of sun and air. This water in Lake Mendota acquires an offensive smell and a disagreeable taste. . . . The products of decomposition of the algae and crustacea of winter and spring remain stored in the deeper water, and undoubtedly the addition of this store of nutritive material to the surface water of the lakes as the thermocline gradually moves downward is one of the factors which occasions the enormous increase of the vegetable plankton in late summer and autumn.

This same idea had been expressed earlier by FitzGerald, (1895). Later Birge cited some studies by Drown on various ponds in Massachusetts, which demonstrated that frequently the content of dissolved oxygen declined rapidly and even totally disappeared below the thermocline. Up to this time Birge had made no chemical studies himself, and hence he could not decide whether the decomposition products or reduced oxygen was the primary factor in eliminating limnetic Crustacea from the deep water of Mendota in summer. Obviously the next step in this developing program was to study the seasonal chemistry of the water.

Birge (1898) was still disturbed by the apparent lack of diel migration in Mendota. In a more detailed investigation of the top 3 m, using a pump and a horizontal tow net, he found that there was indeed a diel migration, but that it was confined to the uppermost 1 or 1½ m, rather than extending over the entire depth of the lake as Francé had found.

The word "thermocline" has been used in the previous discussion. This term was proposed by Birge (1898) as a synonym for the *Sprungschicht* of the German investigators.

Birge (1898) presented a detailed analysis of the annual thermal regime of Mendota for the two years. He noted many phenomena that are now part of our general understanding of the thermal dynamics of lakes—the lowering of the thermocline during summer, the increase in water temperature beneath the ice, the marked rise in temperature of the bottom water during the destruction of thermal stratification in autumn, the variations in position and thickness of the thermocline under various wind stresses, etc. He had already reached a general understanding of the interplay of sun and wind in setting up the thermal structure, as evidenced by the statement that "the warmth of the surface water received from the sun is distributed by the wind through a certain depth of the lake, a depth which is proportional to the violence of the wind and the area of the lake." Through his observations during two full years on Mendota and single visits to neighboring lakes he had reached the conclusion that the temperature of the bottom water in different lakes and its variation from year to year in the same lake is controlled by four factors—(1) the depth of the lake, (2) the area in relation to depth, (3) the shape of the lake and the nature of its surroundings, and (4) the temperature and wind movement during the spring warm-up. Brief summaries of these two major zooplankton studies were also published elsewhere (Birge, 1895*b*, 1897).

Thus, Birge, who only two years earlier had been interested mainly in the systematics of the Cladocera, was now a limnologist interested in "the natural history of an inland lake as 'a unit of environment,' to employ Eigenmann's appropriate phrase." He had already been led into a study of the seasonal thermal regime of Mendota. He realized that the chemistry of the water, especially during periods of stratification, would be important to understand, and he already anticipated the interplay between organisms and environment in the development of this stratification and the control of seasonal cycles of plankton. He was aware of the importance of sun and wind in controlling the seasonal stratification. Here, in essence, is Birge, the limnologist. He and his associates subsequently investigated all these fields, and more besides, constantly keeping in mind the totality of organic and inorganic processes going on in a body of water.

All this was going on while Juday was still a youth in his mid-twenties. He, too, was interested in plankton, and, significantly, his initial approach was concerned with biomass rather than numbers. In Turkey Lake he studied the horizontal and vertical distribution of plankton collected with a vertical tow net and concentrated by centrifuge (Juday, 1897). Most of the plankton occurred in the uppermost 3 m, with very little below 6 m. A similar study, but of lesser scope, was made on Maxinkuckee (Juday, 1902, 1920*c*), in which the plankton was confined almost entirely to the uppermost 12 m, and in which there was positive evidence of upward migration at night. During the Maxinkuckee study Juday used a simple pump for obtaining some deep-water samples from known depths. Because of this background and research interest and because of Birge's frequent contacts with Eigenmann at this time, it seems almost inevitable in retrospect that Juday should have been hired as Biologist in the newly formed Wisconsin Geological and Natural History Survey.

Birge was still interested in the diel migration of the Crustacea, which undoubtedly explains Juday's intensive study of this phenomenon in 30 lakes of southeastern Wisconsin, including Mendota, during his first year with the survey (Juday, 1904*a*). Using a pump and garden hose, Juday sampled at half-hour intervals during the periods encompassing sunset and sunrise, and even throughout the night. He found that although quite a number of species exhibited diel migration, the extent of movement for the same species varied greatly from lake to lake and even from one time to another in the same lake. He believed that light was the chief factor responsible for downward movement, but that other unspecified factors were more responsible for the upward movement. He also confirmed Birge's earlier observations on the very shallow daily readjustments of species in Mendota. Returning to Indiana briefly in the summer of 1901, Juday made similar studies on Winona Lake (Juday, 1903). He found a marked diel migration in *Epischura lacustris, Chaoborus,* and *Leptodora.* Shifts in the distribution of *Diaptomus* and *Cyclops* are possibly obscured by Juday's failure to distinguish species in these genera.

Unfortunately, because of Juday's health, which necessitated his leaving for several years shortly after he had arrived, and because of Birge's increased administrative responsibilities as Acting President of the university, there was a period of relative quiescence in limnology at Wisconsin.

Copepods and other early limnology

One can argue convincingly that Birge was not the first limnologist in Wisconsin. In 1883 C. D. Marsh, who was four years Birge's junior in age, became a Professor of Natural Science at Ripon College, just seven miles from Green Lake. According to his widow (Mrs. Florence W. Marsh, 1938), he had been interested in minute forms of freshwater life since boyhood. It is little wonder then that he was attracted to Green Lake, in spite of the difficulties of horse and buggy travel. Marsh's early trips to Green Lake and to other neighboring lakes awakened an interest in copepods, which he retained throughout his life. He became an authority on this group of microcrustacea, in the same way as Birge on the Cladocera. The two men complemented each other nicely, and they maintained a constant exchange of specimens and information.

Whereas Birge originally was interested in the pure systematics of the Cladocera, Marsh from the beginning was interested in environmental relationships as well. He recorded some depth profiles of Green Lake based on soundings made through the ice some years earlier by C. A. Kenaston, also of Ripon College, and he likewise measured surface and bottom temperatures with a deep-sea thermometer loaned by the U.S. Commissioner of Fish and Fisheries (Marsh, 1892*b*). To his surprise he found both *Pontoporeia* and *Mysis* in the deep water, the first reported instance of this in any of the smaller inland lakes of North America (Marsh, 1892*a*). In August of 1893, before Birge's students began a similar study, Marsh undertook an investigation of the diel migration of plankton Crustacea in Green Lake. A brief report on the early stages of this work was presented in 1894 and a final report in 1898. He used a vertical tow net (he called such a device a "dredge," as did Birge) that could be closed at any depth. There was a marked concentration of entomostraca toward the surface at night, although not all species exhibited such an upward migration. The Miller-Casella deep-sea maximum-minimum thermometer took so long to reach equilibrium that Marsh measured only surface and bottom temperatures, but these observations are presented in graphical form for two entire years. If Marsh had had an instrument that reached thermal equilibrium more rapidly and was more amenable to being operated from a sailboat, it is conceivable he would have anticipated Birge and FitzGerald in describing the annual thermal cycle of a lake.

He was disturbed by the great variability among replicate plankton samples. He considered depth of water as one of the chief controls in the geographic distribution of copepods, and he suggested dividing lakes into shallow and deep lakes

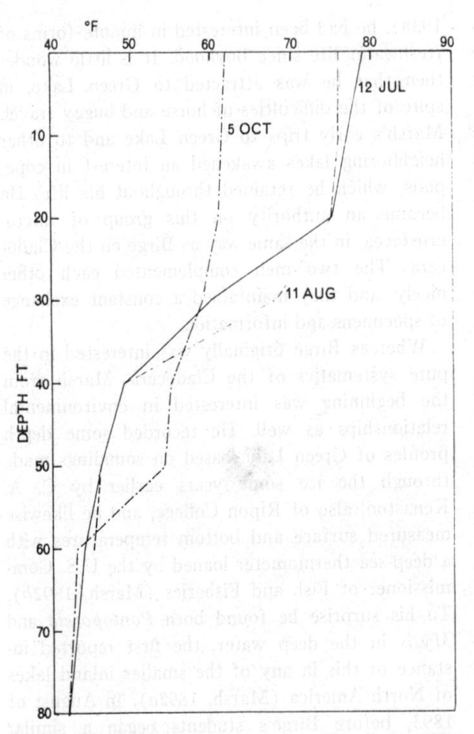

Fig. 1.9.—Thermal stratification in Pine Lake, Wisconsin, in 1879. From data given by Gifford and Peckham, 1882.

on the basis of their copepod faunas, with a dividing line at 40 m. Later (Marsh, 1901) he revised this to 30 or 35 m. This represents one of the very early attempts to recognize different kinds or types of lakes on the basis of organisms that live in them.

When the Wisconsin Geological and Natural History Survey began operations, Marsh was hired to construct a bathymetric map of Green Lake (Marsh, 1899). Subsequently he made a comparative study of the net plankton of Green and Winnebago lakes as representing deep and shallow lakes, respectively, with less intensive observations on about 30 other lakes, including some in the northern forested region (Marsh, 1903). Seasonal curves for the individual species of entomostraca and for the total volume of plankton were given. An interesting observation was that the cladoceran *Eurycercus lamellatus*, which is generally considered a littoral form, bulked large in the summer plankton of Lake Winnebago, forming a significant item in the diet of the sheepshead (*Aplodinotus grunniens*).

Marsh left Ripon College in 1903 to obtain his Ph.D. from the University of Chicago, after which he joined the staff of the U.S. Department of Agriculture, investigating the poisoning of domestic animals by various forage plants. During this period of governmental employment, which lasted 25 years, he continued his interest in the systematics of the copepods. He examined collections from all parts of the United States and published 35 papers on his results, which are listed in a carefully prepared bibliography on pages 57–58 of his 1933 paper. Included are revisions of North American *Diaptomus* (Marsh, 1907) and *Cyclops* (Marsh, 1910), as well as the chapter on copepods in *Fresh-Water Biology* (Marsh, 1918).

Other systematic studies on copepods include those by Juday (1914*b*, 1923*a*, 1925) and Juday and Muttkowski (1915), the latter containing descriptions of two copepods—*Diaptomus pribilofensis* and *Heterocope septentrionalis*—that are widespread in the North American Arctic, as well as the papers by Wright (1927, 1928). In addition Bere (1931*a*, 1935) did some work on copepods parasitic on fish.

A few other early studies deserve mention. One of the most interesting of these is the measurement of subsurface temperatures in Pine Lake by Gifford and Peckham (1882), with a "self-registering thermometer" made by Hicks of London. Although most of the time only the surface and bottom temperatures were measured, on six occasions the temperature was measured at 10-ft intervals to depths as great as 80 ft. Gifford and Peckham themselves made no detailed analysis of the data, but the series for 12 July and 11 August, when plotted, show a very nice sigmoid curve, and the series for 5 October shows a substantial lowering of the thermocline (Fig. 1.9). The bottom water of Pine Lake remained at 41–42° F throughout the season from 11 May to 5 October. The authors found a uniform temperature on 2 December, concerning which they remarked that "probably in a larger lake this condition would not be reached before January."

Another early study was made by Hoy (1872) on the deep-water fauna of Lake Michigan. Based first on an analysis of the stomach contents of deep-water fishes, the study was supplemented by some dredgings on 24 June 1870. Hoy obtained *Mysis relicta* (named *M. diluvianus* by Stimpson) and a new species of cottid, both of which Hoy regarded as marine forms providing evidence of a former direct connection between Lake Michigan and the sea!

Another early study is that of Trelease (1889) on the blooms of blue-green algae in Mendota and Monona. The blooms were not of equal intensity in all years. That of 1882 in Mendota was so bad (chiefly *Microcystis* and *Coelosphaerium*) it prevented boating along the Madison shore. One interesting observation was on the production of horizontal rainbows by refraction or diffraction of light from floating colonies of blue-green algae partially projecting through the smooth water surface. This same phenomenon was noted later by Juday (1916*a*, 1920*b*).

Limnology: 1911–23

Dissolved gases

Ever since the study on the seasonal changes in the vertical distribution of plankton Crustacea and the suggestion that these organisms were eliminated from the bottom water in summer by chemical changes, Birge had wanted to begin a study of lake chemistry. In 1904 a gasometric method for determining O_2, CO_2, and N_2 was finally worked out with the help of two chemists at the university, and this method was subsequently used in 1905 to study the gas content principally of Mendota. Juday returned in August of that year. Whether it was at his instigation or Birge's is not certain, but in the subsequent summers the unmodified Winkler method was used for dissolved oxygen, and titrations to the methyl orange and phenolphthalein endpoints with N/44 Na_2CO_3 and N/44 HCl, as appropriate, were used in determining free CO_2, half-bound CO_2, and fixed CO_2. Gaseous nitrogen was not determined after the early part of 1906 (Birge and Juday, 1911).

I should add, parenthetically, that the concepts of "bound CO_2" and "half-bound CO_2," originally proposed by Seyler (see Juday, Birge, and Meloche, 1935), have led to some confusion. They are based first of all on the assumption that any alkalinity to the methyl orange endpoint results primarily from calcium (or magnesium) carbonate and bicarbonate, which is not true for all surface waters and certainly not true of those hypolimnions in which ammonia accumulates during summer stratification (Ohle, 1952). Second, the interpretation of titration results is made even more difficult by Ohle's (1952) study, which showed that the carbon dioxide equilibrium is really an "inequilibrium," which follows the law of mass action very slowly under certain conditions, and that in hard-water lakes during periods of photosynthesis there can be considerable quantities of colloidal $CaCO_3$ in the water, which do not affect the *p*H or conductivity but do affect the methyl orange alkalinity. Third, most persons today generally report their results as parts per million of methyl orange alkalinity, but since it is not always clear whether N/44 acid (as recommended by Birge and Juday) or N/50 acid (as recommended by the American Public Health Association, 1955) was used in the titrations, the results may not be interpretable by the reader. To the writer it seems that the European practice of reporting methyl orange alkalinity as the number of milliequivalents of acid (acid combining capacity = SBV) required to titrate one liter of water to the methyl orange endpoint is least ambiguous and much to be preferred.

During the five years of study a tremendous amount of information on water chemistry, temperatures, and plankton was accumulated from 156 lakes in the three major lake districts of the state. Juday gradually was given almost complete responsibility for the study, and he took full charge of the field parties in the last two seasons. Birge, in spite of his heavy administrative duties, however, reported that he gave almost daily attention to the work and constantly reviewed the progress of the program. The final report (Birge and Juday, 1911) certainly represents one of the milestones of limnology.

In the meantime, Birge's concepts regarding ecological unity of lakes were being sharpened, and he was revealing himself more and more as a master essayist. His address in 1903 as President of the American Microscopical Society was "The thermocline and its biological significance" (Birge, 1904*b*). In this he described how the depth and extent of warming of a lake are determined by the interplay between the velocity and duration of the wind and the "thermal resistance to mixture" of the water. He distinguished between and illustrated two types of lakes—those with no zooplankton below the thermocline in summer and those with zooplankton in this region. As yet he had no terms for the zones above and below the thermocline, although he sometimes referred to them as the superthermocline and subthermocline. (He proposed the terms epilimnion and hypolimnion in a later paper [Birge, 1910*b*].) In two ideas he was not quite correct according to our current concepts. He considered that only slow water movements occurred in the hypolimnion, and he believed that plankton algae are kept in suspension by vital rather than mechanical means.

In his 1906 essay Birge presented a description of the seasonal change in oxygen content of Mendota, obtained the previous year by the laborious gasometric technique. He compared the conditions in Mendota briefly with those in Green Lake and with two lakes showing a metalimnetic oxygen maximum. "The amount of oxygen in the lower water depends not merely on the length of time that the bottom water is cut off from the external air, but it depends also upon the amount of decomposable material discharged into it by the upper water and on the volume of the lower water, which, in turn, depends on the depth of the lake." In this statement he combined the concepts of morphometry and productivity as controlling the O_2 content of the hypolimnion, which Thienemann (1928) later developed in greater detail. Birge also cited temperature as an important influence on oxygen content through its effect on the rate of decomposition in deep water, and he noted that small lakes have a greater relative contribution of decomposable organic matter from the littoral zone than do large lakes.

As President of the American Fisheries Society, Birge presented an address on "The respiration of an inland lake" (Birge, 1907*b*). Comparing a lake to an organism, or more specifically to the blood of an organism, Birge described the annual cycle of respiration in relation to the thermal cycle—a full inspiration in autumn, a less complete one in spring, none at all under the ice in winter, and only shallow respiration in summer. This essay, along with others, reflects Birge's overall view of a lake as a unit of environment—as an individual with physiological processes analogous to those of an organism.

In the second gas essay Birge (1910*e*) discussed the seasonal stratification of temperature, oxygen, nitrogen, carbon dioxide, and carbonates (expressed as CO_2) of Lake Mendota and then compared these conditions with those of lakes elsewhere in the state. In this essay, which is liberally documented by graphs showing stratification, Birge further developed his ideas regarding the interplay among productivity of plankton, the sun-wind factor, and the volume of deep water in determining the oxygen content of deep water in summer. Because of the large area of Mendota, the resulting low thermocline (and, therefore, small volume of deep water) and high temperature of the bottom water, and the great quantities of decomposable material that rain into the deep water, the oxygen here disappears rapidly and completely. For these reasons, Birge regretted somewhat that Mendota had been selected for such detailed study.

Other lakes showed different conditions. Green Lake had abundant oxygen in deep water throughout the summer (and also contained *Pontoporeia* and *Mysis*, as noted earlier by Marsh [1892*a*]). Some of the lakes exhibited oxygen maxima or minima in the thermocline. Birge noted that although the gas picture varied from lake to lake, certain patterns were emerging. This is further evidence of Birge's increasing interest in lake typology. He pointed out, too, that hard-water lakes have a greater carbon dioxide reserve than soft-water lakes and, therefore, generally have more plankton. As a result, the soft-water lakes in northeastern Wisconsin typically had better oxygen conditions in deep water during stratification than did those in the southeast. Birge related the oxygen content of the hypolimnion to its volume and the quantity of decomposable matter discharged into it.

In another paper from this period Birge (1910*b*) developed the concept of "thermal resistance to mixture" and asserted that the magnitude of this quantity increases the farther the water temperature departs from 4° C. And in the paper in which he proposed the terms *epilimnion* and *hypolimnion* (Birge, 1910*c*), he refused to accept Wedderburn's conclusion that the great vertical excursions of the thermocline in Loch Ness are the result of *temperature seiches* (internal seiches) and not the direct result of wind blowing the warm surface water from one side of the lake to another. Mortimer (1956) gives a more detailed exposition of this deficiency in Birge's otherwise excellent insight into limnological processes.

In the meantime some lesser papers were being published as an outcome of these studies. Juday

(1908*a*) briefly described the work of the Wisconsin Survey, noting among other things that whereas the calcium content of deep water increases during summer stratification, magnesium remains essentially unchanged. He further noted the occurrence of cocoons or cysts of *Cyclops bicuspidatus* in the deep-water sediments of Mendota in summer, an observation recorded in greater detail in a separate paper (Birge and Juday, 1908, which incidentally was the first paper published jointly by these men).

In this latter paper Birge and Juday reported that although the occurrence of *Cyclops* cysts at the surface of the deep-water sediments in Lake Mendota is largely coincident with the period of summer stratification and hence of oxygen depletion, the immature copepods begin to encyst while there is still adequate oxygen at the bottom and begin to excyst before oxygen is restored to the bottom by the fall overturn. The occurrence of these cysts was also noted in several other lakes, two of which had oxygen in the hypolimnion throughout the summer. At this time Birge and Juday did not have any device for sampling the lake bottom on an areal basis. They merely noted that the cysts were abundant.

Juday (1908*b*) was also interested in the effects of oxygen depletion on other animals occurring in deep water. He found eleven kinds of Protozoa and one rotifer that were active when no oxygen was present. *Chaoborus* (*Corethra*) was abundant in the oxygen-free water of the hypolimnion. An ostracod and a fingernail clam remained alive but inactive during the anaerobic period, as shown by aquarium experiments. When the clams were placed in aerated water, they quickly resumed activity. *Tubifex, Limnodrilus,* and a nematode were active under anaerobic conditions also. Wanting to make certain that oxygen was really lacking, Juday used two different titration procedures plus the gasometric procedure previously developed at Wisconsin. All gave negative results. Hence, there was quite a variety of animals that either appeared to remain active during anaerobic conditions lasting about 3 months or else survived these extreme conditions through dormancy and inactivity.

The newly discovered story of oxygen depletion in deep water was also being related to fish distribution. Lake trout were being stocked widely at this time, but with relatively infrequent success. Juday and Wagner (1908) found that Lake Kawaguesaga, in which stocking had been repeatedly unsuccessful, had no oxygen below a depth of 10.5 m, whereas in Trout Lake, in which lake trout were known to occur and were in fact caught in experimental gill nets, oxygen extended to the bottom. Moreover, Juday and Wagner found that perch and crappies placed in water pumped from the oxygenless hypolimnion of Mendota rapidly died. Such relationships as these are common knowledge now, but at the time they opened up new vistas for the fishery biologists.

All these essays and incidental reports were mere previews of the final report (Birge and Juday, 1911), which, in spite of its volume, is scarcely more than a summary of the tremendous amount of work accomplished. In the introduction Birge reviewed the annual cycle of thermal stratification, superposed on which is the utilization and production of O_2 and CO_2 by biological processes in a lake. For the first time in his writings Birge used the terms *zone of photosynthesis* and *zone of decomposition,* although he did not define them as we do today. He related the thickness of the zone of photosynthesis to the color and turbidity of the water, somewhat anticipating the later studies on light transmission. He regarded the supply of carbon dioxide, including the "half-bound" CO_2 in the bicarbonate reserves, as probably the chief controlling factor in productivity. The shape of the lake basin was likewise regarded as an important factor in productivity. The carbon dioxide removed from the epilimnion by the sinking of dead and dying plankton "is only partially replaced from the air or other sources. Since this is the case, the form of the lake basin is of importance in the economy of the lake. A lake with shoal margins, offering a chance for much decomposition above the thermocline, will produce—other things being equal—more plankton than a lake of similar area, but with steep slopes and deep water. In the latter case, almost all matter manufactured near the surface is decomposed in deep water below the region of circulation; and therefore is more or less permanently withdrawn from the possibility of being used again."

Birge returned again and again to likening lakes to organisms. He regarded as one of the chief values of their study the demonstration of the "existence of physiological processes in lakes as complex, as distinct, and as varied as those of one of the higher animals." To understand the gas story thoroughly, information was needed on the quantitative aspects of plankton and of the bacterial processes of decomposition. One of the big unanswered questions was why lakes varied so tremendously in their productivity and their ability to support plankton. The horizons of Birge and Juday had now broadened to include almost all aspects of descriptive limnology. Eventually they and their associates actively investigated many of them, using experimental procedures more and more in the later years of this period.

In the main body of the final report Juday described the annual cycles of oxygen and carbon dioxide stratification in Lake Mendota and a number of other Wisconsin lakes. The utilization of oxygen in the hypolimnion was brought about primarily by the decomposition of phytoplankton raining down from above, and "the rapidity of the decrease of oxygen in the lower water depends chiefly on three factors, the quantity of decomposable material, the temperature of the water, and the volume of water below the thermocline." Instances of supersaturation in the thermocline were associated with decreases of free CO_2, fixed (bound) CO_2, and silica and were properly related to the photosynthetic activities of algae.

Three categories of lakes were described, based on the amounts of fixed CO_2 present: *soft water lakes* with less than 5 cc/L, *medium water lakes* with 5 to 22 cc/L, and *hard water lakes* with more than 22 cc/L. (These classes were later [e.g., Wilson, 1935] redefined in terms of ppm of bound CO_2: 0–10, 10–30, and >30 ppm, respectively.) All lakes examined in northeastern and northwestern Wisconsin were either soft- or medium-water lakes, whereas all lakes of southeastern Wisconsin were hard-water lakes. In most of the medium- and hard-water lakes, the surface waters in summer were alkaline to phenolphthalein through withdrawal of half-bound (bicarbonate) CO_2. No measurements of *p*H were available at this time. Alkalinity and acidity were defined in relation to the phenolphthalein endpoint rather than to *p*H 7 as is done today.

Juday also divided lakes into those that circulate throughout the summer and those that are stratified. The latter were in turn classified into three types on the basis of the oxygen content of the hypolimnion: (1) abundant oxygen throughout summer, (2) disappearance of oxygen from a part of the hypolimnion, (3) disappearance of oxygen from all or nearly all of the hypolimnion. The distribution of plankton, based on pump collections, was related to the various types of oxygen distribution.

Some initial studies were made on metallic ions, on the gases of anaerobic decomposition, and on the greater content of inorganic nitrogen compounds in the hypolimnion in summer.

In a final section Juday listed a variety of unsolved problems, some of which are prophetic of later studies undertaken by the Wisconsin limnologists:

> The sunlight in the zone of photosynthesis offers numerous problems both quantitative and qualitative; the relation between transparency and color of water and the depth to which photosynthesis may extend; the relation of light conditions to the rate of liberation of oxygen by chlorophyl-bearing organisms, and especially the rate of liberation and accumulation of this gas in the excess oxygen stratum; the factors, whether due to light or other causes, which fix a lower limit of depth for the manufacture of excess oxygen. . . .

Hence, we see in this monograph an emerging understanding of the interaction between growth and decomposition of the plankton and the physical factors of thermal stratification and basin morphometry in controlling the chemistry of the hypolimnion. We also see some of the first attempts to classify lakes on the basis of their hardness and the summer oxygen content of the hypolimnion. One of the big advantages of these extensive monographs by Birge and Juday is that they contain great volumes of primary data, which have enabled subsequent limnologists to make new interpretations. Thus, Hutchinson (1938) used the data from Lake Mendota and Green Lake in developing the theory of areal hypolimnetic oxygen deficits. According to his analysis, Green Lake, in spite of its deep-water *Pontoporeia, Mysis,* and whitefish, is actually more productive than Mendota. The large volume of its hypolimnion makes it morphometrically oligotrophic.

Thermal capacity of lakes

Wishing to determine how widely applicable to other lake regions their conclusions on the Wisconsin lakes were, Birge and Juday, with support from the U.S. Bureau of Fisheries, studied the Finger Lakes of New York at various times from 1910 to 1912 (Birge and Juday, 1914). In this and subsequent joint papers Birge was responsible for reporting on the temperature and hydrography, Juday on the dissolved gases and plankton.

Birge was becoming more and more interested in the mechanics of thermal stratification, the means whereby water below the surface becomes warmed, and the total amount of heat a lake has stored at the time of maximum summer conditions. In this paper (Birge and Juday, 1914) and one published the next year (Birge, 1915), Birge compared the heat budgets of first-class American lakes with those of comparable European lakes for which data were available. First-class lakes were defined as those of sufficient size (at least 10 km long and 2 km wide) and depth (mean depth at least 30 m) to permit the lake "to acquire the maximum amount of heat possible under the weather conditions."

Birge noted that on the whole the heat budgets of European lakes were smaller than those of comparable North American lakes, which he explained mainly as the result of the much warmer epilimnions in the American lakes. He noted that "the August temperature conditions in the lakes of central Europe, in general, resemble those of American lakes in June. In the European lakes a relatively thick epilimnion is not formed in early and mid-summer as in ours, and a well-marked epilimnion is hardly developed in these lakes until the surface begins to cool. From this fact comes the statement, not uncommon in European writers on lakes, that the thermocline is a phenomenon which develops during the cooling period of a lake. No student of American lakes would make such a statement, since the epilimnion is fully formed in July, even in a large lake, while in smaller lakes it may be well developed in June or even in May." Because the heat content of the epilimnion can vary greatly during the summer with changes in the weather, Birge proposed later (Birge and Juday, 1920) that for purposes of comparing the heat budgets of American lakes, a mean epilimnial temperature of 23° C be used, rather than the temperature actually measured.

Forel calculated heat budgets from the heat content of a column of water at the position of maximum depth. Halbfass used the entire heat content of a lake. Birge rejected both of these as being unsuitable for comparing lakes and used instead the average amount of heat per square centimeter of lake surface. For this procedure he acknowledged the priority of Wojeikoff in 1902 and Wedderburn in 1910. He also suggested that the best measure of heat income under most conditions is the *wind distributed heat,* which is defined as all the heat in a lake above 4° C. Later (1915) he suggested this be called the *summer heat income.* The total amount of heat stored in a first-class lake in the latitude of Madison at maximum summer conditions represents ⅓ to ½ of the 60,000 cal/cm^2 that is delivered to the lake during summer stratification.

Birge approached the question of stability of stratification through a consideration of the amount of work required for the wind to move the warmed surface water down to the various depths in the lake (Birge, 1916). Schmidt was also considering stability of stratification but from a different approach. Schmidt calculated the amount of work that would be required to distribute all the heat in a lake uniformly and hence bring it back to indifferent equilibrium (homothermy), whereas Birge calculated the amount of work required to produce the existing thermal stratification. Birge regarded his method as more simple and direct, although Schmidt's method is now generally considered more useful (Mortimer, 1956).

To help obtain an estimate of the magnitude of *in situ* absorption of solar radiation, Birge had an instrument (called a pyrlimnometer) constructed in 1912 for measuring solar energy directly at any depth (Birge, 1922). This instrument consisted of a series of 20 iron-constantan thermocouples, the junctions of which were covered with blackened silver discs. The instrument was sensitive to energy inputs as low as 0.006 cal/cm^2 per min. Hence, the initial studies on solar radiation at Wisconsin were based on total energy rather than spectral composition.

With this instrument in hand Birge and Juday revisited the Finger Lakes in 1918 (Birge and Juday, 1921). From measurements of the transmission of solar radiation through the water and assuming that all the subsurface radiation is absorbed and stored, Birge calculated that the maximum warming of the Finger Lakes by direct penetration of solar radiation was about 16% of the summer heat income, which was about the same as in Mendota. In Lake Okoboji (Birge and Juday, 1920) this amounted to about 20%. Because of the rapid rate of absorption of solar radiation, chiefly the infrared, about 95% of the heating by direct absorption is confined to the top 5 m. Curves for the total heat content with relation to depth were partitioned into direct solar heat and wind-distributed heat. Birge and Juday were at-

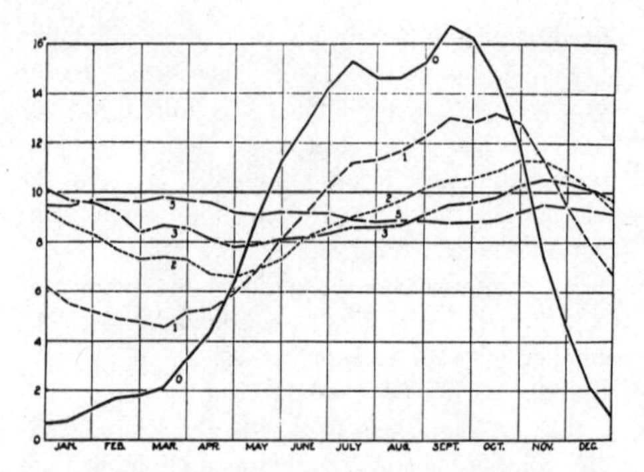

Fig. 1.10.—Mean annual fluctuations of temperature in the sediments of Lake Mendota over a three-year period at the 12-m station. The number associated with each curve is its depth in the sediments. From Birge, Juday, and March, 1928.

tracted to Lake Okoboji by the high deep-water temperatures reported by Tilton, but these were judged to be in error.

Birge and Juday studied the thermal structure of Lake Mendota over a 16-year period. During this time the mean annual heat budget (from winter minimum to summer maximum) was about 23,000–24,000 cal/cm². Two problems that were still unresolved were the amounts of energy absorbed by the sediments each year and the effect of the sediments in warming the water beneath the ice in winter. By means of electrical resistance thermometers (Birge, 1922) Birge and Juday (Birge *et al.*, 1928) measured the mud temperatures at 0.5-m intervals from the surface to 5 m, at water depths of 8, 12, 18, and 23.5 m, throughout the years 1918–20, with scattered observations beginning in 1916 and extending into 1921. They found that the annual energy budget of the sediments was about 2,000 cal/cm², hence roughly 10% of the annual energy budget of the water. Over winter about 650 cal/cm² were given up by the mud to the water, which represented only about one-fourth of the total heat gain beneath the ice. Most of the remainder was ascribed to the penetration of solar radiation through the ice.

Heat transmission varied according to the nature of the sediments. At the 8-m station this amounted to 57% per m, whereas at the deeper stations with more organic sediments it was about 50% per m. The lag in transmission from meter to meter was about one month (Fig. 1.10).

Quantitative plankton

Another monumental study by Birge and Juday was that on the plankton of the Madison lakes, chiefly Mendota (Birge and Juday, 1922). Emphasis in this study was on the amount of plankton present in the lake at a given time, its seasonal variation, and the seasonal variation in its chemical composition, expressed as nitrogen, crude protein, ether extract, carbohydrates, crude fiber, nitrogen-free extract, and ash. The chemical work was done by Schuette of the Chemistry Department, or under his direction, using methods he had reported earlier (Schuette, 1918). The voluminous report, which was written largely by Juday, represented a culmination of his early interests in the biomass aspects of plankton.

All samples were collected at a single station in Mendota, and the data from these were then extrapolated to the entire lake. This involved certain assumptions that were not critically examined. Massive water samples (2,000–38,000 L each, although most were in the 10,000–20,000 L range) were pumped from the various depths in the water column. From 1911 through 1914 only net plankton was studied, but from 1915 through 1917 nannoplankton was sampled as well. The nannoplankton was concentrated by means of a DeLaval centrifuge developed for clarifying oils and varnishes. Tests showed that this centrifuge recovered 98% of the algae and Protozoa, as well as 25–50% of the bacteria. Smaller samples were taken for determining the genera, or in some instances the species, of organisms present, although this aspect of the study is completely subsidiary to the biomass aspects. During this intensive study, more than 2 million liters of water were strained, yielding 1.3 kg dry weight of net plankton for chemical analyses. About half this amount of nannoplankton and detritus was obtained.

Juday reported that "the various forms reach their maximum numbers at different times of the year, some even in winter, so that there is no definite harvest time [as in agriculture] at which this material may be collected and the annual production of it thereby determined. The plankton, therefore, must be considered as a 'standing crop' since it is present at all seasons of the year and since it does not possess any definite period of maturity; in other words, it constitutes a continuous stream of life which presents different degrees of abundance during the course of its annual cycle."

The annual cycle of plankton standing crop or biomass in Mendota exhibited peaks in spring and autumn at the time of the overturns and lesser amounts during summer and winter (Fig. 1.11). Nannoplankton was of much greater relative importance in the spring bloom than in the fall bloom. Diatoms were the dominant organisms at both these times. During a two-year period the standing crop of plankton in Mendota averaged 240 kg/ha expressed as dry organic matter. Live (wet) weight would be approximately 10 times greater. On the average the weight of nannoplankton in the standing crop was about 5 times greater than that of the net plankton. Juday defined nannoplankton as all organisms that slip through the finest bolting silk net, not merely those smaller than 25 μ as Lohmann had originally defined the term.

Birge (1923) discussed the results of this study in more general terms, pointing out that the average standing crop of offshore phytoplankton in Mendota was about 2 tons wet weight per acre and that of the zooplankton about 200 lbs per acre.

Certain minor studies and observations were made in connection with the plankton investigation that were prophetic of future studies by Birge and Juday and their co-workers. Thus, several analyses of the organic nitrogen content of Mendota water before and after centrifugation showed a much higher content of dissolved organic nitrogen than of nannoplankton or net plankton nitrogen, the figures being 758.0, 103.5, and 21.5 mg/m³, respectively. Juday considered and rejected Pütter's theory concerning the uptake of dissolved organic matter in the nutrition of higher organisms, largely because Pütter had neglected the nannoplankton.

Quantitative studies on the role of bacteria in the plankton were begun in 1919. Direct counts gave about 10 times as many bacteria as culture methods, but even the highest numbers obtained (60,000/ml) are much lower than we now know occur in lake waters (Kuznezow, 1959). Hence, Juday's calculations of the biomass of bacteria in Mendota and his opinion as to their importance are low by a corresponding amount.

Juday realized that the big problem for the future was not the measurement of the instantaneous biomass but the rate of turnover, or in other words the productivity. Standing crop merely represents the momentary balance between losses and gains in the population. Without presenting

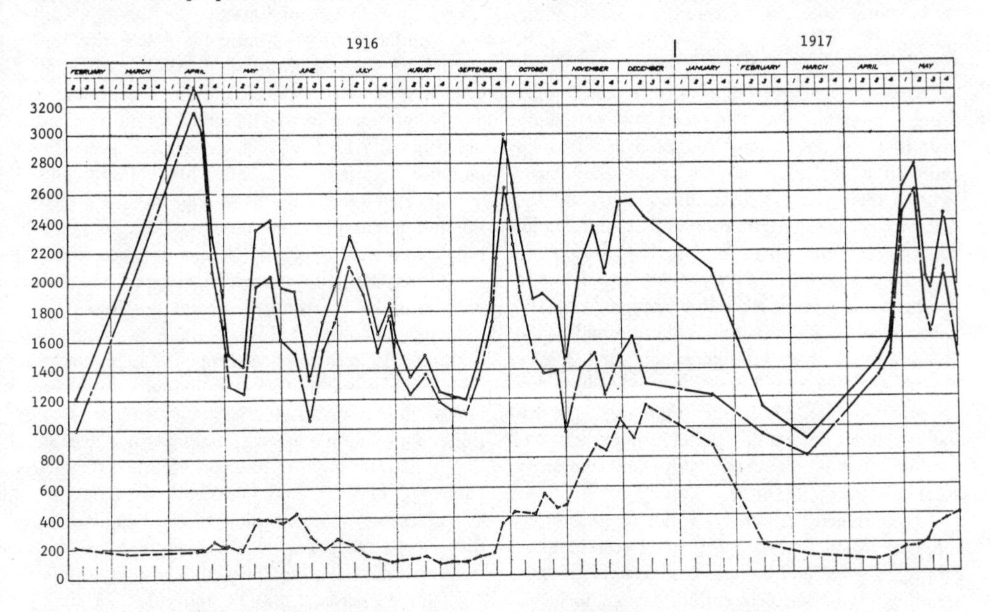

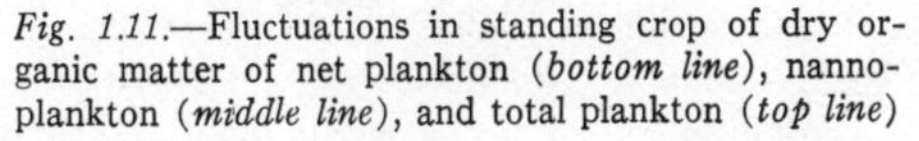
Fig. 1.11.—Fluctuations in standing crop of dry organic matter of net plankton (*bottom line*), nannoplankton (*middle line*), and total plankton (*top line*) in Lake Mendota in 1916–17, in mg per m³ of water. From Birge and Juday, 1922.

any real basis for his opinion, he stated that the mean turnover rate probably "will be found to fall somewhere between one and two weeks during the greater part of the year."

Juday had already developed the plankton trap associated with his name (Juday, 1916*b*), but interestingly this device up to this time had been used only to test the efficiency of the pump used in collecting the massive samples of net and nannoplankton.

In several other studies of this period (Birge and Juday, 1914, 1920, 1921) Juday related the vertical distribution of zooplankton to the chemical and thermal stratification of the lakes. Unfortunately, as in the dissolved gases paper (Birge and Juday, 1911), the organisms were grouped into such large categories (Crustacea, nauplii, rotifers, diatoms, and all other algae) that only the grosser aspects of vertical distribution could be perceived.

In the Lake Okoboji paper (Birge and Juday, 1920) Juday used the term mesolimnion as equivalent to thermocline, and in the plankton monograph (Birge and Juday, 1922) he used the term exclusively for this stratum. Later (e.g., Juday and Birge, 1932) he readopted the term thermocline.

Transmission of solar radiation

One of the fields of limnological research most closely associated with Birge and Juday—mainly with Birge, although most of the papers are coauthored with Juday—and in which they made some of their greatest contributions is that of the transmission of solar radiation and the factors influencing the differences in this transmission observed from lake to lake and from one depth to another in the same lake. Four papers based on field studies in southeastern Wisconsin and at the Trout Lake Laboratory were published jointly (Birge and Juday, 1929*b*, 1930, 1931, 1932; the paper listed as 1929*a* in the references is essentially a summary of the 1929*b* paper), and one significant paper was published on laboratory studies (James and Birge, 1938).

As mentioned previously, Birge became interested in radiant energy early in his career after he had "discovered" the seasonal cycle of thermal stratification in Mendota. In an attempt to measure the penetration of solar energy into lakes, he used a black-bulb thermometer in 1900–1901 to depths of 5 m in Mendota and in 11 other lakes. These results were never published, although they agreed substantially with those presented in the first paper (Birge and Juday, 1929*b*).

In 1912 studies on transmission were initiated using a pyrlimnometer, at first with iron-constantan thermocouples and later with bismuth-silver thermocouples. This instrument converts the entire solar spectrum into energy units. In the first study the radiation was reported as g · cal/cm², but in later studies the energy at any depth was always expressed as a percentage of the incident energy at the surface. Likewise, the spectral composition at any depth was based on the percentage that the energy in each color region, defined on the basis of wave length, comprised of the total energy at that level.

The results of these studies are well known. They showed that almost none of the infrared radiation and very little of the ultraviolet radiation were transmitted through more than 1 m of lake water. Hence, below a depth of 1 m the solar energy is almost entirely in the visible portion of the spectrum. Likewise, because only about half of the energy of solar radiation is in the visible portion of the spectrum, even distilled water transmits only about 47% of the total spectral energy through 1 m of water.

In the lakes reported on in the first paper, the top 10 cm of water absorbed 40–65% of the incident radiation. Birge at this time was still primarily interested in the thermal aspects of solar radiation and their effects on lakes. During the period 15 April to 15 August Mendota receives on the average about 60,000 cal/cm², of which the top meter of water absorbs 74%. Most of this heat is used in evaporation, is returned to the air during the daily warming-cooling cycle, or is distributed by the wind to greater depths.

Below a depth of 1 m, a lake maintains a characteristic transmission through the epilimnion and frequently into the metalimnion, except in highly colored waters. The pyrlimnometer used by Birge and Juday was progressively modified so that its lower limit of usefulness was extended from 1% to 0.1% and eventually to 0.01% or less of the surface radiation. In the hypolimnion they found that there was frequently a reduced transmission, especially near the bottom, resulting from increased turbidity and color. In some of the clearest lakes, however, the characteristic transmission was maintained from a depth of 1 m to the bottom.

In 1930 (Birge and Juday, 1931) the pyrlimnometer was fitted with a series of interchangeable filters so that the spectral composition of the light at different depths in the top 10 m could be studied. These studies demonstrated that "transmission is determined by three main factors: (1) the selective action of water, which is transparent to short wave radiation and opaque to long waves; (2) the selective action of stain which acts more strongly on the short wave radiation and is effective in proportion to the amount and kind present; (3) the action of suspended matter—organic and inorganic—which offers more obstruction to short wave radiation, but is not definitely selective." These studies provided a description of the spectral changes of light with increasing depth in lakes having different levels of dissolved color and turbidity.

In the last published study by Birge on this subject (James and Birge, 1938) measurements of the absorption of monochromatic light of different wave lengths in the laboratory enabled the resolution of the total transmission curve into the three components controlled respectively by the water itself, colors (both dissolved and colloidal), and suspensoids.

Other studies of solar radiation in relation to lakes were made by Pietenpol, Whitney, and Davis. Pietenpol (1918) carried out one of the earliest studies on the selective absorption of visible light by means of a spectrophotometer. Some of his data were subsequently used by Birge and Juday (1931) for comparison. Whitney (1937, 1938*a*) discovered a microstratification in lakes, as indicated by changes in light transmission over short vertical distances. He found that whereas the transmission in the epilimnion was generally uniform, in the metalimnion and hypolimnion the water was stratified alternately into more transparent and less transparent layers, which varied in thickness from only a few centimeters to about two meters. Replicate transmission curves from different places in the same lake showed a marked correspondence, indicating a definite layered condition within the stratified part of the lake. There was some correlation between numbers of bacteria and transparency. "The fact that the greatest changes in transparency usually occur in regions of relatively large temperature change may indicate that a lake acts as a filter or sorting machine in which particles of different sizes, shapes, and densities slowly sink and accumulate at levels of similar densities as determined by temperature and other physical conditions" (Whitney, 1938).

In another study Whitney (1938*c*) extended Birge's measurements on light transmission to intensities as small as 2×10^{-6} of surface light by means of a photoelectric cell and an amplifier. Measurements of light intensity at any depth were made with the thermopile (pyrlimnometer) and with the photocell face up, then with the photocell face down to determine the amount of scattering. The ratios of scattered light to direct light intensity ranged from 1/20 to 1/180. Later papers by Whitney (1941*b*, 1941*c*) represent an expansion of these studies.

Davis (1941) was interested in the loss of light at the surface by reflection from the surface and by scattering upward out of the water by suspensoids. Surface loss did not vary with surface roughness so long as the sun was within 50° of the zenith. Surface loss was somewhat greater for diffuse skies than for clear skies, and the loss of red light was less than for total light at relatively high zenith angles.

Limnology: The last two decades (1924–44)

Much of the limnological work reviewed up to this point resulted principally from the efforts of the two pioneer limnologists, with the help of relatively few associates and assistants. But during the last two decades of the Birge-Juday era, limnology was being pushed forward on broad fronts both in the Madison area and at the newly formed Trout Lake Laboratory. At the University of Wisconsin zoologists, botanists, chemists, bacteriologists, geologists, and others were being drawn more and more into a cooperative program. Each summer at Trout Lake up to a dozen or more assistants and visiting investigators were studying all kinds of limnological problems. During this interval more than 260 papers were published by the Wisconsin group and their associates. The peak came in the late '30's and early '40's, with as many as 34 papers listed for a single year (Juday and Hasler, 1946). During this period the *Transactions of the Wisconsin Academy of Sciences, Arts, and Letters* were dominated by the enormous and varied limnological output of this group of limnologists, sometimes to the virtual exclusion of other studies. Since I cannot hope to do justice to all these papers, I shall have to summarize them briefly by major topics.

Methods and equipment

The Wisconsin limnologists were fortunate in having available Mr. J. P. Foerst, the mechanic and instrument maker in the Department of Physics. Often merely from rough sketches or from suggestions as to the kinds of data needed, Foerst was able to devise items of equipment, many of which are standard items even today for the inland limnologist—Kemmerer water bottle, Juday plankton trap, case for reversing thermometer, Ekman dredge, Foerst centrifuge, etc. These items and others are described by Juday (1916*b*, 1926*c*, 1929) and Birge (1922). After leaving the university, Foerst and his brother, H. M. Foerst, established the Foerst Mechanical Specialties in Chicago for manufacturing these items of equipment.

The Department of Chemistry at the university developed or modified many methods for the analysis of chemical substances in water and in the residues of aquatic plants and animals. Many of the chemical analyses of these materials were made by the staff members themselves (Kemmerer, Schuette, Robinson, Meloche) or by students under their direction. Thus, Schuette (1918), as already mentioned, developed methods for the microanalysis of plankton residues, the results of which formed such an integral part of the plankton monograph (Birge and Juday, 1922). Later Schuette and Hoffman (1922) and Schuette and Alder (1927, 1929*a*, 1929*b*) carried out a series of short studies on the chemical composition of various aquatic plants. Kemmerer and Hallett (1927*a*, 1927*b*, 1927*c*) improved the methods for the microanalysis of ammonia, carbon, and loss on ignition. Taylor (1928) developed new micro methods for the analysis of lake water residues. Black (1929) turned his attention to lake sediments. Robinson and Kemmerer (1930*a*, 1930*b*, 1930*c*) reported on the methods for determining Kjeldahl nitrogen, organic phosphorus, and silica in waters.

V. W. Meloche was one of the very important contributors in the cooperative lake program. Titus and Meloche (1931) described the determination of total phosphorus, and Meloche with various other persons (Meloche and Setterquist, 1933; Meloche *et al.*, 1938; Lohuis *et al.*, 1938; Meloche and Pingrey, 1938; Knudson, Juday, and Meloche, 1940) reported on methods for determining calcium and magnesium and also presented results of studies on the silica, sodium, and potassium contents of lake waters. In addition Meloche with Juday and Birge (Juday, Birge, and Meloche, 1935, 1938, 1941) reported on detailed chemical analyses of lake waters and sediments. These will subsequently be described in somewhat greater detail. Many other chemical, physical, and biological methods were developed or modified, but these were more commonly described in papers of broader scope.

Chemistry of lake waters

Nitrogen compounds and dissolved organic matter.—As already mentioned, some preliminary analyses made during the plankton study (Birge and Juday, 1922) showed that the amount of dissolved organic matter in lake waters considerably exceeded the amount of particulate organic matter. Investigation of this problem was continued in 1922–24 (Domogalla *et al.*, 1925), chiefly on Lake Mendota. After being run through a centrifuge to remove the particulate matter, large samples of water were evaporated at low heat under vacuum. The residues were analyzed for amino and non-amino nitrogen, protein, peptone, and diamino acid nitrogen. The content of inorganic nitrogen as ammonia, nitrite, and nitrate was also studied. Large seasonal changes were observed in all components, with a high in winter and a low in summer. There was a sudden and marked increase in ammonia and nitrate in February. It was felt that these seasonal changes were probably closely related to changing activities of the biota. An indication that protein itself was present in the water was the formation of a precipitate with phosphotungstic acid.

In another study this same year (Peterson *et al.*, 1925) free amino acids were demonstrated in Mendota. Tryptophane, tyrosine, and histidine occurred at mean concentrations of about 13 mg/m^3, cystine at 4 mg/m^3. Arginine was also found. The dissolved nitrogen was separated into 12 components, which accounted for 90% of the total.

Domogalla *et al.* (1926) related the seasonal variations in ammonia and nitrate content of Mendota to the activity of ammonifying, nitrifying, and nitrate-reducing bacteria. Beneath the ice both ammonia and nitrate increased markedly since the dissolved oxygen did not become completely exhausted, whereas during the more severe summer stratification ammonia increased but nitrate decreased. These changes were related to bacterial populations.

Domogalla and Fred (1926) presented graphs and tables showing the annual cycles of ammonia, nitrate, organic nitrogen, soluble phosphorus, and rates of nitrification and nitrate reduction for all five lakes at Madison. Lake Monona at this time was receiving the treated sewage effluent from Madison, which considerably influenced its nitrogen chemistry and that of the two lower lakes—Waubesa and Kegonsa. Domogalla claimed to be able to demonstrate the direct utilization of ammonia by algae in Monona.

Birge and Juday (1927*a*; the paper listed as 1927*b* contains essentially the same information, and the paper published the previous year is a brief preliminary report of the same results) were concerned with the origin and significance of the large amounts of dissolved organic matter in fresh water. The average nitrogen to carbon ratio of this material was 1 to 15, whereas that of the plankton in Mendota was 1 to 7.1 and in two other Madison lakes about the same. The ratio in rooted aquatics was 1 to 17, but Birge and Juday calculated that the annual production of rooted aquatics in Mendota was scarcely one-third of the mean standing crop of dissolved organic matter and, hence, could scarcely serve as its source. They concluded that the plankton was the chief source, and that the disparity in the nitrogen to carbon ratios resulted from a more ready utilization of the nitrogen components in the material.

Birge and Juday (1927*b*) divided lakes into two classes according to the major source of the dissolved organic matter: those with the major source within the lake itself (chiefly plankton) were called *autotrophic,* whereas those with a considerable portion from outside the lake as extractives from soils and peat were called *allotrophic.* Evidence indicated that the quantity of dissolved organic matter was quite constant for a given lake, varying relatively little with depth or time. Hence, as in mineral analyses, a single sample of water should be adequate to define this characteristic.

Birge and Juday considered that the dissolved organic matter has about the same potential food value as plankton itself, since the proportions of free amino, peptide, and non-amino nitrogen are about the same, and since five free amino acids were already demonstrated to occur. However, even though the quantity of this material is much greater than that of particulate organic matter, its great dilution would present a problem in its utilization, at least by animals. Many of the net algae are too large to be eaten by the zooplankton, but the quantity of nannoplankton and find detritus is many times the weight of all Crustacea and rotifers and, hence, should provide adequate particulate food for them. Birge and Juday indicated that the whole problem of the nutrition of the plankton was very much unsettled.

The last paper (Birge and Juday, 1934) presented data from 529 lakes in the northeastern lake district. Whereas in the lakes of southeastern Wisconsin reported previously, there was little detectable outside influence on the amount of dissolved organic matter, the northeastern lakes exhibited the entire spectrum from clear seepage lakes to very dark-colored drainage lakes. In the clearest lakes, including Lake Michigan, the dissolved organic content was about six times that of the plankton, from which Birge concluded that "the plankton is the primary source of about 6 times its own weight of dissolved organic matter in the waters of those lakes which derive little or no organic matter from outside sources." He showed, however, that this simple ratio could not be used in partitioning the dissolved organic matter of higher carbon lakes into external and internal sources. As noted elsewhere, the summer standing crops of plankton in these soft-water lakes were only a fourth to a half as great as those in the hard-water lakes of southeastern Wisconsin.

Within a radius of 25 mi of the Trout Lake Station there are more than 500 lakes. Juday and Birge conducted a general chemical, physical, and biological survey of these lakes, many of them being visited only once but some of them being visited regularly in successive summers. A complete set of field determinations comprised 19 different chemical, physical, and biological items. Except for the few lakes that were singled out for special study, the remainder lose their identity in the subsequent reports; they are merely tallied in the appropriate range of the particular character being considered. In their early years Birge and Juday were concerned with the lake as a unit of environment. At the

Trout Lake Laboratory they were concerned with all the lakes in a particular region and the frequency distribution of these lakes over the total ranges of the particular variables being considered. The resulting papers lack the satisfying unity of the ecosystem approach, and one feels that many significant interrelationships may have been overlooked as a result.

One of the main contributions of this series of studies was to show the lake in relation to its watershed. The isolated *seepage lakes* (those without surface inlet or outlet) as a statistical group tended to be quite different in nearly all characteristics studied from the *drainage lakes* (those that are in direct surface connection with their watershed).

Transparency, color, and specific conductance (Juday and Birge, 1933).—Transparency measured in the northeastern lakes with a 10-cm Secchi disc ranged from 0.3 m to 13.6 m. In general, seepage lakes had higher transparencies than drainage lakes, which is related to the higher dissolved color of the latter. Color was more important than the amount of plankton in influencing transparency. The dissolved color, which ranged as high as 340 ppm on the Pt-Co scale, remained fairly constant for two or more months during the summer. Some lakes showed about the same color from surface to bottom, whereas others showed a marked increase toward the bottom. In a few of the latter this resulted from iron rather than vegetable stain. There was a distinct positive relationship between dissolved color and organic carbon in the water, as would be expected. The carbon to nitrogen ratio increased with increasing color.

Specific conductance varied from 6 to 132 micromhos (reciprocal megohms), which is on the whole much less than that in the lakes of southeastern Wisconsin. The specific conductance was quite characteristic of a given lake, tending to remain approximately the same from one year to the next. Specific conductance showed a strong positive relationship with fixed CO_2 and with total Ca and Mg, these being the dominant cations in the water. In some of the lakes there was no appreciable increase in specific conductance and fixed CO_2 in the hypolimnion over the summer, whereas in others there was. The role of biological activity in modifying the specific conductance was not discussed in sufficient detail.

Mineral content.—Juday, Birge, and Meloche (1938) reported on analyses for silicon, iron, manganese, calcium, magnesium, fluoride, chloride, sulfate, and ammonia, nitrite, and nitrate nitrogen in more than 500 lakes of northeastern Wisconsin. Ranges and frequency distributions were given for each of these substances, as well as the distribution with depth and the relation to the general chemical conditions in the lakes. Stratification was related largely to biological activity of various types. An interesting result was that whereas manganese was seldom found in surface waters, iron in quantities up to 2 mg/L occurred in the surface waters of 74 lakes. The paper is valuable mainly for the description of the mineral content of a whole lake region, without giving much insight into the dynamics of particular lakes.

Lohuis *et al.* (1938) analyzed the sodium and potassium contents of lakes in northeastern Wisconsin. Drainage lakes contained more total dissolved solids and also more total Na and K than seepage lakes, but the latter had a higher proportion of Na and K. The total amount of alkali, reported as potassium, generally ranged between 2 and 4 ppm. In drainage lakes sodium and potassium were approximately equal in abundance, whereas in seepage lakes there was generally less sodium than potassium.

Meloche *et al.* (1938) studied the silica content of Mendota during the period of 1935–36 and related this to the abundance of diatoms. During summer stratification there was a distinct increase in silica in the hypolimnion. The increase in diatoms in fall was related to the upward transport of silica during fall turnover. An incidental feature of the paper is that it contains a description of the Kemmerer water sampler as currently constructed.

Phosphorus.—The results of the first paper (Juday *et al.*, 1927) were included in the second (Juday and Birge, 1931), which reported the phosphorus content of 479 lakes in northeastern Wisconsin. The mean summer content of soluble phosphorus was 0.003 mg/L and of organic phosphorus (obtained by subtracting soluble from total phosphorus) was 0.020 mg/L, although the modal value of the organic phosphorus was 0.014 mg/L. Contrary to what would be expected, there was not always a decrease in soluble phosphorus over summer, even in stratified lakes. Stratified lakes with little plankton exhibited essentially no increase in hypolimnetic phosphorus during summer stratification, whereas those richer in plankton generally showed an increased content of soluble phosphorus, and frequently of organic phosphorus as well, in the deep water. This paper, as most others of this chemical series, was concerned less with interpretation of results than with their descriptive presentation. The main value is in showing the range of phosphorus contents in lake waters of one compact lake district.

Dissolved oxygen and oxygen consumed (Juday and Birge, 1932).—The oxygen content of the surface waters of more than 500 lakes in northeastern Wisconsin averaged only 82% of saturation. This was surprising but unexplained. It checked with 88% saturation recorded in this lake district in 1907–1910, and it contrasted with the average condition of 100% saturation in the lakes of the northwestern and southeastern lake districts. Juday and Birge compared the unmodified Winkler method with the Rideal-Stewart modification, and, finding no significant difference between these methods, they continued to use the standard Winkler because of its greater convenience.

Juday and Birge considered the applicability of lake types to the lakes of northeastern Wisconsin. Defining oligotrophy and eutrophy on the basis of the late summer oxygen content of the hypolimnion, they accepted the extremes of the trophic series but not the intermediate stages proposed by Naumann. Because of great differences between the brown-colored lakes of northeastern Wisconsin and those of western Europe, they rejected dystrophy as representing a separate type of lake.

In their discussion of oxygen deficits they considered the absolute and actual deficits for the entire lake, not just for the hypolimnion. Various analyses were made of percentage saturation for entire lakes and of the hypolimnion to epilimnion ratios as proposed by Thienemann. Unfortunately, the concept of areal hypolimnetic deficit was only getting started in Norway at this time, so that the analyses made are not too significant in terms of modern limnology.

Oxygen consumed was measured not by biochemical-oxygen-demand methods but by hot acid digestion with potassium permanganate. An average of about 40% of the total organic carbon in the water was oxidized by this procedure. The total amount of oxygen consumed varied directly with the dissolved color and organic carbon in the water.

Eh, pH, and carbon dioxide.—As already mentioned, Juday and Birge in their early work considered a water sample alkaline or acid according to its color reaction to phenolphthalein. Their first paper involving the measurement of *p*H was that by Juday, Fred, and Wilson (1924), in which they followed changes in *p*H with season and depth in Lake Mendota for a two-year period from 1919 to 1922. Single series of observations were obtained in summer from six other lakes in southeastern Wisconsin. They observed that in summer particularly, and to a somewhat lesser extent under the ice in winter, the upper waters had a substantially higher *p*H than the lower waters. They ascribed the summer differences to the photosynthetic activity of algae towards the surface and to the decomposition of organic matter in the hypolimnion.

Prior to 1932 all their *p*H measurements were made by colorimetric comparison. In that year, however, Freeman *et al.* (1933) devised a calomel-quinhydrone cell for measuring *p*H *in situ*. A similar cell was used for measuring the *p*H of water samples hauled to the surface. The two methods gave essentially the same results, and both were considered superior to the calculation of *p*H from titration data for free and bound CO_2.

In an extensive paper on carbon dioxide and *p*H, Juday, Birge, and Meloche (1935) reviewed the basis for the concepts of free, half-bound, and bound CO_2. In this paper the effects of watershed on controlling the chemistry of lakes were clearly apparent. Seepage lakes tended to have lower *p*H, more free CO_2, and less bound CO_2 in their surface waters than did drainage lakes. Although *p*H generally decreased toward the bottom of a lake, associated with increases in free CO_2 and bound CO_2, Lake Mary exhibited an increase in *p*H—a phenomenon Yoshimura had designated by the term dichotomous stratification. Lake Mary is now known to be quite definitely meromictic.

Allgeier *et al.* (1941) measured the oxidation-reductional potential (*Eh*) by means of an *in situ* probe. Values ranged from +0.512 to +0.077 volt in the water to a low value of −0.140 volt in the mud. These values were higher than those obtained by previous investigators. Lakes with abundant oxygen in deep water exhibited little reduction in *Eh* toward the bottom, whereas lakes with reduced oxygen and brown-water lakes (they used the term dystrophic here as a general descriptive term) generally exhibited a marked reduction. The investigators showed that in addition to dissolved oxygen ferrous iron and hydrogen sulfide helped control the *Eh* levels, and they suggested that organic reducing systems were probably active as well.

Sediments

The brief study by Black (1929) represented the first investigation in Wisconsin of the chemistry of lake sediments. Black analyzed surficial sediments of several Wisconsin lakes and three Alaskan lakes for SiO_2, Fe_2O_3, Al_2O_3, CaO, MgO, P_2O_5, SO_4, CO_2, and organic carbon. The results, presented in tables, were only partially related to the depositional environments. Black noted, as Birge and Juday (1911) had earlier, that relatively little Mg occurs in the sediments even of hard-water lakes because of the high solubility of $MgCO_3$ relative to $CaCO_3$. Two brief experiments were conducted on gas production by Mendota sediments at room temperature and in a refrigerator. More than half of the gas produced anaerobically was methane.

Steiner and Meloche (1935) were concerned with the carbohydrate fraction in the organic analyses of Birge and Juday (1934), because to the extent that this material is ligneous, it would not be readily available as food to organisms. They found that lignin comprised 30 to 48% of the organic matter of lake sediments, 18% of net plankton, and 10 to 20% of nannoplankton. The lignin content was generally higher in sediments of northeastern lakes than of southeastern.

Juday, Birge, and Meloche (1941) put out a second extensive paper on the chemistry of lake sediments, involving 18 lakes in northeastern Wisconsin and 3 in southeastern Wisconsin. The same substances were measured as in the paper by Black and by the same methods, except that in the sediments of soft-water lakes, the content of organic matter was approximated by loss on ignition. Sediments of the northeastern lakes had an average organic content three times greater than that of the southeastern lakes. This was related to the higher lignin content reported by Steiner and Meloche and also to the greater abundance of mud bacteria in the southeastern lakes, the unstated assumption being that the organic matter of the sediments is more readily decomposed in the hard-water southeastern lakes. Results of the analyses were given in tabular form.

W. H. Twenhofel, a geologist at the University of Wisconsin, was a well-known authority on sediments and sedimentation. He began investigating lake sediments as an aid to understanding the depositional environment of freshwater rocks. He was interested in transport of materials, as indicated by mechanical analysis, in the diagenesis of sediments with the potential production of petroleum, and in other geological problems. His early studies were based on surficial samples collected with an Ekman dredge and on partial cores obtained with a weight-driven pipe 6 ft long of Twenhofel's own design. Although this sampler was sometimes driven more than 6 ft into the sediments, it never collected 6 ft of material. Because the stratigraphic relationships were inexact, Twenhofel analyzed only the top, middle, and bottom of each core. In later studies Twenhofel and his associates obtained complete cores with a Jenkin corer from several northeastern lakes. On one occasion the Wilson sampler was used.

The first study was on Mendota (Twenhofel, 1933) and the second on Monona (Twenhofel, 1937). In both these lakes there was a black, soupy sediment at the top (which Twenhofel first called "ooze" and then "sludge"), with lighter-colored and firmer sediments below. Twenhofel thought this represented a diagenetic series—that the deeper, light-colored sediments were like the sludge when first deposited, but that they subsequently suffered a loss of organic content by biological activity. More recently Murray (1956) has related this almost abrupt change in sediment type in these lakes to the influence of man in the region, bringing about an increased importation of clastics and a decline in hypolimnetic oxygen, resulting in the deposition of ferrous sulfide in the sediments. Neither geologist established any chronology for the sediments of these lakes.

A similar study was made on Devils Lake in southern Wisconsin (Twenhofel and McKelvey, 1939), which is a very soft-water lake in a quartzite basin. As expected, SiO_2 was the chief inorganic constituent of the sediments, and there was little or no Ca and Mg. The organic content was estimated at 15–20% of the dry weight. The nitrogen content of the sediments varied from

0.018–0.88%. Bacterial analyses run by Janice Stadler showed smaller total counts than in Mendota. Chloroform and ether extracts yielded 1–9 lbs "oil" per ton dry weight of sediments.

Attention was also turned to Crystal Lake in the northeastern lake district (Twenhofel and Broughton, 1939). This lake was selected because of its extremely soft water and absence of dissolved color. A large part of the paper is devoted to the mechanical analysis of the sediments. A few cores that penetrated into sand indicated that the maximum thickness of the sediments was about 3 m, which gave a mean annual accumulation of 0.25 mm/yr based upon a time interval of 10,000 years. Lignin comprised about half the organic matter. The numbers of bacteria were small (Carpenter, 1939), being only about 1/1000 those of Mendota muds. Aerobic and facultative bacteria decreased markedly with depth.

Conger (1939), a diatom specialist, studied the diatoms in the cores from Crystal Lake. He found a much greater diversity than expected, 85 species in all, most of which were epiphytic and benthonic. Planktonic species were generally in the minority. The flora was more diverse at the bottom of the cores, and there was evidence that the depositional environment of the early sediments was less acid and richer in nutrients than subsequently. Several middle samples were dominated by *Fragilaria construens,* indicating rich "blooms" of this species. In some places pine pollen and other types comprised more than 50% of the mass of the sediments. Conger's study is interesting in being one of the first to utilize diatoms in interpreting previous ecological conditions in a lake.

Little Long Lake in Vilas County was selected for study because of complete enclosure by forest, absence of any organized drainage, and the very dark brown color of its water (Twenhofel and McKelvey, 1942). About half of the dry weight of the sediments was organic matter, of which lignin comprised 40–69%. At the surface the sediments consisted of more than 90% water, whereas at greater depths they were more compact. The SiO_2 content was low, and there was no Ca or Mg. Bacterial counts were likewise low. The maximum thickness of sediments, as measured by a spud sampler, was only about 10 ft.

The next two studies (Twenhofel *et al.*, 1942, 1944) were frustrating for Twenhofel in that the sediments of the lakes investigated consisted mainly of a greenish-yellow gel (algal gyttja) composed of floccules or sphaerules about 2 mm in diameter. Attempts at removing the organic matter by H_2O_2 or concentrated H_2SO_4 were unsuccessful, and hence the mechanical analyses were virtually meaningless. Twenhofel concluded that the sediment accumulated as floccules, which had incorporated pollen grains, diatoms, sponge spicules, and small amounts of clastics, and that they probably were not transported at all after primary deposition. In addition to the ordinary collection of surface samples with an Ekman dredge, complete cores were collected with a Jenkin sampler. As a result, information was obtained regarding progressive changes in sediment chemistry, as well as in the total thickness of the sediments. These latter were 30.3 ft in Grassy Lake, 4.7 m in Nebish, 5.9 m in Little John, and 42.5 ft in Allequash.

The sediments were approximately the same from surface to bottom, indicating to Twenhofel that there had been little diagenesis. The water content of these sediments was generally 90% or even higher, so that the 30 ft of sediment in Grassy Lake would represent a postglacial accumulation of only 1.5 ft dry sediment in the deepest part of the lake and scarcely more than 2 in. in the shoreward portions. Except at the bottom, the organic content of the sediments averaged 65–70%, with lignin comprising about a third. Using the Dumas method, Twenhofel obtained a nitrogen content of about 4% in the sediments of Grassy Lake.

Twenhofel did not observe any horizontal structuring or stratification in the sediments of any of the lakes he studied. Without any definite evidence except observation of the activity of oligochaetes in sediment samples in bottles, he attributed this to the activities of organisms, rather than to redeposition by currents. Unfortunately no chronologies were established for these cores, although pollen diagrams were later constructed for a number of localities in northeastern Wisconsin, including Crystal Lake and a nearby bog (Wilson and Cross, 1941). As a minor matter of interest, in his last paper (Twenhofel *et al.*, 1944) Twenhofel was willing to abandon the term "sludge" in favor of gyttja.

In a general paper Twenhofel and McKelvey (1941) described the conditions of sedimentation in freshwater lakes, the types of sediments occurring in them and the types of rocks that would result, the presumed diagenesis after deposition, and the general absence of stratification in freshwater sediments of their experience.

In another general paper Conger (1942) discussed the various factors that may control the production of diatomaceous deposits in lakes. The northeastern lake district is especially good for investigating this problem, since the sediments of some of them consist of 60–75% SiO_2, chiefly diatoms. Five such lakes, including Crystal Lake, were listed.

Studies on the pollen chronology of Wisconsin lake and bog sediments during this period were conducted chiefly by Wilson and Potzger and their associates. In chronological sequence these are Hansen (1937), Wilson (1937), Wilson and Galloway (1937), Wilson (1938), Wilson and Webster (1942*a*, 1942*b*), Potzger (1942), Potzger and Keller (1942), Potzger and Richards (1942), Potzger (1943), Wilson and Webster (1944). No summary will be attempted of these papers here, since in most instances they have little to do with interpretation of earlier limnological conditions. The paper of Wilson (1932) might also be mentioned, as representing a description of the site which has since resulted in the firm correlation by radiocarbon dating between the Two Creeks interstadial of North America and the Alleröd of Europe.

Bacteria

Birge and Juday early recognized the importance of bacteria in controlling the chemical conditions in lakes, in providing sources of particulate food for the water-filtering zooplankters, and in bringing about the regeneration of plant nutrients. Their early studies on numbers of bacteria in Mendota were continued by Fred *et al.* (1924) and Snow and Fred (1926). Bere (1933) was one of the early persons to make direct microscopic counts of bacteria in waters. Studies of this type were extended by Stark and McCoy (1938) to the lakes of northeastern Wisconsin.

Studies on the numbers of bacteria and their function and physiology in bottom sediments were carried out by Allgeier *et al.* (1932, 1934), Williams and McCoy (1934, 1935), Henrici and McCoy (1938), Carpenter (1939), and Erikson (1941). Miscellaneous studies were carried out by Hardman and Henrici (1939), Stadler and ZoBell (1939), Zobell (1940), and Zobell and Stadler (1940*a*, 1940*b*).

During this interval quite a number of persons received Masters' and Ph.D. degrees in bacteriology on problems directly related to limnology. These include Yvette Hardman, Mary A. Jansky, Dorothy E. Kinkel, and William H. Stark. No attempt was made to compile a total list of such persons and the titles of their dissertations.

Benthos of Lake Mendota

In 1913 at Juday's suggestion, Muttkowski (1918) commenced a study of the littoral benthos of Mendota. The littoral zone was defined as extending down to 7 m. The offshore limit of the rooted aquatics at a depth of 5 or possibly 6 m was generally co-extensive with a shell zone, marking the lower limit of wave action. Quantitative samples were collected from areas of 1 m^2 marked out on the bottom, and qualitative samples were collected from greater depths by means of a special rake and net. Quantitative results were expressed only as numbers of individuals per m^2. Nearly all the immature insects were raised to adults, enabling positive identification of species. Hence, distribution, habitat preferences, and phenology were described in terms of species rather than genera or families, as is commonly done in such studies.

Muttkowski presented a very nice discussion of the phenology of the population growth and emergence of the various insect species over the course of the year. Much of the population dynamics is associated, probably causally, with the seasonal growth of attached algae and higher aquatics. Early in the spring *Cladophora* dominates the plant community until a water temperature of about 16°C is reached, after which the rooted aquatics gain the ascendancy. *Sialis* makes a massive shoreward migration in early May from fairly deep water. The new brood of larvae remain in the littoral zone until fall overturn, when they migrate downward into the aphytal zone. Muttkowski related the distribution of the various species to depth and habitat and recorded the apparent optimum conditions.

In the summer of 1916 Muttkowski extended his studies to deeper water by means of the Ekman dredge, but he did not publish his results. Juday continued these studies until August 1918, concentrating mainly on the deep-water zone within the 20-m contour. His paper (Juday, 1921*b*) included Muttkowski's results as well.

Besides merely counting numbers of individuals in each dredge haul, Juday obtained mean wet weights and dry weights of individual organisms from which standing crops could be calculated. These amounted to 697 kg/ha wet weight (excluding *Pisidium*) in the deep-water zone and 360 kg/ha in the intermediate zone (7–20 m). *Chaoborus* dominated in the deep-water zone, and *Chironomus* in the intermediate zone.

Only 6 kinds of macro-invertebrates were found in the deep-water zone: *Limnodrilus, Tubifex, Pisidium idahoense, Corethra punctipennis, Chironomus tentans,* and *Protenthes choreus. Limnodrilus* outnumbered *Tubifex* about 4 to 1, and together they constituted about 58 kg/ha live weight. *Pisidium,* as already noted, becomes dormant during the long anaerobic period of summer stratification. *Chaoborus* (*Corethra*) exhibits a very marked population maximum in winter and almost disappears from the deep-water sediments during emergence time in summer.

Higher aquatic plants

The first study of rooted aquatic plants in Wisconsin was that of Denniston (1922) on Lake Mendota, based on samples collected in 1912. Distribution along the shore seemed to be governed by bottom type and degree of exposure, whereas depth distribution was believed governed by light. There was little or no zonation parallel to shore.

Two of the most significant early studies on the quantitative aspects of aquatic plants were those by Rickett (1921, 1924) on Mendota and Green Lake. Rickett carefully collected all the plants within a reference frame measuring ½ m on a side that was lowered to the bottom. In water no more than 2–3 m deep, simple diving sufficed, whereas at greater depths a diving hood was employed. In Green Lake Rickett was impressed by the marked decline in visual light intensity near the lower limit of the rooted plants: at 7 m the impression was that of fairly bright sunlight, whereas almost darkness prevailed at 10 m. Had SCUBA been available at that time, Rickett undoubtedly would have become a devotee.

Of the 21 species present in Mendota, *Vallisneria spiralis* made up about one-third of the total biomass and the 6 species of *Potamogeton* together about one-half. *Cladophora* was important early in the year. In the 26% of the lake bottom occupied by aquatic plants, the biomass averaged 2,019 kg dry weight per ha. Rickett arbitrarily set up three depth zones: 0–1 m, 1–3 m, and greater than 3 m. The percentages of the total biomass in each of these zones were 30%, 45%, and 25%, respectively. The lower limit of plant growth (*Ceratophyllum*) was generally about 5 m, although in a few localities it extended to 6.5 m.

In Green Lake there were 27 species of macrophytes, of which *Chara* constituted about 50% of the total dry weight. The plants extended to greater depths than in Mendota—to 8 m along the average shore and up to 10 m on shallow slopes, although only to 4 or 5 m on very steep slopes. The mean crop in the plant zone (29% of the lake bottom) was 1,780 kg dry weight per ha, distributed among the depth zones by the percentages 9%, 40%, and 50%, respectively. The greater percentage in deep water is related to greater light transmission in Green Lake than in Mendota and perhaps also to the colder temperatures at these depths. *Chara, Myriophyllum,* and *Ceratophyllum* were abundant in the deep water. Birge provided the information that 1% levels of surface light in summer occur at about 8 m in Green Lake and 4 m in Mendota.

On the basis of a brief visit to the northeastern lake district, Fassett (1930) described a number of growth forms of aquatics, which in addition to the now-recognized categories of submerged, floating leaf, and emergent plants included a fourth category of short, stiff rosettes. He further noted that in Crystal Lake (a very transparent lake) the bottom was bare from 6 to 15 m, but that at greater depths there was an almost continuous carpet of *Fontinalis flaccida* and *Drepanocladus fluitans.*

Studies comparable to those of Rickett were carried out in the northeastern lake district by Wilson (1935, 1937, 1941) and by Potzger and Van Engel (1942). The results of these studies, summarized in Table 1.1, show that the soft-water and medium-hard-water lakes of northern Wisconsin have much smaller annual crops than do the hard-water lakes of the southeast. The number of species is comparable, or even greater, as is the percentage of the total lake bottom

TABLE 1.1

Summer crops of higher aquatic plants in Wisconsin lakes[a]

Lake	Area (ha)	Max. depth (m)	Color (ppm)	Bound CO_2 (ppm)	No. spp.[b] aquatics	% Lake bottom occupied	Max. depth occupied (m)	Total dry wt. (kg)	Av. crop on occupied bottom (g dry wt./m²)
Southeastern lakes									
Mendota	3940	26	14	34	20	26	5(6.5)	2,100,000	202
Green	2972	72	5	38	27	29	8(10)	1,527,900	178
Northeastern lakes									
Little John	67.2	6	18	12	13	31	3	112	0.52
Muskellunge	372	21	8	10	31	52	7	883	0.45
Silver	87.2	19	4	17	15	23	6	17	0.08
Trout	1583	35	6	20	36+	27	6.5	321	0.07
Sweeney	63.5	6	52	18	27	30	2.3	332	1.73
Weber	15.6	13	1	2	8	34	5	894	16.8

[a] Sources of data are the following: *Mendota:* Rickett (1921); *Green:* Rickett (1924); *Little John, Muskellunge, Silver:* Wilson (1935); *Trout:* Wilson (1941); *Sweeney:* Wilson (1937); *Weber:* Potzger and Van Engel (1942). Data on color and bound CO_2 for certain lakes were obtained from Juday and Birge (1933) and Juday, Birge, and Meloche (1935). Morphometric data for Weber Lake (Juday and Birge, 1941) were used for recalculating some of the data in Potzger and Van Engel (1942).

[b] Includes only *Chara* and *Nitella* among the algae. In Mendota *Cladophora* is important in spring and early summer. In Green Lake Rickett lists 6 species of algae other than charophytes that may be abundant at times. *Nostoc* is locally abundant in Lake Muskellunge.

colonized by the plants. The differences in production are a reflection of the basic differences in the lakes. The lakes of the two districts differ not only in mineral content of the water but also in bottom type. The northeastern lakes, located in a sandy outwash plain, frequently have sandy bottoms with organic sediments occurring some distance offshore. The sandy sediments are frequently colonized by rosette-type plants (*Isoetes,* etc.), which have relatively little dry-weight biomass.

Another major difference between the two lake regions is that in the northern lakes studied, Zone I (0–1 m water depth) had from 60 to 76% of the total biomass, except in Weber Lake (studied by Potzger and Van Engel) where 29% of the total biomass occurred in Zone I.

Wilson intended to use the same methods as Rickett, but he soon decided it was too uncomfortable working at the cold temperatures near the lower limits of the colonization zones. He adopted, therefore, a lightened Petersen dredge, which sampled 625 cm² of bottom. Potzger and Van Engel, on the other hand, maintained that the dredge Wilson used was not so efficient as the standard Petersen dredge, which was considerably heavier and sampled 729 cm² of bottom. Comparison showed that the standard Petersen dredge collected four to five times as much material as did the modified dredge. However, even increasing Wilson's crop estimates by this factor still leaves them about two orders of magnitude smaller than those of the southern lakes.

Wilson, Potzger, and Van Engel were concerned with communities or associations of plants in relation to soil type and depth zone. Wilson was also interested in the successional relationships of the communities, from primitive lakes with inorganic soils to advanced lakes with organic soils. In a review paper (Wilson, 1939*a*) he attempted to relate the distribution and abundance of aquatic plants to various environmental factors. Light was of particular importance in depth distribution. Some species had their lower limit at 70% of surface light, whereas others extended down to 2% (Wilson, 1941).

Fishes

While he was on the zoology faculty of the University of Wisconsin, A. S. Pearse was interested in the faunistic and general ecological relationships in lakes. His studies of this period included a number on the food of fishes (Pearse, 1915, 1918, 1921*a*, 1921*b*, 1924), the habits of the black crappie (Pearse, 1919) and the yellow perch (Pearse and Achtenberg, 1920), a study on the chemical composition of fishes (Pearse, 1925), and a study on the general ecology of lake fishes

(Pearse, 1934). In later years Couey (1935) and Nelson and Hasler (1942) also investigated the food habits of fishes.

Juday and Birge had been receiving research support from the U.S. Bureau of Fisheries for some years, but with the establishment of the Trout Lake Laboratory they also began receiving support from the Wisconsin Conservation Department for conducting studies on growth rates, food habits, general life histories, and management of fishes, including creel censuses. Studies of this kind led to the Ph.D. degree for most of the later students associated with Juday, as described in a subsequent section.

This phase of aquatic investigations in Wisconsin may be considered to have started with the study by Wright (1929) on the growth of the rockbass. From this time on there was a steady succession of papers concerned with various aspects of fishery biology and management. Included are a number of papers by Ralph Hile on the cisco (Hile, 1936*a*, 1936*b*, 1938) and the rockbass (Hile, 1941, 1942, 1943), and one by Hile and Deason (1934) on the whitefish. Hile was also interested in the bathymetric distribution of fishes in response to chemical and thermal stratification (Hile and Juday, 1941).

Many of the other papers in fishery biology of this period were written either by Juday, based on data gathered by W.P.A. workers, or by students who received their Ph.D.'s under Juday's supervision (Schneberger, Schloemer, Bennett, Spoor, and Frey). Of these papers only those based in whole or in part on Ph.D. theses are listed in the bibliography here. The remainder may be obtained from the bibliography by Juday and Hasler (1946).

One important paper of this period described the method developed by Schnabel (1938) for estimating the size of a fish population in a lake by marking and recapturing. Another interesting note by Woodbury (1942) ascribed the mortality of some fish to oxygen embolism, developed when the fish moved out of a zone of supersaturation.

Photosynthesis, productivity, community structure

In the last decade of his productive life as a limnologist, Juday began turning his attention to the measurement of the rates of energy fixation and the subsequent utilization of this energy within the trophic structure of the ecosystem. These are central problems in all phases of ecology today. Juday and his associates carried out some pioneer studies in this area and made some fundamental contributions. The big background of information on the light climate, chemical regime, and standing crops of plankton and other habit communities in Wisconsin lakes was put to good use in these studies.

In the early studies cultured algae or concentrations of naturally occurring algae and higher aquatics were placed in light and dark bottles and suspended at various depths in lakes of contrasting color and transparency. Photosynthesis was computed then from changes in the oxygen content of the bottles.

Schomer (1934) selected three lakes with contrasting amounts of dissolved color for experiments beginning in 1932. He placed active tips of *Elodea, Ceratophyllum,* and *Chara* in light and dark bottles, and he did the same with cultures of *Coccomyxa* and *Chlorella.* In general, maximum photosynthesis was at the surface on dull days and below the surface on bright days. The compensation level was inversely related to dissolved color, being at 10–15 m in the clearest lake and at 1–2 m in the darkest lake. Schomer considered the larger plants to be of less value in such experiments than algae, and in all subsequent experiments algae were used.

The results of 1933 were described by Schomer and Juday (1935) and in a somewhat abbreviated version by Juday and Schomer (1935). Using the same two species of cultured algae as in the previous year, the authors calculated the oxygen production per million cells per 3-hour exposure period in the middle of the day and related this to the quantity of radiation received at each level as measured by the pyrlimnometer. Again they found that the depth of the maximum rate of photosynthesis and the depth of the compensation level were directly related to the transparency of the water. Maximum percentage utilization of the light occurred at depths where energies of 1.2 to 8 cal/cm^2 per 3 hours were being received, which was roughly 1% of the surface radiation in the two less transparent lakes and 4% in the most transparent. In all lakes the percentage of available energy utilized by the algae increased with depth, reaching values as high as 11% in the darkest lake. Efficiencies at the surface were unrealistically low, because they were calculated from total radiation, not merely from that in the visible portion of the spectrum.

In Crystal Lake the compensation level for the cultured algae occurred at 16.5 m, although Juday (1934*a*) had earlier pointed out that three species of mosses thrive in this lake at depths between 18 and 20 m. Maximum production of oxygen in all the series was 0.528 mg/million cells per 3 hours.

Curtis and Juday (1937) wished to determine if naturally occurring algae had responses similar to those observed for the cultured algae. Accordingly they ran experiments comparable to those of the previous years with cultured *Chlorella* as a control against *Chlorella* from sponges and colonies of *Ophridium,* two species of *Anabaena* that were available in relatively pure composition, *Gloeothece, Gloeotrichia,* a filamentous diatom growing in shallow water, and *Spirogyra.* Dissolved color in the lakes used ranged from 0 to 364 ppm on the Pt-Co scale. The filamentous diatom was the only species that always had its maximum photosynthesis at the surface. The others had their maxima below the surface during intense illumination.

Curtis and Juday conducted one of the very early bioassays of productivity in this study. They suspended cultured *Chlorella* cells in filtered Trout Lake water having a total CO_2 content (free, half-bound, and bound) of 38.2 mg/L and in water from Crystal Lake having a CO_2 content of only 5.5 mg/L. Both series of bottles, including dark bottles, were then suspended in parallel in Trout Lake. The algae in the Trout Lake water produced 2½ times as much oxygen as those in the Crystal Lake water, which represented a utilization of 40% of the total CO_2 in the Crystal Lake water but only 15% of that in the Trout Lake water. Available CO_2 may have been a limiting factor, although the authors admitted that other unmeasured factors might have been equally important.

The following year (1936), W. M. Manning, a plant physiologist in the Department of Botany at Wisconsin, joined the summer program at Trout Lake with a much more sophisticated approach toward productivity (Manning, Juday, and Wolf, 1938*b*). *Chlorella* was used as in previous years, plus *Cladophora,* two species of *Anabaena,* and active tips of *Potamogeton, Vallisneria,* and *Sagittaria.* Records of surface radiation were made with a recording solarimeter, and of subsurface radiation with the pyrlimnometer. In all instances the photosynthesis accomplished was plotted as a function of light intensity (ergs/cm²/sec) rather than as a function of depth in the water. The curves are similar in shape, naturally, with maximum photosynthesis generally at a value less than surface illumination. *Cladophora,* which was collected from surface waters, was the only plant studied that exhibited maximum photosynthesis at the surface on bright days. In the three higher aquatic plants there was no definite evidence of light adaptation over the natural depth ranges of the species. Manning also determined quantum efficiencies of the various algae used, finding considerably lower values than had been reported previously, although the values for *Chlorella* obtained in the field agreed with those he had previously determined in the laboratory (Manning *et al.,* 1938; Manning, Juday, and Wolf, 1938*a*). Later he also demonstrated that non-chlorophyll pigments can play a part in photosynthesis (Dutton and Manning, 1941).

The first study of the amount and distribution of chlorophyll in lakes was carried out by Kozminski (1938) in 1937. On the basis of vertical series from 17 lakes in the Trout Lake region, Kozminski attempted to establish five types of chlorophyll distribution, which subsequently were shown by Riley and by Manning and Juday (1941) to have no typological significance. The chlorophyll content of the various samples ranged as high as 386 mg/m³. Kozminski predicted that the mean chlorophyll content of the trophogenic zone would turn out to be a good index of lake productivity, since in the lakes studied this showed a general correlation with trophic level inferred from other parameters.

Most surprising was the high concentration of chlorophyll in the hypolimnion of some lakes at depths far below the 1% level. Some of this was inactive chlorophyll, probably resulting from the sinking of dead phytoplankton out of the zone of production, but some was definitely active. Kozminski suggested a relationship between depth of lake and the development of such a concentration zone, as influenced by the availability of regenerated plant nutrients at depths where there was still at least a trace of light. This explanation anticipated a similar one proposed later by Gessner (1949).

Manning and Juday (1941) attempted to obtain an estimate of primary productivity in an entire lake from the depth distribution of chlorophyll, the production of 7 mg of oxygen per hour per 1 mg of chlorophyll at optimum light, the variation in this value with light intensities above and below the optimum, and the measured light intensities in the lake at meter depth intervals and at hour time intervals throughout the day. The productivity varied from 14 to 44 kg glucose/ha per day for the seven lakes studied. The authors showed that in two lakes the deep-water concentration of chlorophyll found by Kozminski resulted from settling of phytoplankton. In Scaffold Lake, on the other hand, samples of water from 10 m, which were anaerobic and far below the 1% level of light (light intensity $< 10^{-5}$ at 6 m), actively photosynthesized at normal light intensities on being aerated. The organism involved was later reported to be *Pelogloea bacillifera* (Dutton and Juday, 1944).

Perhaps the most important study in this series was the last one (Juday, Blair, and Wilda, 1943), in which the daily productivity for an entire lake was determined from continuous records of dissolved oxygen at several depths, as measured by dropping mercury electrodes (Manning, 1940). Temperature was also recorded continuously at these depths, the dropping mercury electrodes were periodically checked by the Winkler method, light at all depths was measured with the pyrlimnometer, chlorophyll content of the water was determined, and light and dark bottle series of the lake water itself were run to determine rates of respiration and to obtain an independent check on productivity calculated from the *in situ* oxygen changes. Productivities measured by the dropping mercury electrodes ranged from 13 to 34 kg O_2/ha per day. Respiration ranged from 48 to 71% of the oxygen produced on clear and partly cloudy days to over 200% on one cloudy day.

Rates of production measured by the light and dark bottle series varied with length of exposure, in general the shorter exposure times having higher rates. Juday related this largely to increase in bacteria, based on observations by Fred *et al.* (1924) that bacteria in Mendota water samples in bottles increased 10 to 20 fold when suspended at depth for 8 days, and on similar observations by Stark *et al.* (1938). Accordingly, he recommended that the exposure time for light-dark bottle series based on oxygen changes should not exceed 48 hours. Rates of turnover were calculated from the glucose production per day in relation to the standing crop of centrifuged plankton.

Juday recommended that *in situ* oxygen values, measured continuously or at close intervals and supplemented by measurements of dark-bottle respiration, would suffice to give a general picture of the metabolic changes occurring in a lake. This paper anticipated by some years more recent attempts to determine productivity from diel *in situ* changes in oxygen, carbon dioxide, and *p*H.

In two other papers of this period Juday attempted to set up an energy budget for Mendota (Juday, 1940) and to investigate the relationships between various components of the standing crop of organic matter (Juday, 1942).

The mean annual solar radiation reaching Mendota over a 28-year period amounted to 118,872 cal/cm². The physical energy budget consisted of melting ice (3,500 cal), annual heat budget of water (24,200 cal), annual heat budget of bottom sediments (2,000 cal), evaporation (29,300 cal), losses at surface (28,500 cal), and conduction, convection, and radiation (31,000 cal). The basis for estimating the last term was the least satisfactory.

Assuming that the phytoplankton has a turnover rate of one week, assuming that it uses in its own metabolism one-third of the energy fixed, and making allowances for respiration rates of the various other components of the biota, Juday calculated that the biological energy utilization amounted to about 1% of the total energy actually entering the lake. Related to light energy alone, this percentage would be about twice as great.

In the later paper Juday (1942) compared the summer standing crops in two soft-water lakes in northeastern Wisconsin (Weber, Nebish) with those in two hard-water lakes in southeastern Wisconsin (Mendota, Green). The various components of the biota considered were phytoplankton, higher aquatics, zooplankton, benthos, and fish (only in the two northern lakes), plus dissolved organic matter. The community structure in each of the northern lakes was illustrated graphically by a triangle, similar to the well-known Eltonian pyramids. Although plant production was much smaller in the two northern lakes, the ratio of the standing crop of plants to that of benthos and zooplankton suggested a greater efficiency among the animals of soft-water lakes. In all four lakes the standing crop of dissolved organic matter was greater than that of the living organisms. In the hard-water lakes the dissolved organic matter was relatively more abundant than in the soft-water lakes.

Weber Lake is particularly interesting because

attempts were made during the 1930's to increase its production by the addition of fertilizers. The quantities of fertilizers added are summarized by Potzger and Van Engel (1942), and a brief progress report on the experiment is contained in Juday and Schloemer (1938*a*). In each of the four years 1932-35 commercial mineral fertilizers were added to the lake, without any appreciable effects on the standing crop. In 1936 the addition of 3,000 lbs of soybean meal resulted in a marked increase in everything except fish and dissolved organic matter, and this effect persisted through 1938 without the addition of any more fertilizer. In 1939 the addition of 2,000 lbs of cottonseed meal further increased the standing crops of phytoplankton, bottom flora, zooplankton, and benthos. This increase persisted through 1940, but the standing crops declined in 1941. The data on the standing crops in these various years were summarized by Juday (1942). The conclusion was reached that productivity in the northern lakes is limited not so much by mineral nitrogen and phosphorus as by the general CO_2 supply and perhaps by various essential organic factors in the water.

Biotal and general biological studies

The major studies of this general category are listed in Table 1.2.

Miscellaneous studies

Cole (1921) investigated the respiratory mechanisms of the benthic animals living under anaerobic conditions in summer. As a result of experimental oxidation of guaiacum or benzidene under conditions of darkness, he claimed that decomposing plant tissues give off small amounts of atomic oxygen, even under anaerobiosis, and that this oxygen is utilized by the animals.

Pennak (1939, 1940) studied the psammolittoral organisms in sandy beaches of 15 Wisconsin lakes. He obtained extensive data on the quantitative horizontal and vertical distribution of the various organisms and on the chemistry of the psammolittoral environment. Pennak paid special attention to the tardigrades, copepods, and rotifers, with only incidental observations on the other groups of organisms.

One other subject that should be mentioned is the cultural eutrophication of the Madison lakes. For a long time the sewage effluent from Madison was emptied into Lake Monona. This created such large blooms of algae that attempts were made to control the algae with copper sulfate beginning in 1918. Domogalla (1935) reviewed the outcome of this treatment. In 1926 a new sewage treatment plant was constructed, the effluent of which was discharged into the Yahara River just above Lake Waubesa. Subsequently, Lakes Waubesa and Kegonsa were treated with $CuSO_4$ along with Monona.

The long-term effects of accumulation of copper in lake sediments have caused concern to biologists. However, Mackenthun and Cooley (1952) showed by experiments that the present levels of copper in the sediments of Monona are not appreciably toxic even to *Pisidium* over a 60-day period.

The already productive lakes at Madison were being so seriously affected by the sewage plant

TABLE 1.2
Summary of major taxonomic, distributional, and life history studies on aquatic organisms during the Birge-Juday era

Group	References (listed in bibliography)
Algae	Smith (1916*a*, 1916*b*, 1918*b*, 1920, 1924); Prescott (1944)
Higher aquatic plants	Fassett (1940, 1957)
Protozoa	Juday (1919); Noland (1925*a*, 1925*b*); Noland and Finley (1931)
Sponges	Jewell (1935, 1939); Neidhoefer (1938, 1940)
Rotifers	Harring and Myers (1922, 1924, 1926, 1927); Myers (1930); Edmondson (1940)
Hirudinea	Bere (1931*b*)
Mollusks	Baker (1914, 1924, 1928); Morrison (1929, 1932*a*, 1932*b*)
Cladocera	Birge (1879, 1892, 1893, 1910*d*, 1918*a*); Woltereck (1932)
Copepoda	Marsh (1893, 1907, 1910, 1918, 1933); Lehmann (1903); Birge and Juday (1908); Juday (1914*b*, 1923*a*, 1925); Juday and Muttkowski (1915); Wright (1927, 1928)
Malacostraca	Marsh (1892*a*); Holmes (1909); Jackson (1912); Juday and Birge (1927); Creaser (1932)
Water mites	Marshall (1903, 1914, 1921, 1929, 1930, 1931–40)
Insects	Vorhies (1905, 1909); Dickinson (1936)
Fishes	Wagner (1908, 1911); Pearse (1934); Greene (1927, 1935)

effluent that the Governor of Wisconsin in 1941 appointed a committee to investigate this problem. The committee found in an intensive two-year study that the effluent from the Madison sewage disposal plant contributed 75% of the total inorganic nitrogen and 88% of the total inorganic phosphorus entering Lake Waubesa. The blooms of algae in this lake were so heavy as a result that the water from Waubesa was the major source of both nitrogen and phosphorus in Lake Kegonsa. These investigations were reviewed briefly by Sawyer (1947). Because of these findings, the sewage effluent from Madison is now diverted by conduit around the lakes, entering the Yahara River below Lake Kegonsa.

Training of graduate students

In spite of the large overall limnological output during this era, relatively few students received graduate degrees under the direct supervision of Birge or Juday. During the late 1800's Birge supervised a number of Bachelor's theses (those of Olson, Harder, and Merrill having already been mentioned) and a few Masters' theses, including those of Julius Nelson and Ruth Marshall. Birge never had any Ph.D. students.

Juday, on the other hand, supervised the Ph.D. research of 13 students between 1928 and 1940. Most of the later dissertations were concerned with growth rates of fishes in Wisconsin, which reflected the need Juday felt for studying fishes as one of the biotal components of the total ecosystem and also reflected the continuing financial support from the U.S. Bureau of Fisheries and the Wisconsin Conservation Department for "practical" investigations. All these dissertations are listed below; those which have been published may be found in the references.

1928 Edward Joseph Wimmer. A study of two limestone quarry pools. (Wimmer, 1929)

1928 Stillman Wright. Studies in aquatic biology.
I. A chemical and plankton study of Lake Wingra. 35 p.
II. A revision of the South American species of *Diaptomus.* (Wright, 1927)
III. A contribution to the knowledge of the genus *Pseudodiaptomus.* (Wright, 1928)

1929 Abraham H. Wiebe. Productivity of fish ponds. I. The plankton. 95 + ii p. Published as Investigations of plankton production in fish ponds. (Wiebe, 1930)

1930 Willis L. Tressler. Limnological studies of Lake Wingra. 35 + v p. Published by W. L. Tressler and B. P. Domogalla as Limnological studies of Lake Wingra. (Tressler and Domogalla, 1931)

1931 J. P. E. Morrison. A report on the Mollusca of the northeastern Wisconsin lake district. (Morrison, 1932*a*)
Studies on the life history of *Acella haldemani* ("Desh." Binney). (Morrison, 1932*b*)

1932 Ruby Bere. The bacterial content of some Wisconsin lakes. 27 p.
The effect of freezing on the number of bacteria in ice and water from Lake Mendota. 18 p.
Copepods parasitic on fish of the Trout Lake region, with descriptions of two new species. (Bere, 1931*a*)

1933 Edward Schneberger. The growth of yellow perch (*Perca flavescens* Mitchill) from Nebish, Silver and Weber lakes in Vilas County, Wisconsin. 73 p. (Schneberger, 1935)

1936 William A. Spoor. The age and growth of the sucker, *Catostomus commersonii* (Lacépède), in Muskellunge Lake, Vilas County, Wisconsin. 90 p. (Spoor, 1938)

1937 Arthur D. Hasler. The physiology of digestion of plankton Crustacea.
I. Some digestive enzymes of *Daphnia.* (Hasler, 1935)
II. Further studies on the digestive enzymes of A. *Daphnia* and *Polyphemus,* B. *Diaptomus* and *Calanus.* 13 + ii p. (Hasler, 1937)

1938 Robert W. Pennak. The ecology of the psammolittoral organisms of some Wisconsin lakes, with special reference to the Tardigrada, Copepoda, and Rotatoria. 180 p. (Pennak, 1940)

1939 George W. Bennett. Limnological investigations in Wisconsin and Nebraska.
I. The limnology of some gravel pits near Louisville, Nebraska. 63 p.
II. The growth of the large mouthed black bass, *Huro salmoides* (Lacépède), in the waters of Wisconsin. (Bennett, 1937)
III. Growth of the small-mouthed black bass, *Micropterus dolomieu* (Lacépède), in Wisconsin waters. (Bennett, 1938)

1939 Clarence L. Schloemer. The age and rate of growth of the bluegill, *Helioperca macrochira* (Rafinesque). 113 p.

1940 David G. Frey. Growth and ecology of the carp, *Cyprinus carpio* Linnaeus, in four lakes of the Madison Region, Wisconsin. 248 p. Partially published. (Frey, 1942)

During the period from 1920 through 1940, when the program of cooperative lake studies was being rapidly expanded, students in other departments were likewise obtaining Ph.D.'s based on limnological research. The departments most directly involved were Bacteriology, Botany, Chemistry, and Geology. No attempt was made to search out all the pertinent dissertations of this period, although the total number probably equals or even exceeds that of Juday's students.

References

This list of references includes all the limnological publications of Birge and Juday, some of which are not listed in the previous bibliography by Juday and Hasler (1946). Although the present list is extensive and covers quite completely the limnological output of this period, it does not include many papers on the growth, ecology, and management of fishes or on the parasites of fishes and other organisms. Some of the lesser taxonomic papers have also been omitted. These may be found in the listing by Juday and Hasler. Finally, a few of the papers listed in the present bibliography have not been summarized in the text because of space limitations.

ALLGEIER, R. J., B. C. HAFFORD, AND CHANCEY JUDAY. 1941. Oxidation-reduction potentials and *p*H of lake waters and of lake sediments. Trans. Wisconsin Acad. Sci. Arts Lett., **33**: 115–133.

ALLGEIER, R. J., W. H. PETERSON, AND CHANCEY JUDAY. 1934. Availability of carbon in certain aquatic materials under aerobic conditions of fermentation. Intern. Rev. Hydrobiol., **30**: 371–378.

ALLGEIER, R. J., W. H. PETERSON, CHANCEY JUDAY, AND E. A. BIRGE. 1932. The anaerobic fermentation of lake deposits. Intern. Rev. Hydrobiol., **26**: 444–461.

AMERICAN PUBLIC HEALTH ASSOCIATION. 1955. Standard methods for the examination of water, sewage, and industrial wastes. 10th ed. APHA, AWWA, FSIWA. Amer. Pub. Health Assoc., New York. 522 + xix p.

BAKER, F. C. 1914. The molluscan fauna of Tomahawk Lake, Wisconsin, with special reference to its ecology. Trans. Wisconsin Acad. Sci. Arts Lett., **17**: 200–246, pl. 11–17.

———. 1924. The fauna of the Lake Winnebago region; a quantitative and qualitative survey with special reference to the Mollusca. Trans. Wisconsin Acad. Sci. Arts Lett., **21**: 109–146.

———. 1928. The fresh water Mollusca of Wisconsin. I. Gastropoda. II. Pelecypoda. Wisconsin Geol. Nat. Hist. Surv., Bull. 70, 507 + xx p., pl.–28; 495 + vi p., pl. 29–105.

BENNETT, G. W. 1937. The growth of the large-mouthed black bass in the waters of Wisconsin. Copeia, 1937(2): 104–118.

———. 1938. Growth of the small-mouthed black bass in Wisconsin waters. Copeia, 1938(4): 157–170.

BERE, RUBY. 1931*a*. Copepods parasitic on fish of the Trout Lake region, with descriptions of two new species. Trans. Wisconsin Acad. Sci. Arts Lett., **26**: 427–436.

———. 1931*b*. Leeches from the lakes of northeastern Wisconsin. Trans. Wisconsin Acad. Sci. Arts Lett., **26**: 437–440.

———. 1933. Numbers of bacteria in inland lakes of Wisconsin as shown by the direct microscopic method. Intern. Rev. Hydrobiol., **29**: 248–263.

———. 1935. Further notes on the occurrence of parasitic copepods on fish of the Trout Lake region, with a description of the male of *Argulus biramosus*. Trans. Wisconsin Acad. Sci. Arts Lett., **29**: 83–88.

BIRGE, E. A. 1878. On Crustacea cladocera collected at Cambridge, Mass., 1876, and at Madison, Wis., 1877. Ph.D. Thesis, Harvard Univ.

———. 1879. Notes on Cladocera. Trans. Wisconsin Acad. Sci. Arts Lett., **4**: 77–112.

———. 1881. Notes on Crustacea in Chicago water supply, with remarks on the formation of the carapace. Chicago Med. J. and Exam., **44**(6): 584–590, 1 pl.

———. 1892. Notes and list of Crustacea Cladocera from Madison, Wisconsin. Trans. Wisconsin Acad. Sci. Arts Lett., **8**: 379–398.

———. 1893. Notes on Cladocera, III. Trans. Wisconsin Acad. Sci. Arts Lett., **9**: 275–317.

———. 1894. A report on a collection of Cladocera, mostly from Lake St. Clair, Michigan, p. 45–47, 1 table. *In* J. E. Reighard. A biological examination of Lake St. Clair. Bull. Michigan Fish Comm., **4**: 1–60.

———. 1895*a*. Cladocera [of Turkey Lake, Indiana]. Proc. Indiana Acad. Sci., **5**: 244–246.

———. 1895*b*. On the vertical distribution of the pelagic crustacea of Lake Mendota, Wis., during July, 1894. Biol. Centr., **15**: 353–355.

———. 1897. The vertical distribution of the limnetic crustacea of Lake Mendota. Biol. Centr., **17**: 371–374.

———. 1898. Plankton studies on Lake Mendota. II. The Crustacea of the plankton from July, 1894, to December, 1896. Trans. Wisconsin Acad. Sci. Arts Lett., **11**: 274–448.

———. 1901*a*. The cone net. J. Appl. Microscop., **4**: 1405–1407.

———. 1901*b*. Report of the Limnological Commission. Trans. Amer. Microscop. Soc., **22**: 193–196.

———. 1904*a*. Report on the Cladocera, p. 149–153, pl. 25. *In* H. B. Ward, A biological reconnaissance of some elevated lakes in the Sierras and Rockies. Studies Zool. Lab., Univ. Nebraska, No. 60, p. 127–154.

———. 1904*b*. The thermocline and its biological significance. Trans. Amer. Microscop. Soc., **25**: 5–33, pl. 1, 2.

———. 1906. Gases dissolved in the waters of Wisconsin lakes. Trans. Amer. Fish. Soc., 1906: 143–163.

———. 1907*a*. The oxygen dissolved in the waters of Wisconsin lakes. Rept. Wisconsin Comm. Fish., 1907: 119–140. (This is the same article as Birge, 1906.)

———. 1907*b*. The respiration of an inland lake. Trans. Amer. Fish. Soc., 1907: 223–241.

———. 1908*a*. The respiration of an inland lake. Popular Sci. Month., **72**: 337–351. (This is the same article as Birge, 1907*b*.)

———. 1908*b*. [Phyllopoda] p. 203–205. *In* S. E. Meek, The zoology of lakes Amatitlan and Atitlan, Guatemala, with special reference to ichthyology.

Field Columbian Museum, Zool. Ser., **7**(6): 159–206.

———. 1910*a*. The apparent sinking of ice in lakes. Science, N.S., **32**: 81–82.

———. 1910*b*. An unregarded factor in lake temperatures. Trans. Wisconsin Acad. Sci. Arts Lett., **16**(2): 989–1004, pl. 64, 65.

———. 1910*c*. On the evidence for temperature seiches. Trans. Wisconsin Acad Sci. Arts Lett., **16**(2): 1005–1016, pl. 66.

———. 1910*d*. Notes on Cladocera, IV. Trans. Wisconsin Acad. Sci. Arts Lett., **16**(2): 1017–1066.

———. 1910*e*. Gases dissolved in the waters of Wisconsin lakes. Bull. U.S. Bur. Fish., **28**: 1273–1294.

———. 1913. Absorption of the sun's energy by lakes. Science, **38**: 702–704.

———. 1915. The heat budgets of American and European lakes. Trans. Wisconsin Acad. Sci. Arts Lett., **18**(1): 166–213. (Fig. 1 separate.)

———. 1916. The work of the wind in warming a lake. Trans. Wisconsin Acad. Sci. Arts Lett., **18**(2): 341–391.

———. 1918*a*. The water fleas (Cladocera), p. 676–740. *In* H. B. Ward and G. C. Whipple, Fresh-water biology. John Wiley & Sons, New York.

———. 1918*b*. Coefficients of absorption in various lake waters. Notes on the foregoing paper [Pietenpol, q.v.]. Trans. Wisconsin Acad. Sci. Arts Lett., **19**(1): 580–593.

———. 1922. A second report on limnological apparatus. Trans. Wisconsin Acad. Sci. Arts Lett., **20**: 533–552, pl. 39, 40.

———. 1923. The plankton of the lakes. Trans. Amer. Fish. Soc., **52**: 118–130.

———. 1929. Fish and their food. Trans. Amer. Fish. Soc., **59**: 188–194.

———. 1936. Biology of Lake Mendota. Tech. Club of Madison, 1936: 11–12.

———. 1938. Note [re C. D. Marsh], p. 541–543. *In* Mrs. Florence W. Marsh, Professor C. Dwight Marsh and his investigations of lakes. Trans. Wisconsin Acad. Sci. Arts Lett., **31**: 535–543.

Birge, E. A., and Chancey Juday. 1908. A summer resting stage in the development of *Cyclops bicuspidatus* Claus. Trans. Wisconsin Acad. Sci. Arts Lett., **16**: 1–9.

———. 1911. The inland lakes of Wisconsin. The dissolved gases of the water and their biological significance. Wisconsin Geol. Nat. Hist. Surv., Bull. 22, 259 + x p.

———. 1914. A limnological study of the Finger Lakes of New York. Bull. U.S. Bur. Fish., **32**: 525–609, pl. 111–116.

———. 1920. A limnological reconnaissance of West Okoboji. Univ. Iowa Studies Nat. Hist., **9**(1): 1–56, 1 pl.

———. 1921. Further limnological observations on the Finger Lakes of New York. Bull. U.S. Bur. Fish., **37**: 210–252.

———. 1922. The inland lakes of Wisconsin. The plankton. I. Its quantity and chemical composition. Wisconsin Geol. Nat. Hist. Surv., Bull. 64, 222 + ix p.

———. 1926. The organic content of lake water. Proc. Natl. Acad. Sci., **12**: 515–519.

———. 1927*a*. Organic content of lake water. Bull. U.S. Bur. Fish., **42**: 185–205.

———. 1927*b*. The organic content of the water of small lakes. Proc. Amer. Phil. Soc., **66**: 357–372.

———. 1929*a*. Penetration of solar radiation into lakes, as measured by the thermopile. Bull. Natl. Research Council, **68**: 61–76.

———. 1929*b*. Transmission of solar radiation by the waters of inland lakes. Trans. Wisconsin Acad. Sci. Arts Lett., **24**: 509–580.

———. 1930. A second report on solar radiation and inland lakes. Trans. Wisconsin Acad. Sci. Arts Lett., **25**: 285–335.

———. 1931. A third report on solar radiation and inland lakes. Trans. Wisconsin Acad. Sci. Arts Lett., **26**: 383–425.

———. 1932. Solar radiation and inland lakes. Fourth Report. Observations of 1931. Trans. Wisconsin Acad. Sci. Arts Lett., **27**: 523–562.

———. 1934. Particulate and dissolved organic matter in inland lakes. Ecol. Monogr., **4**: 440–474.

Birge, E. A., Chancey Juday, and H. W. March. 1928. The temperature of the bottom deposits of Lake Mendota; a chapter in the heat exchanges of the lake. Trans. Wisconsin Acad. Sci. Arts Lett., **23**: 187–231, tables 17–20 in appendix.

Birge, E. A., O. A. Olson, and H. P. Harder. 1895. Plankton studies on Lake Mendota. I. The vertical distribution of the pelagic crustacea during July, 1894. Trans. Wisconsin Acad. Sci. Arts Lett., **10**: 421–484, pl. 7–10.

Birge, E. A., and W. H. Rich. 1927. Observations on Karluk Lake, Alaska. Ecology, **8**: 384.

Black, C. S. 1929. Chemical analysis of lake deposits. Trans. Wisconsin Acad. Sci. Arts Lett., **24**: 127–133.

Black, R. F. 1959. Friends of the Pleistocene [report of meeting]. Science, **130**: 172–173.

———. 1960. The "Driftless area" of Wisconsin was glaciated. Program 1960 Ann. Meetings, Geol. Soc. Amer.: 59.

Bordner, J. S. 1939. Inventory of northern Wisconsin lakes. Wisconsin State Planning Board, Div. Land Econ. Inventory, Bull. 5, 64 p.

Brooks, J. L. 1959. Cladocera, p. 587–656. *In* W. T. Edmondson (ed.), Fresh-water biology. 2nd ed. John Wiley & Sons, New York.

Brooks, J. L., G. L. Clarke, A. D. Hasler, and L. E. Noland. 1951. Edward Asahel Birge (1851–1950). Arch. Hydrobiol., **45**: 235–243.

Broughton, W. A. 1941. The geology, ground water and lake basin seal of the region south of the Muskellunge Moraine, Vilas County, Wisconsin. Trans. Wisconsin Acad. Sci. Arts Lett., **33**: 5–20.

Carpenter, P. L. 1939. Bacterial counts in the muds of Crystal Lake—an oligotrophic lake of northern Wisconsin. J. Sediment. Petrol., **9**: 3–7.

Chamberlin, T. C. 1894. Glacial phenomena of North America, p. 724–774. *In* James Geikie, The great ice age and its relation to the antiquity of man. 3rd ed. D. Appleton, New York.

———. 1895. Classification of American glacial deposits. J. Geol., **3**: 270–277.

Chase, W. J., and L. E. Noland. 1927. The history and hydrography of Lake Ripley (Jefferson County, Wisconsin). Trans. Wisconsin Acad. Sci. Arts Lett., **23**: 179–186.

Cole, A. E. 1921. Oxygen supply of certain animals living in water containing no dissolved oxygen. J. Exptl. Zool., **33**: 293–320.

Coleman, F. H. 1941. Supplementary climatic notes for Wisconsin, p. 1199–1200. *In* Climate and man, Yearbook of Agriculture. U.S. Dept. Agr.

Conger, P. S. 1939. The contribution of diatoms to the sediments of Crystal Lake, Vilas County, Wisconsin. Amer. J. Sci., **237**: 324–340, pl. 1–2.

———. 1942. Accumulation of diatomaceous deposits. J. Sediment. Petrol., **12**(2): 55–66.

Couey, F. M. 1935. Fish food studies of a number of northeastern Wisconsin lakes. Trans. Wisconsin Acad. Sci. Arts Lett., **29**: 131–172.

Creaser, E. P. 1932. The decapod crustaceans of Wisconsin. Trans. Wisconsin Acad. Sci. Arts Lett., **27**: 321–338.

Curtis, J. T., and Chancey Juday. 1937. Photosynthesis of algae in Wisconsin lakes. III. Observations of 1935. Intern. Rev. Hydrobiol., **35**: 122–133.

Davis, F. J. 1941. Surface loss of solar and sky radiation by inland lakes. Trans. Wisconsin Acad. Sci. Arts Lett., **33**: 83–93.

Denniston, R. H. 1922. A survey of the larger aquatic plants of Lake Mendota. Trans. Wisconsin Acad. Sci. Arts Lett., **20**: 495–500. (Table 2 folded in.)

Dickinson, W. E. 1936. The mosquitoes of Wisconsin. Bull. Pub. Museum, Milwaukee, **8**(3).

Domogalla, B. P. 1935. Eleven years of chemical treatment of the Madison lakes: Its effect on fish and fish foods. Trans. Amer. Fish. Soc., **65**: 115–121.

Domogalla, B. P., and E. B. Fred. 1926. Ammonia and nitrate studies of lakes near Madison, Wisconsin. J. Amer. Soc. Agron., **18**: 897–910.

Domogalla, B. P., E. B. Fred, and W. H. Peterson. 1926. Seasonal variations in the ammonia and nitrate content of lake waters. J. Amer. Water Works Assoc., **15**: 369–385.

Domogalla, B. P., Chancey Juday, and W. H. Peterson. 1925. The forms of nitrogen found in certain lake waters. J. Biol. Chem., **63**: 269–285.

Dutton, H. J., and Chancey Juday. 1944. Chromatic adaptation in relation to color and depth distribution of freshwater phytoplankton and large aquatic plants. Ecology, **25**: 273–282.

Dutton, H. J., and W. M. Manning. 1941. Evidence for carotenoid-sensitized photosynthesis in the diatom *Nitzschia closterium*. Amer. J. Botan., **28**: 516–526.

Edmondson, W. T. 1940. The sessile Rotatoria of Wisconsin. Trans. Amer. Microscop. Soc., **59**: 433–459.

Enteman, Minnie M. 1900. Variations in the crest of *Daphnia hyalina*. Amer. Nat., **34**: 879–890.

Erikson, Dagny. 1941. Studies on some lake-mud strains of *Micromonospora*. J. Bacteriol., **41**: 277–300.

Fassett, N. C. 1930. The plants of some northeastern Wisconsin lakes. Trans. Wisconsin Acad. Sci. Arts Lett., **25**: 157–168.

———. 1940. A manual of aquatic plants. McGraw-Hill, New York. 382 + vii p.

———. 1957. A manual of aquatic plants. 2nd ed. Revision appendix by Eugene C. Ogden. Univ. Wisconsin Press, Madison. 405 + ix p.

Fenneman, N. M. 1902. On the lakes of southeastern Wisconsin. Wisconsin Geol. Nat. Hist. Surv., Bull. 8, 178 + xv p.

———. 1938. Physiography of eastern United States. McGraw-Hill, New York. 714 + xiii p.

Field, J. B., C. A. Elvehjem, and Chancey Juday. 1943. A study of the blood constituents of carp and trout. J. Biol. Chem., **148**: 261–269.

Field, J. B., Lynn L. Gee, C. A. Elvehjem, and Chancey Juday. 1944. The blood picture in furunculosis induced by *Bacterium salmonicida* in fish. Arch. Biochem., **3**: 277–284.

FitzGerald, Desmond. 1895. The temperature of lakes. Trans. Amer. Soc. Civil Engineers, **34**(756): 67–109; Discussion, 110–114.

Flanigon, T. H. 1942. Limnological observations on three lakes in eastern Vilas County, Wisconsin. Trans. Wisconsin Acad. Sci. Arts Lett., **34**: 167–175.

Forbes, S. A. 1890. Preliminary report upon the invertebrate animals inhabiting Lakes Geneva and Mendota, Wisconsin, with an account of the fish epidemic in Lake Mendota in 1884. Bull. U.S. Fish. Comm., **8**: 473–487, pl. 72–74.

Francé, R. H. 1894. Zur Biologie des Planktons. Biol. Centr., **14**(2): 33–38.

Fred, E. B., F. C. Wilson, and Audrey Davenport. 1924. The distribution and significance of bacteria in Lake Mendota. Ecology, **5**: 322–339.

Freeman, Stephen, V. W. Meloche, and Chancey Juday. 1933. The determination of the hydrogen ion concentration of inland lake waters. Intern. Rev. Hydrobiol., **29**: 346–359.

Frey, D. G. 1942. Studies on Wisconsin carp. I. Influence of age, size, and sex on time of annulus formation by the 1936 year class. Copeia, 1942(4): 214–223.

———. 1955. The Winona Lake Biological Station. Amer. Inst. Biol. Sci. Bull., **5**(3): 20–22.

Fries, Carl, Jr. 1938. Geology and ground water of the Trout Lake region, Vilas County, Wisconsin. Trans. Wisconsin Acad. Sci. Arts Lett., **31**: 305–322.

Frye, J. C., and H. B. Willman. 1960. Classification of the Wisconsinan Stage in the Lake Michigan glacial lobe. Illinois State Geol. Surv., Circ. 285, 16 p.

Gessner, Fritz. 1949. Der Chlorophyllgehalt im See und seine photosynthetische Valenz als geophysikalisches Problem. Schweiz. Z. Hydrol., **11**: 378–410.

Gifford, Elizabeth M., and G. W. Peckham. 1882. Temperature of Pine, Beaver and Okauchee lakes, Waukesha County, Wisconsin, extending from May to December, 1879; also particulars of depths of

Pine Lake. Trans. Wisconsin Acad. Sci. Arts Lett., **5**: 273–275.

GREENBANK, JOHN. 1945. Limnological conditions in ice-covered lakes, especially as related to winter-kill of fish. Ecol. Monogr., **15**: 343–392.

GREENE, C. W. 1927. An ichthyological survey of Wisconsin. Papers Michigan Acad. Sci. Arts Lett., **7**: 299–310.

———. 1935. The distribution of Wisconsin fishes. Wisconsin Conserv. Comm. 235 p.

HANSEN, H. P. 1937. Pollen analyses of two Wisconsin bogs of different age. Ecology, **18**: 136–148.

HARDMAN, YVETTE. 1941. The surface tension of Wisconsin lake waters. Trans. Wisconsin Acad. Sci. Arts Lett., **33**: 395–404.

HARDMAN, YVETTE, AND A. T. HENRICI. 1939. Studies of freshwater bacteria. V. The distribution of *Siderocapsa treubii* in some lakes and streams. J. Bacteriol., **37**: 97–104, pl. 1.

HARRING, H. K., AND F. J. MYERS. 1922. The rotifers of Wisconsin. Trans. Wisconsin Acad. Sci. Arts Lett., **20**: 553–662, pl. 41–61.

HARRING, H. K., AND F. J. MYERS. 1924. The rotifer fauna of Wisconsin. II. A revision of the notommatid rotifers, exclusive of the Dicranophorinae. Trans. Wisconsin Acad. Sci. Arts Lett., **21**: 415–549, pl. 16–43.

HARRING, H. K., AND F. J. MYERS. 1926. The rotifer fauna of Wisconsin. III. A revision of the genera Lecane and Monostyla. Trans. Wisconsin Acad. Sci. Arts Lett., **22**: 315–423, pl. 8–47.

HARRING, H. K., AND F. J. MYERS. 1927. The rotifer fauna of Wisconsin. IV. The Dicranophorinae. Trans. Wisconsin Acad. Sci. Arts Lett., **23**: 667–808, pl. 23–49.

HASLER, A. D. 1935. The physiology of digestion of plankton crustacea. I. Some digestive enzymes of Daphnia. Biol. Bull., **68**: 207–214.

———. 1937. The physiology of digestion in plankton crustacea. II. Further studies on the digestive enzymes of (A) Daphnia and Polyphemus; (B) Diaptomus and Calanus. Biol. Bull., **72**: 290–298.

HATHAWAY, E. S. 1927. The relation of temperature to the quantity of food consumed by fishes. Ecology, **8**: 428–434.

———. 1928. Quantitative study of the changes produced by acclimatization in the tolerance of high temperatures by fishes and amphibians. Bull. U.S. Bur. Fish., **43**(2): 169–192.

HENRICI, A. T., AND ELIZABETH MCCOY. 1938. The distribution of heterotrophic bacteria in the bottom deposits of some lakes. Trans. Wisconsin Acad. Sci. Arts Lett., **31**: 323–361. (Fig. 1 separate.)

HILE, RALPH. 1936*a*. Age and growth of the cisco, *Leucichthys artedi* (Le Sueur), in the lakes of northeastern Highlands, Wisconsin. Bull. U.S. Bur. Fish., **48**: 211–317.

———. 1936*b*. Summary of investigations on the morphometry of the cisco, *Leucichthys artedi* (Le Sueur), in the lakes of the northeastern Highlands, Wisconsin. Papers Michigan Acad. Sci. Arts Lett., **21**: 619–634, pl. 62.

———. 1938. Morphometry of the cisco, *Leucichthys artedi* (Le Sueur), in the lakes of the Northeastern Highlands, Wisconsin. Intern. Rev. Hydrobiol., **36**: 57–130.

———. 1941. Age and growth of the rock bass. *Ambloplites rupestris* (Rafinesque), in Nebish Lake, Wisconsin. Trans. Wisconsin Acad. Sci. Arts Lett., **33**: 189–337.

———. 1942. Growth of the rock bass, *Ambloplites rupestris* (Rafinesque), in five lakes of northeastern Wisconsin. Trans. Amer. Fish. Soc., **71**: 131–143.

———. 1943. Mathematical relationship between the length and the age of the rock bass, *Ambloplites rupestris* (Rafinesque). Papers Michigan Acad. Sci. Arts Lett., **28**: 331–341.

HILE, RALPH, AND H. J. DEASON. 1934. Growth of the whitefish, *Coregonus clupeaformis* (Mitchill), in Trout Lake, northeastern Highlands, Wisconsin. Trans. Amer. Fish. Soc., **64**: 231–237.

HILE, RALPH, AND CHANCEY JUDAY. 1941. Bathymetric distribution of fish in lakes of the northeastern Highlands, Wisconsin. Trans. Wisconsin Acad. Sci. Arts Lett., **33**: 147–187.

HOLMES, S. J. 1909. Description of a new subterranean amphipod from Wisconsin. Trans. Wisconsin Acad. Sci. Arts Lett., **16**(1): 77–80.

HOY, P. R. 1872. Deep-water fauna of Lake Michigan. Trans. Wisconsin Acad. Sci. Arts Lett., **1**: 98–101.

HUTCHINSON, G. E. 1938. On the relation between the oxygen deficit and the productivity and typology of lakes. Intern. Rev. Hydrobiol., **36**: 336–355.

JACKSON, H. H. T. 1912. A contribution to the natural history of the amphipod, *Hyalella knickerbockeri* Bate. Bull. Wisconsin Nat. Hist. Soc., **10**: 49–60.

JAMES, H. R., AND E. A. BIRGE. 1938. A laboratory study of the absorption of light by lake waters. Trans. Wisconsin Acad. Sci. Arts Lett., **31**: 1–154.

JEWELL, MINNA E. 1935. An ecological study of the fresh-water sponges of northeastern Wisconsin. Ecol. Monogr., **5**: 461–504.

———. 1939. An ecological study of the fresh-water sponges of Wisconsin. II. The influence of calcium. Ecology, **20**: 11–28.

JUDAY, CHANCEY. 1896. Hydrographic map of Turkey Lake. Proc. Indiana Acad. Sci. (Frontispiece), 1895.

———. 1897. The plankton of Turkey Lake. Proc. Indiana Acad. Sci., 1896: 287–296.

———. 1902. The plankton of Lake Maxinkuckee, Indiana. Trans. Amer. Microscop. Soc., **24**: 61–62.

———. 1903. The plankton of Winona Lake. Proc. Indiana Acad. Sci., 1902: 120–133.

———. 1904*a*. The diurnal movement of plankton crustacea. Trans. Wisconsin Acad. Sci. Arts Lett., **14**: 534–568.

———. 1904*b*. Fishes of Boulder County, Colorado. Univ. Colorado Studies, **2**(2): 113–114.

———. 1905. List of fishes collected in Boulder County, Colorado, with description of a new species of Leuciscus. Bull. U.S. Bur. Fish., **24**: 223–227.

———. 1906*a*. The food of the trout of the Kern River Region. Bull. U.S. Bur. Fish., **25**: 43–49.

———. 1906*b*. Ostracoda of the San Diego region. I.

Halocypridae. Univ. California Publl. Zool., **3**(2): 13–38, pl. 3–7.
———. 1907*a*. Cladocera of the San Diego region. Univ. California Publ. Zool., **3**(10): 157–158.
———. 1907*b*. Notes on Lake Tahoe, its trout and trout-fishing. Bull. U.S. Bur. Fish., **26**: 133–146.
———. 1907*c*. Ostracoda of the San Diego region. II. Littoral forms. Univ. California Publ. Zool., **3**(9): 135–156, pl. 18–20.
———. 1907*d*. Studies on some lakes in the Rocky and Sierra Nevada Mountains. Trans. Wisconsin Acad. Sci. Arts Lett., **15**(2): 781–793, pl. 48–50.
———. 1907*e*. A study of Twin Lakes, Colorado, with especial consideration of the food of the trouts. Bull. U.S. Bur. Fish., **26**: 147–178, pl. 3.
———. 1908*a*. Resumé of the recent work on lakes by the Wisconsin Geological and Natural History Survey. Intern. Rev. Hydrobiol., **1**: 240–242.
———. 1908*b*. Some aquatic invertebrates that live under anaerobic conditions. Trans. Wisconsin Acad. Sci. Arts Lett., **16**: 10–16.
———. 1908*c*. [Copepoda] p. 205. *In* S. E. Meek, The zoology of lakes Amatitlan and Atitlan, Guatemala, with special reference to ichthyology. Field Columbian Museum, Zool. Ser., **7**(6): 159–206.
———. 1910. Some European biological stations. Trans. Wisconsin Acad. Sci. Arts Lett., **16**(2): 1257–1277.
———. 1913. Air in the depths of the ocean. Science, N.S., **38**: 546–547.
———. 1914*a*. The inland lakes of Wisconsin. II. The hydrography and morphometry of the lakes. Wisconsin Geol. Nat. Hist. Surv., Bull. 27, 137 + xv p.
———. 1914*b*. A new species of Diaptomus. Trans. Wisconsin Acad. Sci. Arts Lett., **17**(2): 803–805.
———. 1915. Limnological studies on some lakes in Central America. Trans. Wisconsin Acad. Sci. Arts Lett., **18**: 214–250.
———. 1916*a*. Horizontal rainbows on Lake Mendota. Monthly Weather Rev., **44**: 65–67.
———. 1916*b*. Limnological apparatus. Trans. Wisconsin Acad. Sci. Arts Lett., **18**(2): 566–592, pl. 34–38.
———. 1919. A freshwater anaërobic ciliate. Biol. Bull., **36**: 92–95.
———. 1920*a*. The Cladocera of the Canadian Arctic Expedition, 1913–18. Rept. Canadian Arctic Exped., 1913–18, Vol. 7: Crustacea, Part H: Cladocera, p. 3*E*–8*E*.
———. 1920*b*. Horizontal rainbows. Science, N.S., **51**: 188.
———. 1920*c*. The plankton, p. 105–110. *In* B. W. Evermann and H. W. Clark, Lake Maxinkuckee, physical and biological survey. Vol. II. Indiana Dept. Conserv., Indianapolis.
———. 1921*a*. Observations on the larvae of *Corethra punctipennis* Say. Biol. Bull., **40**: 271–286.
———. 1921*b*. Quantitative studies of the bottom fauna in the deeper waters of Lake Mendota. Trans. Wisconsin Acad. Sci, Arts Lett., **20**: 461–493.
———. 1922. Limnological observations on Lake George. A biological survey of Lake George, New York. New York Conserv. Comm., 1921: 37–51.
———. 1923*a*. An interesting copepod from the Finger Lakes, New York. Science, **58**: 205.
———. 1923*b*. The water-fleas, p. 16–17. *In* A scientific survey of Turners Lake, Isle-au-Haut, Maine, New York State Museum. Published privately.
———. 1924*a*. The productivity of Green Lake, Wisconsin. Verh. intern. Verein. Limnol., **2**: 357–360.
———. 1924*b*. Summary of quantitative investigations on Green Lake, Wisconsin. Intern. Rev. Hydrobiol., **12**: 1–12.
———. 1925. *Senecella calanoides*, a recently described fresh-water copepod. Proc. U.S. Natl. Museum, **66**(Article 4): 1–6.
———. 1926*a*. Freshwater Cladocera from southern Canada. Canadian Field-Nat., **40**: 99–100.
———. 1926*b*. Sand flotation on lakes. Science, N.S., **64**: 138.
———. 1926*c*. A third report on limnological apparatus. Trans. Wisconsin Acad. Sci. Arts Lett., **22**: 299–314.
———. 1927. Freshwater Cladocera from the east shore of Hudson and James bays. Canadian Field-Nat., **41**: 130–131.
———. 1929. Limnological methods. Arch. Hydrobiol., **20**: 517–524.
———. 1934*a*. The depth distribution of some aquatic plants. Ecology, **15**: 325.
———. 1934*b*. Growth of game fish. Field and Stream, December, 1934: 7–72.
———. 1935. Chemical composition of large aquatic plants. Science, **81**: 273.
———. 1937. Trout Lake. The Limnological Laboratory. The Biologist, **18**: 177–182.
———. 1938*a*. Fish records for Lake Wingra. Trans. Wisconsin Acad. Sci. Arts Lett., **31**: 533–534.
———. 1938*b*. Wisconsin lakes and fish investigations. Progr. Fish-Cult., **39**: 18–21.
———. 1940. The annual energy budget of an inland lake. Ecology, **21**: 438–450.
———. 1942. The summer standing crop of plants and animals in four Wisconsin lakes. Trans. Wisconsin Acad. Sci. Arts Lett., **34**: 103–135.
———. 1943. The utilization of aquatic food resources. Science, **97**(2525): 456–458.
JUDAY, CHANCEY, AND G. W. BENNETT. 1935. The growth of game fish in Wisconsin waters. Mimeographed report, 13 p.
JUDAY, CHANCEY, AND E. A. BIRGE. 1927. *Pontoporeia* and *Mysis* in Wisconsin lakes. Ecology, **8**: 445–452.
———. 1930. The highland lake district of northeastern Wisconsin and the Trout Lake limnological laboratory. Trans. Wisconsin Acad. Sci. Arts Lett., **25**: 337–352.
———. 1931. A second report on the phosphorus content of Wisconsin lake waters. Trans. Wisconsin Acad. Sci. Arts Lett., **26**: 353–382.
———. 1932. Dissolved oxygen and oxygen consumed in the lake waters of northeastern Wisconsin. Trans. Wisconsin Acad. Sci. Arts Lett., **27**: 415–486.
———. 1933. The transparency, the color and the specific conductance of the lake waters of northeastern Wisconsin. Trans. Wisconsin Acad. Sci. Arts Lett., **28**: 205–259.
———. 1941. Hydrography and morphometry of some

northeastern Wisconsin lakes. Trans. Wisconsin Acad. Sci. Arts Lett., **33**: 21–72.

JUDAY, CHANCEY, E. A. BIRGE, G. I. KEMMERER, AND R. J. ROBINSON. 1927. Phosphorus content of lake waters of northeastern Wisconsin. Trans. Wisconsin Acad. Sci. Arts Lett., **23**: 233–248.

JUDAY, CHANCEY, E. A. BIRGE, AND V. W. MELOCHE. 1935. The carbon dioxide and hydrogen ion content of the lake waters of northeastern Wisconsin. Trans. Wisconsin Acad. Sci. Arts Lett., **29**: 1–82.

JUDAY, CHANCEY, E. A. BIRGE, AND V. W. MELOCHE. 1938. Mineral content of the lake waters of northeastern Wisconsin. Trans. Wisconsin Acad. Sci. Arts Lett., **31**: 223–276.

JUDAY, CHANCEY, E. A. BIRGE, AND V. W. MELOCHE. 1941. Chemical analyses of the bottom deposits of Wisconsin lakes. II. Second report. Trans. Wisconsin Acad. Sci. Arts Lett., **33**: 99–114.

JUDAY, CHANCEY, J. M. BLAIR, AND E. F. WILDA. 1943. The photosynthetic activities of the aquatic plants of Little John Lake, Vilas County, Wisconsin. Amer. Midland Nat., **30**: 426–446.

JUDAY, CHANCEY, E. B. FRED, AND F. C. WILSON. 1924. The hydrogen ion concentration of certain Wisconsin lake waters. Trans. Amer. Microscop. Soc., **43**: 177–190.

JUDAY, CHANCEY, AND A. D. HASLER. 1946. List of publications dealing with Wisconsin limnology 1871–1945. Trans. Wisconsin Acad. Sci. Arts Lett., **36**: 469–490.

JUDAY, CHANCEY, CLARENCE LIVINGSTON, AND HUBERT PEDRACINE. 1938. A census of the fish caught by anglers in Lake Waubesa in 1937. Mimeographed report, 7 p.

JUDAY, CHANCEY, AND V. W. MELOCHE. 1943. Physical and chemical evidence relating to the lake basin seal in certain areas of the Trout Lake region of Wisconsin. Trans. Wisconsin Acad. Sci. Arts Lett., **35**: 157–174.

JUDAY, CHANCEY, AND R. A. MUTTKOWSKI. 1915. Entomostraca of St. Paul Island, Alaska. Bull. Wisconsin Nat. Hist. Soc., **13**: 23–31.

JUDAY, CHANCEY, W. H. RICH, G. I. KEMMERER, AND ALBERT MANN. 1932. Limnological studies of Karluk Lake, Alaska, 1926–30. Bull. U.S. Bur. Fish., **12**: 407–436.

JUDAY, CHANCEY, AND C. L. SCHLOEMER. 1936. Growth of game fish in Wisconsin waters. Fourth report. Mimeographed report, 17 p.

JUDAY, CHANCEY, AND C. L. SCHLOEMER. 1938*a*. Effect of fertilizers on plankton production and on fish growth in a Wisconsin lake. Progr. Fish-Cult., **40**: 24–27.

JUDAY, CHANCEY, AND C. L. SCHLOEMER. 1938*b*. Growth of game fish in Wisconsin waters. Fifth report. Mimeographed report, 26 p.

JUDAY, CHANCEY, AND EDWARD SCHNEBERGER. 1930. Growth studies of game fish in Wisconsin waters. Mimeographed report, 7 p.

JUDAY, CHANCEY, AND EDWARD SCHNEBERGER. 1933. Growth studies of game fish in Wisconsin waters. Second report. Mimeographed report, 10 p.

JUDAY, CHANCEY, AND H. A. SCHOMER. 1935. The utilization of solar radiation by algae at different depths in lakes. Biol. Bull., **69**: 75–81.

JUDAY, CHANCEY, AND L. E. VIKE. 1938. A census of the fish caught by anglers in Lake Kegonsa. Trans. Wisconsin Acad. Sci. Arts Lett., **31**: 527–532.

JUDAY, CHANCEY, AND GEORGE WAGNER. 1908. Dissolved oxygen as a factor in the distribution of fishes. Trans. Wisconsin Acad. Sci. Arts Lett., **16**: 17–22.

KEMMERER, GEORGE, J. F. BOVARD, AND W. R. BOORMAN. 1923. Northwestern lakes of the United States: Biological and chemical studies with reference to possibilities in production of fish. Bull. U.S. Bur. Fish., **39**: 51–140.

KEMMERER, GEORGE, AND L. T. HALLETT. 1927*a*. An improved method of organic microcombustion. Ind. Engr. Chem., **19**: 173–176.

KEMMERER, GEORGE, AND L. T. HALLETT. 1927*b*. Improved micro-Kjeldahl ammonia distillation apparatus. Ind. Engr. Chem., **19**: 1295–1296.

KEMMERER, GEORGE, AND L. T. HALLETT. 1927*c*. Micro determination of carbonate carbon. Ind. Engr. Chem., **19**: 1352–1354.

KNUDSON, H. W., CHANCEY JUDAY, AND V. W. MELOCHE. 1940. Silicomolybdate method for silica. Ind. Engr. Chem., Anal. Ed., **12**: 270–273.

KNUDSON, H. W., V. W. MELOCHE, AND CHANCEY JUDAY. 1940. Colorimetric analysis of a two-component color system. Ind. Engr. Chem., Anal. Ed., **12**: 715–718.

KOZMINSKI, ZYGMUNT. 1938. Amount and distribution of the chlorophyll in some lakes of northeastern Wisconsin. Trans. Wisconsin Acad. Sci. Arts Lett., **31**: 411–438.

KUZNEZOW, S. I. 1959. Die Rolle der Mikroorganismen im Stoffkreislauf der Seen. (Translated from the Russian by Alfred Pochmann.) VEB Deutscher Verlag der Wissenschaften, Berlin. 301 + x p.

LAPHAM, I. A. 1876. Oconomowoc Lake, and other small lakes of Wisconsin, considered with reference to their capacity for fish-production. Trans. Wisconsin Acad. Sci. Arts Lett., **3**: 31–36.

LATHBURY, ALISON, AND R. A. BRYSON. 1958. Studies of the physiographic features of Lake Mendota. I. Sublacustrine gullies. Unpublished.

LEHMANN, HARRIET. 1903. Variations in form and size of *Cyclops brevispinosus* Herrick and *Cyclops americanus* Marsh. Trans. Wisconsin Acad. Sci. Arts Lett., **14**(1): 279–298, pl. 30–33.

LOHUIS, DELMONT, V. W. MELOCHE, AND CHANCEY JUDAY. 1938. Sodium and potassium content of Wisconsin lake waters and their residues. Trans. Wisconsin Acad. Sci. Arts Lett., **31**: 285–304.

MACKENTHUN, K. M., AND H. L. COOLEY. 1952. The biological effect of copper sulphate treatment on lake ecology. Trans. Wisconsin Acad. Sci. Arts Lett., **41**: 177–187.

MANNING, W. M. 1938. Photosynthesis. J. Phys. Chem., **42**(6): 815–854.

———. 1940. A method for obtaining continuous records of dissolved oxygen in lake waters. Ecology, **21**: 509–512.

———. 1943. Physical factors influencing the accuracy of the dropping mercury electrode in measurements of photochemical reaction rates. Trans. Wisconsin Acad. Sci. Arts Lett., **35**: 221–233.

Manning, W. M., and R. E. Juday. 1941. The chlorophyll content and productivity of some lakes in northeastern Wisconsin. Trans. Wisconsin Acad. Sci. Arts Lett., **33**: 363–394.

Manning, W. M., Chancey Juday, and Michael Wolf. 1938*a*. Photosynthesis in *Chlorella*. Quantum efficiency and rate measurements in sunlight. J. Amer. Chem. Soc., **60**: 274–278.

Manning, W. M., Chancey Juday, and Michael Wolf. 1938*b*. Photosynthesis of aquatic plants at different depths in Trout Lake, Wisconsin. Trans. Wisconsin Acad. Sci. Arts Lett., **31**: 377–410.

Manning, W. M., J. F. Stauffer, B. M. Duggar, and Farrington Daniels. 1938. Quantum efficiency of photosynthesis in *Chlorella*. J. Amer. Chem. Soc., **60**: 266–274.

Marsh, C. D. 1892*a*. On the deep-water Crustacea of Green Lake. Trans. Wisconsin Acad. Sci. Arts Lett., **8**: 211–213.

———. 1892*b*. Notes on the depth and temperature of Green Lake. Trans. Wisconsin Acad. Sci. Arts Lett., **8**: 214–218, pl. 6.

———. 1893. On the Cyclopidae and Calanidae of central Wisconsin. Trans. Wisconsin Acad. Sci. Arts Lett., **9**: 189–224, pl. 3–6.

———. 1894. On the vertical distribution of pelagic Crustacea in Green Lake, Wisconsin. Amer. Nat., **28**: 807–809.

———. 1898. On the limnetic Crustacea of Green Lake. Trans. Wisconsin Acad. Sci. Arts Lett., **11**: 179–224, pl. 5–14.

———. 1899. Hydrographic map of Green Lake. Wisconsin Geol. Nat. Hist. Surv., Map No. 7.

———. 1901. The plankton of fresh water lakes. Trans. Wisconsin Acad. Sci. Arts Lett., **13**: 163–187.

———. 1903. The plankton of Lake Winnebago and Green Lake. Wisconsin Geol. Nat. Hist. Surv., Bull. 12, 94 + vi p.

———. 1907. A revision of North American species of Diaptomus. Trans. Wisconsin Acad. Sci. Arts Lett., **15**: 381–516, pl. 15–28.

———. 1910. A revision of the North American species of Cyclops. Trans. Wisconsin Acad. Sci. Arts Lett., **16**(2): 1067–1134, pl. 72–81.

———. 1918. Copepoda, p. 741–789. *In* H. B. Ward and G. C. Whipple, Fresh-water biology. John Wiley & Sons, New York.

———. 1933. Synopsis of the calanoid crustaceans, exclusive of the Diaptomidae, found in fresh and brackish waters, chiefly of North America. Proc. U.S. Natl. Museum, **82**(18): 1–58, pl. 1–24.

Marsh, Mrs. Florence W. [Mrs. C. D.]. 1938. Professor C. Dwight Marsh and his investigation of lakes. Trans. Wisconsin Acad. Sci. Arts Lett., **31**: 535–543.

Marshall, Ruth. 1903. Ten species of Arrenuri belonging to the subgenus Megalurus Thon. Trans. Wisconsin Acad. Sci. Arts Lett., **14**: 145–172, pl. 14–18.

———. 1914. Some new American water mites. Trans. Wisconsin Acad. Sci. Arts Lett., **17**(2): 1300–1304, pl. 92–93.

———. 1921. New American water mites of the genus Neumania. Trans. Wisconsin Acad. Sci. Arts Lett., **20**: 205–213, pl. 2–4.

———. 1929. The morphology and developmental stages of a new species of Piona. Trans. Wisconsin Acad. Sci. Arts Lett., **24**: 401–404.

———. 1930. The water mites of the Jordan Lake region. Trans. Wisconsin Acad. Sci. Arts Lett., **25**: 245–253.

———. 1931–40. Preliminary list of the Hydracarina of Wisconsin. Part I. The red mites. Trans. Wisconsin Acad. Sci. Arts Lett., **26**: 311–319 (1931); Part II, Trans. Wisconsin Acad. Sci. Arts Lett., **27**: 339–357 (1932); Part III, Trans. Wisconsin Acad. Sci. Arts Lett., **28**: 37–61 (1933); Part IV, Trans. Wisconsin Acad. Sci. Arts Lett., **29**: 273–297 (1935); Part V, Trans. Wisconsin Acad. Sci. Arts Lett., **30**: 225–251 (1937); Part VI, Trans. Wisconsin Acad. Sci. Arts Lett., **32**: 135–165 (1940).

Marshall, W. S., and N. C. Gilbert. 1905. Notes on the food and parasites of some fresh-water fishes from the lakes at Madison, Wis. Rept. U.S. Bur. Fish., 1904: 513–522.

Martin, Lawrence. 1932. The physical geography of Wisconsin. 2nd ed. Wisconsin Geol. Nat. Hist. Surv., Bull. 36, 608 + xxiii p.

Meloche, V. W., G. Leader, L. Safranski, and Chancey Juday. 1938. The silica and diatom content of Lake Mendota water. Trans. Wisconsin Acad. Sci. Arts Lett., **31**: 363–376.

Meloche, V. W., and Katherine Pingrey. 1938. The estimation of magnesium in lake water residues. Trans. Wisconsin Acad. Sci. Arts Lett., **31**: 277–283.

Meloche, V. W., and T. Setterquist. 1933. The determination of calcium in lake water and lake water residues. Trans. Wisconsin Acad. Sci. Arts Lett., **28**: 291–296.

Merrill, Harriet Bell. 1893. The structure and affinities of *Bunops scutifrons* Birge. Trans. Wisconsin Acad. Sci. Arts Lett., **9**(2): 319–342, pl. 14–15.

Morrison, J. P. E. 1929. A preliminary list of the mollusca of Dane County, Wisconsin. Trans. Wisconsin Acad. Sci. Arts Lett., **24**: 405–425.

———. 1932*a*. A report on the mollusca of the northeastern Wisconsin lake district. Trans. Wisconsin Acad. Sci. Arts Lett., **27**: 359–396.

———. 1932*b*. Studies on the life history of *Acella haldemani* ("Desh." Binney). Trans. Wisconsin Acad. Sci. Arts Lett., **27**: 397–414.

Mortimer, C. H. 1956. An explorer of lakes, p. 165–211. *In* G. C. Sellery, E. A. Birge. Univ. Wisconsin Press, Madison.

Murray, R. C. 1956. Recent sediments of three Wisconsin lakes. Bull. Geol. Soc. Amer., **67**: 883–910.

Muttkowski, R. A. 1918. The fauna of Lake Mendota. A qualitative and quantitative survey with special reference to the insects. Trans. Wisconsin Acad. Sci. Arts Lett., **19**(1): 374–482.

Myers, F. J. 1930. The rotifer fauna of Wisconsin. V. The genera Euchlanis and Monommata. Trans.

Wisconsin Acad. Sci. Arts Lett., **25**: 353–413, pl. 10–26.

NEEDHAM, J. G., *et al.* 1941. A symposium on hydrobiology. Univ. Wisconsin Press, Madison. 405 + ix p.

NEESS, J. C., AND W. W. BUNGE, JR. 1956. An unpublished manuscript of E. A. Birge on the temperature of Lake Mendota. I. Trans. Wisconsin Acad. Sci. Arts Lett., **45**: 193–238.

NEESS, J. C., AND W. W. BUNGE, JR., 1957. An unpublished manuscript of E. A. Birge on the temperature of Lake Mendota. II. Trans. Wisconsin Acad. Sci. Arts Lett., **46**: 31–89.

NEIDHOEFER, J. R. 1938. *Carterius tenosperma* (Potts), a species of fresh-water sponge new to Wisconsin. Trans. Amer. Microscop. Soc., **57**: 82–84.

———. 1940. The fresh-water sponges of Wisconsin. Trans. Wisconsin Acad. Sci. Arts Lett., **32**: 177–197, pl. 1–27.

NELSON, M. M., AND A. D. HASLER. 1942. The growth, food, distribution and relative abundance of the fishes of Lake Geneva, Wisconsin, in 1941. Trans. Wisconsin Acad. Sci. Arts Lett., **34**: 137–148.

NOLAND, L. E. 1925*a*. Factors influencing the distribution of fresh water ciliates. Ecology, **6**: 437–452.

———. 1925*b*. A review of the genus Coleps with descriptions of two new species. Trans. Amer. Microscop. Soc., **44**: 3–13.

———. 1945. Chancey Juday. Limnol. Soc. Amer., Spec. Publ. 16: 1–3.

NOLAND, L. E., AND H. E. FINLEY. 1931. Studies on the taxonomy of the genus Vorticella. Trans. Amer. Microscop. Soc., **50**: 81–123.

NOLAND, W. E. 1950. The hydrography, fish, and turtle population of Lake Wingra. Trans. Wisconsin Acad. Sci. Arts Lett., **40**(2): 5–58.

O'DONNELL, D. J. 1943. The fish population in three small lakes in northern Wisconsin. Trans. Amer. Fish. Soc., **72**: 187–196.

OHLE, WALDEMAR. 1952. Die hypolimnische Kohlendioxyd-Akkumulation als produktionsbiologischer Indikator. Arch. Hydrobiol., **46**(2): 153–285.

OLIVE, E. W. 1905. Notes on the occurrence of *Oscillatoria prolifica* (Greville) Gomont in the ice of Pine Lake, Waukesha County, Wisconsin. Trans. Wisconsin Acad. Sci. Arts Lett., **15**(1): 124–134.

PEARSE, A. S. 1915. On the food of the small shore fishes in the waters near Madison, Wisconsin. Bull. Wisconsin Nat. Hist. Soc., **13**: 7–22.

———. 1918. The food of the shore fishes of certain Wisconsin lakes. Bull. U.S. Bur. Fish., **35**: 245–292.

———. 1919. Habits of the black crappie in inland lakes of Wisconsin. Rept. U.S. Comm. Fish. for 1918. Appendix 3. Bur. Fish. Documents 867: 1–16.

———. 1921*a*. Distribution and food of the fishes of Green Lake, Wis., in summer. Bull. U.S. Bur. Fish., **37**: 253–272.

———. 1921*b*. The distribution and food of the fishes of three Wisconsin lakes in summer. Univ. Wisconsin Studies in Sci., No. 3, 61 p.

———. 1924. Amount of food eaten by four species of fresh-water fishes. Ecology, **5**: 254–258.

———. 1925. The chemical composition of certain fresh-water fishes. Ecology, **6**: 7–16.

———. 1934. Ecology of lake fishes. Ecol. Monogr., **4**: 475–480.

PEARSE, A. S., AND HENRIETTA ACHTENBERG. 1920. Habits of yellow perch in Wisconsin lakes. Bull. U.S. Bur. Fish., **36**: 293–366, pl. 83.

PENNAK, R. W. 1939. The microscopic fauna of the sandy beaches, p. 94–106. *In* Problems in lake biology. Amer. Assoc. Advance. Sci., Publ. No. 10.

———. 1940. Ecology of the microscopic metazoa inhabiting the sandy beaches of some Wisconsin lakes. Ecol. Monogr., **10**: 537–615.

PETERSON, W. H., E. B. FRED, AND B. P. DOMOGALLA. 1925. The occurrence of amino acids and other organic nitrogen compounds in lake waters. J. Biol. Chem., **63**: 287–295.

PIETENPOL, W. B. 1918. Selective absorption in the visible spectrum of Wisconsin lake waters. Trans. Wisconsin Acad. Sci. Arts Lett., **19**(1): 562–579.

POTZGER, J. E. 1942. Pollen spectra from four bogs on the Gillen Nature Reserve, along the Michigan-Wisconsin state line. Amer. Midland Nat., **28**: 501–511.

———. 1943. Pollen study of five bogs in Price and Sawyer counties, Wisconsin. Butler Univ. Botan. Studies, **6**: 50–64.

POTZGER, J. E., AND C. O. KELLER. 1942. A pollen study of four bogs along the southern border of Vilas County, Wisconsin. Trans. Wisconsin Acad. Sci. Arts Lett., **34**: 149–166.

POTZGER, J. E., and RUTH R. RICHARDS. 1942. Forest succession in the Trout Lake, Vilas County, Wisconsin area: A pollen study. Butler Univ. Botan. Studies, **5**: 179–189.

POTZGER, J. E., AND W. A. VAN ENGEL. 1942. Study of the rooted aquatic vegetation of Weber Lake, Vilas County, Wisconsin. Trans. Wisconsin Acad. Sci. Arts Lett., **34**: 149–166.

PRESCOTT, G. W. 1944. New species and varieties of Wisconsin Algae. Farlowia, **1**(3): 349–385.

RAGOTZKIE, R. A. 1960. Compilation of freezing and thawing dates for lakes in North Central United States and Canada. Tech. Rept. No. 3, NR 387–022, Nonr 1202 (07).

RICKETT, H. W. 1920. A quantitative survey of the flora of Lake Mendota. Science, N.S., **52**(1357): 641–642.

———. 1921. A quantitative study of the larger aquatic plants of Lake Mendota. Trans. Wisconsin Acad. Sci. Arts Lett., **20**: 501–527. (Tables 4–6 folded in.)

———. 1924. A quantitative study of the larger aquatic plants of Green Lake, Wisconsin. Trans. Wisconsin Acad. Sci. Arts Lett., **21**: 381–414, 2 pl.

ROBINSON, R. J., AND GEORGE KEMMERER. 1930*a*. Determination of organic phosphorus in lake waters. Trans. Wisconsin Acad. Sci. Arts Lett., **25**: 117–121.

ROBINSON, R. J., AND GEORGE KEMMERER. 1930*b*. The determination of Kjeldahl nitrogen in natural waters. Trans. Wisconsin Acad. Sci. Arts Lett., **25**: 123–128.

ROBINSON, R. J., AND GEORGE KEMMERER. 1930*c*. Determination of silica in mineral waters. Trans.

Wisconsin Acad. Sci. Arts Lett., **25**: 129–134.

SAWYER, C. N. 1947. Fertilization of lakes by agricultural and urban drainage. J. New England Water Works Assoc., **61**: 109–127.

SCHLOEMER, C. L. 1936. The growth of the muskellunge, *Esox masquinongy immaculatus* (Garrard), in various lakes and drainage areas of northern Wisconsin. Copeia, 1936(4): 185–193.

———. 1938. A second report on the growth of the muskellunge, *Esox masquinongy immaculatus* (Garrard), in Wisconsin waters. Trans. Wisconsin Acad. Sci. Arts Lett., **31**: 507–512.

SCHLOEMER, C. L., AND RALPH LORCH. 1942. The rate of growth of the wall-eyed pike, *Stizostedion vitreum* (Mitchill), in Wisconsin's inland waters with special reference to the growth characteristics of the Trout Lake population. Copeia, 1942(4): 201–211.

SCHNABEL, ZOE EMILY. 1938. The estimation of the total fish population of a lake. Amer. Math. Monthly, **45**: 348–352.

SCHNEBERGER, EDWARD. 1935. Growth of the yellow perch (*Perca flavescens* Mitchill) in Nebish, Silver and Weber lakes, Vilas County, Wisconsin. Trans. Wisconsin Acad. Sci. Arts Lett., **29**: 103–130.

SCHOMER, H. A. 1934. Photosynthesis of water plants at various depths in the lakes of northeastern Wisconsin. Ecology, **15**: 217–218.

SCHOMER, H. A., AND CHANCEY JUDAY. 1935. Photosynthesis of algae at different depths in some lakes of northeastern Wisconsin. I. Observations in 1933. Trans. Wisconsin Acad. Sci. Arts Lett., **29**: 173–193.

SCHUETTE, H. A. 1918. A biochemical study of the plankton of Lake Mendota. Trans. Wisconsin Acad. Sci. Arts Lett., **19**: 594–613.

SCHUETTE, H. A., AND HUGO ALDER. 1927. Notes on the chemical composition of some of the larger aquatic plants of Lake Mendota. II. Vallisneria and Potamogeton. Trans. Wisconsin Acad. Sci. Arts Lett., **23**: 249–254.

SCHUETTE, H. A., AND HUGO ALDER. 1929*a*. Notes on the chemical composition of some of the larger aquatic plants of Lake Mendota. III. *Castalia odorata* and *Najas flexilis*. Trans. Wisconsin Acad. Sci. Arts Lett., **24**: 135–139.

SCHUETTE, H. A., AND HUGO ALDER. 1929*b*. A note on the chemical composition of Chara from Green Lake, Wisconsin. Trans. Wisconsin Acad. Sci. Arts Lett., **24**: 141–145.

SCHUETTE, H. A., AND ALICE E. HOFFMAN. 1922. Notes on the chemical composition of some of the larger aquatic plants of Lake Mendota. I. Cladophora and Myriophyllum. Trans. Wisconsin Acad. Sci. Arts Lett., **20**: 529–531.

SELLERY, G. C. 1956. E. A. Birge. Univ. Wisconsin Press, Madison. 221 + vii p.

SMITH, G. M. 1916*a*. A monograph of the algal genus *Scenedesmus* based upon pure culture studies. Trans. Wisconsin Acad. Sci. Arts Lett., **18**(2): 422–530, pl. 25–38.

———. 1916*b*. A preliminary list of algae found in Wisconsin lakes. Trans. Wisconsin Acad. Sci. Arts Lett., **18**(2): 531–565.

———. 1918*a*. The vertical distribution of Volvox in the plankton of Lake Monona. Amer. J. Botan., **5**: 178–185.

———. 1918*b*. A second list of algae found in Wisconsin lakes. Trans. Wisconsin Acad. Sci. Arts Lett., **19**(1): 614–654, pl. 10–15.

———. 1920. Phytoplankton of the inland lakes of Wisconsin. I. Myxophyceae Phaeophyceae, Heterokonteae, and Chlorophyceae exclusive of the Desmidiaceae. Wisconsin Geol. Nat. Hist. Surv., Bull. 57, 243 p., pl. 1–51.

———. 1924. Phytoplankton of the inland lakes of Wisconsin. II. Desmidiaceae. Wisconsin Geol. Nat. Hist. Surv., Bull. 57, 227 p., pl. 52–88.

SNOW, LETITIA M., AND E. B. FRED. 1926. Some characteristics of the bacteria of Lake Mendota. Trans. Wisconsin Acad. Sci. Arts Lett., **22**: 143–154, pl. 4.

SPOOR, W. A. 1938. Age and growth of the sucker, *Catostomus commersonii* (Lacépède), in Muskellunge Lake, Vilas County, Wisconsin. Trans. Wisconsin Acad. Sci. Arts Lett., **31**: 457–505, pl. 1.

STADLER, JANICE, AND C. E. ZOBELL. 1939. Evidence for the aerobic decomposition of lignin by lake bacteria. J. Bacteriol., **38**: 115.

STARK, W. H., AND ELIZABETH MCCOY. 1938. Distribution of bacteria in certain lakes of northern Wisconsin. Centr. Bakteriol. Parasitenk. C., **96**: 201–209.

STARK, W. H., JANICE STADLER, AND ELIZABETH MCCOY. 1938. Some factors affecting the bacterial population of fresh-water lakes. J. Bacteriol., **36**: 653–654.

STEINER, JOHN, AND V. W. MELOCHE. 1935. A study of ligneous substances in lacustrine materials. Trans. Wisconsin Acad. Sci. Arts Lett., **29**: 389–402.

TAYLOR, F. H. L. 1928. A complete systematical analysis of lake water residues by a new micro method. Unpublished thesis, Univ. Wisconsin, 57 p.

THIENEMANN, AUGUST. 1928. Der Sauerstoff im eutrophen und oligotrophen Seen. Die Binnengewässer, Bd. 4, 175 p.

THWAITES, F. T. 1929. Glacial geology of part of Vilas County, Wisconsin. Trans. Wisconsin Acad. Sci. Arts Lett., **24**: 109–125.

TITUS, LESLIE, AND V. W. MELOCHE. 1931. Note on the determination of total phosphorus in lake water residues. Trans. Wisconsin Acad. Sci. Arts Lett., **26**: 441–444.

TITUS, LESLIE, AND V. W. MELOCHE. 1933. A microextractor. Ind. Engr. Chem., Anal. Ed., **5**: 286–288.

TRELEASE, WILLIAM. 1889. The "working" of Madison lakes. Trans. Wisconsin Acad. Sci. Arts Lett., **7**: 121–129, pl. 10.

TRESSLER, W. L., AND B. P. DOMOGALLA. 1931. Limnological studies of Lake Wingra. Trans. Wisconsin Acad. Sci. Arts Lett., **26**: 331–351.

TWENHOFEL, W. H. 1933. The physical and chemical characteristics of the sediments of Lake Mendota, a fresh water lake of Wisconsin. J. Sediment. Petrol., **3**: 68–76.

———. 1937. The bottom sediments of Lake Monona, a fresh-water lake of southern Wisconsin. J. Sediment. Petrol., **7**: 67–77.

TWENHOFEL, W. H., AND W. A. BROUGHTON. 1939. The sediments of Crystal Lake, an oligotrophic lake in Vilas County, Wisconsin. Amer. J. Sci., **237**: 231–252.

TWENHOFEL, W. H., S. L. CARTER, AND V. E. MCKELVEY. 1942. The sediments of Grassy Lake, Vilas County, a large bog lake of northern Wisconsin. Amer. J. Sci., **240**: 529–546.

TWENHOFEL, W. H., AND V. E. MCKELVEY. 1939. Sediments of Devils Lake, a eutrophic-oligotrophic lake of southern Wisconsin. J. Sediment. Petrol., **9**: 105–121.

TWENHOFEL, W. H., AND V. E. MCKELVEY. 1941. Sediments of fresh-water lakes. Bull. Amer. Assoc. Petrol. Geol., **25**: 826–849.

TWENHOFEL, W. H., AND V. E. MCKELVEY. 1942. The sediments of Little Long (Hiawatha) Lake, Wisconsin. J. Sediment. Petrol., **12**: 36–50.

TWENHOFEL, W. H., V. E. MCKELVEY, S. A. CARTER, AND HENRY NELSON. 1944. The sediments of four woodland lakes, Vilas County, Wisconsin. I and II. Amer. J. Sci., **242**: 19–44, 85–104.

VOIGT, L. P. 1958. Wisconsin Lakes. Wisconsin Conserv. Dept., Publ. 218–58, 35 p.

VORHIES, C. T. 1905. Habits and anatomy of the larva of the caddis-fly, *Platyphylax designatus,* Walker. Trans. Wisconsin Acad. Sci. Arts Lett., **15**(1): 108–123, pl. 7–8.

———. 1909. Studies on the Trichoptera of Wisconsin. Trans. Wisconsin Acad. Sci. Arts Lett., **16**: 647–738, pl. 52–61.

WAGNER, GEORGE. 1908. Notes on the fish fauna of Lake Pepin. Trans. Wisconsin Acad. Sci. Arts Lett., **16**(1): 23–37.

———. 1911. The cisco of Green Lake, Wisconsin. Bull. Wisconsin Nat. Hist. Soc., **9**: 73–77.

WELCH, P. S. 1944. Chancey Juday (1871–1944). Ecology, **25**(3): 271–272.

WHITNEY, L. V. 1937. Microstratification of the waters of inland lakes in summer. Science, N.S., **85**: 224–225.

———. 1938*a*. Microstratification of inland lakes. Trans. Wisconsin Acad. Sci. Arts Lett., **31**: 155–173.

———. 1938*b*. Continuous solar radiation measurements in Wisconsin lakes. Trans. Wisconsin Acad. Sci. Arts Lett., **31**: 175–200.

———. 1938*c*. Transmission of solar energy and the scattering produced by suspensoids in lake waters. Trans. Wisconsin Acad. Sci. Arts Lett., **31**: 201–221.

———. 1941*a*. A multiple electromagnetic water sampler. Trans. Wisconsin Acad. Sci. Arts Lett., **33**: 95–97.

———. 1941*b*. A general law of diminution of light intensity in natural waters and the percent of diffuse light at different depths. J. Opt. Soc. Amer., **31**: 714–722.

———. 1941*c*. The angular distribution of characteristic diffuse light in natural waters. J. Marine Research, **4**: 122–131.

WIEBE, A. H. 1930. Investigations of plankton production in fish ponds. Bull. U.S. Bur. Fish., **46**: 137–176.

WILLIAMS, F. T., AND ELIZABETH MCCOY. 1934. On the role of microorganisms in the precipitation of calcium carbonate in the deposits of fresh water lakes. J. Sediment. Petrol., **4**: 113–126.

WILLIAMS, F. T., AND ELIZABETH MCCOY. 1935. The microflora of the mud deposits of Lake Mendota. J. Sediment. Petrol., **5**: 31–36.

WILSON, L. R. 1932. The Two Creeks forest bed, Manitowoc County, Wisconsin. Trans. Wisconsin Acad. Sci. Arts Lett., **27**: 31–46.

———. 1935. Lake development and plant succession in Vilas County, Wisconsin. I. The medium hard water lakes. Ecol. Monogr., **5**: 207–247.

———. 1937. A quantitative and ecological study of the larger aquatic plants of Sweeney Lake, Oneida County, Wisconsin. Bull. Torrey Botan. Club, **64**: 199–208.

———. 1938. The postglacial history of vegetation in northwestern Wisconsin. Rhodora, **40**: 137–175.

———. 1939*a*. Rooted aquatic plants and their relation to the limnology of freshwater lakes, p. 107–122. *In* Problems in lake biology. Amer. Assoc. Advance. Sci., Publ. No. 10.

———. 1939*b*. A temperature study of a Wisconsin peat bog. Ecology, **20**: 432–433.

———. 1941. The larger aquatic vegetation of Trout Lake, Vilas County, Wisconsin. Trans. Wisconsin Acad. Sci. Arts Lett., **33**: 135–146, 2 pl.

WILSON, L. R., AND A. T. CROSS. 1941. A study of the plant microfossil succession in the bottom deposits of Crystal Lake, Vilas County, Wisconsin, and the peat of an adjacent bog. Amer. J. Sci., **241**: 307–315.

WILSON, L. R., AND E. F. GALLOWAY. 1937. Microfossil succession in a bog in northern Wisconsin. Ecology, **18**: 113–118.

WILSON, L. R. AND RUTH M. WEBSTER. 1942*a*. Fossil evidence of wider post-Pleistocene range for butternut and hickory in Wisconsin. Rhodora, **44**: 409–414.

WILSON, L. R., AND RUTH M. WEBSTER. 1942*b*. Microfossil studies of three northcentral Wisconsin bogs. Trans. Wisconsin Acad. Sci. Arts Lett., **34**: 177–193.

WILSON, L. R., AND RUTH M. WEBSTER. 1944. Fossil evidence of wider post-Pleistocene range for butternut and hickory in Wisconsin—a reply. Rhodora, **46**: 149–155.

WIMMER, E. J. 1929. A study of two limestone quarry pools. Trans. Wisconsin Acad. Sci. Arts Lett., **24**: 363–399.

WOLTERECK, RICHARD. 1932. Races, associations and stratification of pelagic daphnids in some lakes of Wisconsin and other regions of the United States and Canada. Trans. Wisconsin Acad. Sci. Arts Lett., **27**: 487–521.

WOODBURY, L. A. 1942. A sudden mortality of fishes accompanying a supersaturation of oxygen in Lake Waubesa, Wisconsin. Trans. Amer. Fish. Soc., **71**: 112–117.

WRIGHT, STILLMAN. 1927. A revision of the South American species of *Diaptomus.* Trans. Amer. Microscop. Soc., **46**: 73–121.

———. 1928. A contribution to the knowledge of

the genus *Pseudodiaptomus*. Trans. Wisconsin Acad. Sci. Arts Lett., **23**: 587–599.

——. 1929. A preliminary report on the growth of the rock bass, *Ambloplites rupestris* (Rafinesque), in two lakes of northern Wisconsin. Trans. Wisconsin Acad. Sci. Arts Lett., **24**: 581–595.

ZoBell, C. E. 1940. Some factors which influence oxygen consumption by bacteria in lake water. Biol. Bull., **78**: 388–402.

ZoBell, C. E., and Janice Stadler. 1940*a*. The effect of oxygen tension on the oxygen uptake of lake bacteria. J. Bacteriol., **39**: 307–322.

ZoBell, C. E., and Janice Stadler. 1940*b*. The oxidation of lignin by lake bacteria. Arch. Hydrobiol., **37**: 163–171.

THE DEVELOPMENT OF ASSOCIATION AND CLIMAX CONCEPTS

E. Lucy Braun

16

THE DEVELOPMENT OF ASSOCIATION AND CLIMAX CONCEPTS

Their Use in Interpretation of the Deciduous Forest [1]

E. Lucy Braun

Ecology is defined as the study of life in relation to environment. It is concerned with what *is* and *why*. The ecological study of a single organism frequently meets with almost insurmountable obstacles because environment, if broken down into its component parts, is no longer environment. The ecological study of a complex piece of vegetation—a forest for example—presents innumerable problems, for the complexity of the environment—with its interacting factors—is multiplied by the complexity of the community; it does not yield to laboratory analysis. The methods of the physiologist fail.

Attempts at mathematical interpretation are for the most part static, and they often obscure rather than reveal the true picture of vegetation. Vegetation is dynamic—an ever-changing complex now appearing quiescent and in complete equilibrium with the habitat, now displaying obvious evidence of change. I believe that failure to recognize the dynamic aspects of vegetation is a primary cause of differing concepts. A dynamic approach is essential to interpretation of our eastern forests.

Observation of natural changes in vegetation long ago resulted in the concept of succession; recognition of quiescent phases led to the climax concept. The association concept arose from the need for designating so-called units of vegetation.

Concepts have changed through the years; they should change. As Cooper

[1] A paper delivered at Storrs, Conn., August 28, 1956, as part of the Ecological Society's symposium, "Approaches to Interpretation of the Eastern Forests of North America."

stated (1926), "a periodic inspection of foundations is most desirable." Concepts are bound to change with progress from local or intensive study to broad and extensive study. This is the reason for change with each individual worker; it has in part been the reason for change in concepts through the past 50 to 60 years of study of vegetation. Geographic location of studies also is a factor in the development of concepts.

Emphasis of the concepts of association, succession, and climax dates from the work of Henry C. Cowles around the beginning of the century (1899, 1901). Communities were recognized and at first called plant societies; later, plant associations. These were seen to be in equilibrium, more or less, with habitat; however, habitats change through the years, forcing change in vegetation. Changes were recognized as biotic, topographic, and climatic—the biotic due to reactions of the vegetation (and other accompanying life), the topographic to erosional and depositional forces, the climatic to the long-range variation in climate (Cowles, 1911; Clements, 1916, 1928). All groups of forces are at work everywhere, but locally one or another group may seem to determine the nature of vegetational change; all have been operative in determining regional vegetation.

Awareness of change led to the succession concept, for decades the guiding principle in ecological study and, I believe, the most fundamental of all concepts relating to the study of vegetation.

Succession is vegetational change—the *gradual* replacement of populations of species by other populations, i.e., of one community by another. The resulting plant succession (or sere) is not a series of steps or stages—no serious student of succession (a process) has ever claimed that a succession is made up of "discrete units." However, within any complete or partial sere there may be recognizable communities, communities sufficiently distinct from one another to justify naming them, usually for their dominants. Succession diagrams naming such "stages" place lines or sometimes arrows leading from one to another, which are intended to indicate gradual change. The whole sere is a continuously but gradually changing complex—the changes forced by biotic, topographic, or climatic factors. It is dynamic. It is a sequence in time, not in space, although spatial relations are often indicative of succession. Given time enough, the process of succession may lead to the establishment of a climax community, a community in which further change awaits major environmental change. It must be recognized that disruptive forces may prevent such progress, that although vegetational change is taking place, successions may be fragmentary and hardly recognizable.

Let us now apply these ideas to a concrete bit of forest vegetation—as the young student or the local worker must do. I select for this the primary forests of the Illinoian Till Plain of southwestern Ohio (Braun, 1936). Reconnaissance observation reveals a number of communities, among them those dominated by pin oak, by white oak, by beech. Each has certain characteristics and, according to the nomenclature so far suggested, may be called an association. Are they distinct and separate units, or are they related one to another? The primary pin oak association is normally an open forest of large trees occupying poorly drained areas wet or swampy in spring. Pin oak is an intolerant species—in spots between the large old trees may be seen younger individuals of this species, but in shady spots beneath the canopy formed by groups of pin oaks, other species occur, shagbark hickory and perhaps an occasional white oak or beech. As long as open spots remain, ecesis of pin oak is possible, but when the canopies of primary trees and older reproduction join, further ecesis of pin oak ceases. Gradually the older pin oak trees die out and the older individuals of other species of the understory replace them. Vegetational change is taking place, succession is in progress. Replacement takes place individual by individual—there is no jump from pin oak dominance to dominance by another species. Wherever a pin oak started in an open spot it may persist along with the new entrants which are changing the habitat by their deeper shade and their different leaf litter. After many forest generations and centuries of occupancy, a forest in which white oak is dominant may come into being. The old white oak forest, with trees 3 to 4 feet in diameter, again reveals evidence of change. Beech trees of various ages are interspersed, but white oak is absent from lower layers. Gradual dying out of white oak and growing to maturity of beech leads to the establishment of a beech forest. Beech, the most tolerant species of the area, is able to reproduce and perpetuate itself. Succession has been taking place because of biotic forces at work and has apparently ended in an unchanging or very slowly changing community—a climax of some sort.

Within any *single* forest tract the distribution of earlier and later developmental stages appears to be related to almost imperceptible topographic differences—depressions and swells differing but a few inches in elevation. But comparison with another tract may disclose that "earlier" stages are lacking and that "later" ones may occupy a greater variety of sites. The whole tract appears as a mixed forest, its unlike component communities often small in extent. It is a pattern of intergrading developmental and climax communities, not a climax pattern due to micro-relief (as suggested by Whittaker, 1953, p. 45). When analyzed by statistical methods of recent years, with quadrats laid out arbitrarily in a grid pattern (as has been done), with all trees above a certain diameter counted, and with basal areas determined, evidence of unlike communities within the tract is obscured and evidence of succession is lost. With this loss, the most fundamental concepts of vegetation are endangered.

Only after repeated field surveys of such tracts and only after the processes at work are thoroughly understood and the relationships of parts of the forest determined, could statistical methods be applied. Continuous transects, with vegetation related in so far as possible to environment, supply the most desirable method of study. Once the community relations have been determined, isolated samples of community types from other areas may be used to amplify concepts.

What are the concepts which can be illustrated by a dynamic approach to such a forest study as the above? (1) An association concept—for here are more or less well-defined communities, each with definite dominants and seemingly more or less in equilibrium with the habitat. (2) A succession concept, based on evidence of change, of lack of accord of canopy and understory, and of forces, here largely biotic, at work directing the changes. (3) A climax concept, for this succession appears to end in a self-perpetuating community. Will these concepts as first developed stand the test of application to other unlike sites in the geographic region and of application in other geographic regions? In part, yes; in part, no.

In unlike sites of the same geographic area, the communities are different, the causal factors of succession may be different, and topographic change rather than biotic direct the course of development. And still, communities are evident—let us for the present call them associations; succession is evident, and a climax is reached—but not a beech climax. At once we question our climax concept. The beech forest of our example is dependent on a particular site—an undissected plain in an immature topography. With the first development of drainage lines, changes begin again and development goes on. This beech forest is a physiographic climax. The climax of mature topographic sites is different and more enduring.

Soon we realize that no two pieces of vegetation are exactly alike, even though they may have the same dominants. Nichols (1923) realized this and proposed the association concrete (applied to a piece of vegetation) and the association abstract—the abstract concept gained by familiarity with many pieces of similar vegetation. Gleason (1926), recognizing unlikeness, suggested the individualistic concept.

In other geographic areas, we find again that concepts must be revised. The seemingly important associations of our local area are not found; a multiplicity of new ones occur. Succession may be less evident—not because succession is not or has not taken place, but because development of primary vegetation has progressed farther, or in some instances because development has been interrupted by frequent disturbance. Climax communities may be found which bear a marked resemblance to those of mature topography of our local area. Such climax communities begin to assume greater importance in our thoughts. We begin to understand Clements' views, which resulted from extensive work. We realize that our association concept is open to question, that there is a tremendous difference between the pin oak association, or the white oak association, or even the beech association, and the climax association which we find recurring again and again over a vast geographic area and as a result of many unlike lines of succession. Clements (1916) proposed the term *associes* for developmental units and retained *association* only for climax units. Even with this restriction, we must realize that all climax units are not equivalent, that there are different kinds of climaxes, chief of which are physiographic climaxes and regional climaxes.

Shall the term *association* be used only for regional climax communities or for all sorts of climax communities, physiographic and regional? Shall it be thus restricted, or shall it be used for all relatively stable communities even though developmental? Or shall the term be restricted to the abstract? The local worker usually shuns the restricted usage and prefers to designate all well-marked communities as associations. (I am not here concerned with division of associations or associes into consociations, consocies, etc.) Extensive work is sure to modify this view, for the great regional associations are not comparable to the local associations. If association is to be used for all, then the regional should be called major associations. In fact, such emphasis is often desirable even when a broad association concept is used.

The validity of an association concept (and of other community units) has been questioned (Whittaker, 1951). I believe everyone will agree that communities are recognizable in the field, that we do see patches of forest to which we can give names, because of dominants. The objection is based on the fact that communities do not have definite limits, but intergrade with one another; that the species which seem to characterize them extend into other communities, although probably in different proportions; that no two communities are exactly alike; that vegetation (barring abrupt site differences) is continuous though differing from place to place.

The climax concept, too, is questioned, perhaps because of rigidity of interpretation in some schools of thought (Whittaker, 1953). The monoclimax and polyclimax ideas do not help. To me, a monoclimax concept appears impossible (although I have been surprised to find that I am by some considered a supporter of the idea); equally, a polyclimax concept seems questionable. I adhere to the idea that there is more than one kind of climax. I do not refer to the complex Clementsian units—as serclimax, disclimax, postclimax, etc.—which I prefer not to recognize. Extensive investigation reveals the recurrence, on mesic although not exactly similar sites, of climax communities closely similar through wide geographic areas. Intensive or local studies usually reveal the existence of other climaxes, stable communities related to other, sometimes extreme, sites. The first group of communities are representative of what I prefer to call the regional climax—an abstract concept. The second are physiographic climaxes, or, if you prefer, topographic, or edaphic, climaxes. This should not be understood to imply that regional climaxes are not influenced by topography—environment operates on all. Regional climaxes are commonly, but not always correctly, referred to as climatic climaxes. The relative prominence of these climax groups and of the developmental communities in any region depends upon the age of the region (geologically speaking) and the degree of advancement of the erosion cycle. Much of the vegetation of old areas (or of mature topography), if undisturbed, has reached stability, and a mosaic or pattern of intergrading climax communities is seen. In young areas, areas of immature topography, succession is still active and much of the vegetation is not climax. Convergence of seres toward a regional climax is best illustrated in such areas. And it is in such areas that the possibility of further development or change is indicated; that is, the stability of the regional climax is open to question.

It should be realized that the terminology which has developed and the concepts implied are a result of the ecologist's need to classify (see Cooper, 1926). In the few areas where undisturbed forest of any extent still remains, patterns of vegetation can be determined and the relationships of communities can be investigated. Throughout almost all the Eastern Deciduous Forest, only small fragments of primary forest remain; no patterns are discernible. The need to classify remains, and the isolated sample must be assigned to a place in the system of classification, although it has no place in a concrete transect of vegetation. Such difficulties do not arise in local studies.

An understanding of the structure of our eastern forests is not dependent on agreement as to concepts. It is dependent on the recognition of development, of the existence of types of communities, and of climax; of climax as a long-term expression of regional factors, both climatic and local, stable as far as human life span is concerned, stable in relation to developmental communities of comparable life form, but nevertheless changing in accord with changes in regional factors.

Our forest vegetation has its roots in the early evolutionary history of angiosperms, their rise and their geographic spread as continental history permitted. The great deployment of deciduous forest (the eoclimax of Clements) came in Tertiary time. Related climaxes with the same general climatic features occur in widely separated geographic areas, in eastern America and eastern Asia, for example. These comprise a Clementsian panclimax, whose parts originated from the ancestral Tertiary forest. In America, that climax believed to be most like the ancestral Tertiary forest is the Mixed Mesophytic Forest climax. Its similarity to the mixed forest of China has been pointed out by Chaney, who has seen both the forests of China and the mixed mesophytic forests of the Cumberland Mountains, the area of best development of Mixed Mesophytic Forest.

Differentiation of the Deciduous Forest has resulted because of environmental changes related to progress of erosion cycles and to climatic changes. Awareness of and understanding of succession in its broadest aspects help to unravel the complex pattern of eastern forest vegetation.

Twenty years ago, in the Cumberland Mountains, great areas of the Kentucky slopes of Big Black Mountain and of the northwest slope of Pine Mountain were clothed with almost unbroken primary forest, affording, by the continuous transect method, opportunity to study the relationship of communities, and hence of community types (Braun, 1935a, 1940, 1942). Gradations related to altitude and to moisture (suggested by topographic position, insolation, etc.) as well as to substratum occurred. The whole displayed a multitude of segregates. The most mesic communities are the most highly mixed communities, including in the canopy a great number of tree species, some with wide ecological amplitude, which are species with many biotypes, and hence a variety of ecotypes, as Dr. Whittaker ably explained in a recent paper on the vegetation of the Great Smoky Mountains (1956). Partly because these very mesic mixed communities often contain a few red oak, white oak, and chestnut trees, Dr. Whittaker has compared them with transition forests of the Smokies, where *Castanea* is classed as "subxeric." However, in the Cumberland Mountains and on the Cumberland Plateau, *Castanea* has a greater variety of ecotypes and may be associated with river birch and sweet gum on river bottoms, with sweet gum, pin oak, and beech on swamp flats; that is, it varies from hydromesic to subxeric, but reaches greatest abundance in submesic and subxeric sites, where it may be codominant with sugar maple and tuliptree in the former, or with chestnut oak in the latter. Other species, also, have wider ecological amplitudes in the Cumberlands than in the Smokies.

To return briefly to concepts—all the mesic mixed communities have much in common, in species and in luxuriance and coverage of herbaceous vegetation. In a scheme of classification, all could be grouped under the broad term mixed mesophytic, that is, as belonging to the Mixed Mesophytic association. But because no two species are alike in physiological requirements, gradation in environment results in segregation of the mixture. For these interrelated similar communities the term association-segregate was proposed as a dynamic term suggesting response to environmental forces at work (Braun, 1935). It was apparent that gradations led from the mesic communities readily classified as belonging to the mixed mesophytic type to less mesic transitional types and finally to subxeric types on ridges. It was also apparent that a more or less distinct break could be found where influence of canopy (late leafing of oak and chestnut and slow decomposition of their leaf litter) was reflected in composition of undergrowth. This break occurs between what may be classed as mixed mesophytic communities and oak-chestnut communities.

In the Cumberland Mountains, mixed mesophytic communities occupy not only the coves but much of the slopes, except convexities and ridges. Nowhere else are such communities better developed. This I believe is because of (1) the long period of continued occupancy of an area of mature topography; (2) the very deep soils which can develop where underlying strata are horizontal or nearly so (in marked contrast to conditions in the Great Smoky Mountains and Blue Ridge); (3) the high annual precipitation, equably distributed and without the occurrence of summer droughts, such as the Southern Appalachians are subject to (see maps by Thornthwaite, 1941). It became apparent that here was the center of best development (not center of dispersal) of mixed mesophytic forest. Extensive travel disclosed the prevalence of mixed mesophytic forest communities over a considerable geographic area, which I later designated as the Mixed Mesophytic Forest region. Beyond this region—in any direction—such communities are more limited in extent. Other climax types prevail. On the basis of prevalence of a particular climax type (and prominence of other community types) other forest regions were designated (Braun, 1950).

The prevalence of any particular climax type may be due to the control of climate; it may be related to the history of erosion cycles and past climates; it may be due to state of development of climax communities; or, more likely, to a combination of all.

Prevalence of a climax type should not be construed as uniformity of vegetation. Great diversity of vegetation is a characteristic of all areas of diverse topography. And each area may have its own peculiar vegetation types as well as some common to other areas of the region. In the intensive study of any local area, all community types should be distinguished, and the status (climax, developmental, primary, secondary) be determined. In an extensive approach, the need for classification is as evident as is the need for classifying species into genera and families—the number of community types would doubtless be in the thousands.

The scheme of classification most used by American ecologists in connection with the Eastern Deciduous Forest is division into what have been called climax associations. For many years a threefold concept prevailed which recognized three climax associations: beech-maple, oak-chestnut, and oak-hickory (Clements, 1916, 1928). In 1916, a fourth climax, mixed mesophytic, was proposed for a mesic climax type with several dominants, among which beech was most important (Braun, 1916). Expansion of the concept became necessary, for beech is not always an important species, in fact is not always present. Two species, *Tilia heterophylla* and *Aesculus octandra,* are the most characteristic species of the mixed mesophytic association (Braun, 1947, 1950). This does not mean that other species cannot be thought of as characteristically present in mixed mesophytic communities, but that these two are essentially confined to such communities and do not occur in any other climax association than mixed mesophytic.

If the four major climax associations mentioned above are recognized, then a geographic region in which the mixed mesophytic association prevails can be distinguished as the Mixed Mesophytic Forest region; this occupies a central position. A region in which an oak-chestnut association prevails occupies a large area bordering it to the east and southeast and is called the Oak-Chestnut Forest region. Farther west, centering in the Ozarks, is the Oak-Hickory Forest region. The Beech-Maple Forest region lies to the north;

here a beech-maple or maple-beech climax prevails. In addition to these four climaxes, a maple-basswood is now generally recognized (Oosting, 1948; Braun, 1950).

The extent of these regions, and I believe the prevalence of the climaxes which characterize them, is a result of past history as well as present climate. That is why I like to think of these climaxes as regional rather than climatic. Evidence points to mid-Tertiary occupancy of much of the eastern half of the United States by mixed forest. Base leveling and climatic change curtailed its extent, leaving remnants, relics, in favorable sites, and giving advantage to less mesic types—just as moisture and slope differences in the Cumberland Mountains cause segregation into types. Glaciation further curtailed the mixed forest. Post-Wisconsin migrations have repopulated the glaciated area, including that part here designated as the Beech-Maple region. Here, more than anywhere else, the importance of succession is evident. Here also the question arises as to the permanence of this classic association. There is evidence that where dissection is beginning, beech-maple dominance is being broken by the entrance of additional mesic species (Braun, 1950, pp. 523, 527). Again, question is raised as to the climatic character of this regional climax.

The associations here distinguished are not concrete pieces of vegetation—they are abstract concepts to which communities and community types can be assigned. The species (or species populations) which make up these communities seldom have narrow habitat limits; mostly they are complex species or species made up of a number of ecotypes and hence may range through a number of different community types. A species which in one place may indicate subxeric conditions, in another may grow in hydromesic sites; hence conclusions reached in one intensive study cannot be carried over into another. Considering the four major associations, two are mesic, two submesic to subxeric. The area where the Mixed Mesophytic climax prevails is separated from the area where the Oak-Hickory climax prevails by a broad belt which I call the Western Mesophytic Forest region, characterized by a mosaic of climaxes including rich mixed mesophytic communities, various less mesic types, as beech-chestnut, white oak–black oak–tuliptree, and oak-hickory (of varying composition), pine communities, cedar barrens, and prairies. No existing gradient can account for the distribution of this motley assortment. Late Tertiary, Pleistocene, and post-Pleistocene climates, topography, and migrations must be considered. The distribution of mixed mesophytic communities in this region cannot be explained in relation to precipitation alone; rather, they appear to have persisted (with changes of course) from the more widespread mixed Tertiary forest, to have persisted where topography has been continuously favorable, and can and did, in drier times, partly compensate for reduced precipitation.

To summarize—I think of the Eastern Deciduous Forest (ignoring secondary forest) as made up of a vast number of intergrading communities. Some of the communities are developmental, others climax in nature, some are very local in occurrence and extent, others recur frequently over considerable geographic extent. The developmental communities give clues to trends. Those climax communities which are limited to specific sites, more or less extreme for their region, illustrate the great range of types possible in a region. Those climax types which recur again and again, represented by similar though individually different communities, are the ones which can be classified into the major associations of the Deciduous Forest and are the ones which best illustrate history and development of that forest through the long ages of its existence.

LITERATURE CITED

BRAUN, E. L. 1916. The physiographic ecology of the Cincinnati region. Ohio Biol. Surv. Bull. 7.

———. 1935. The undifferentiated deciduous forest climax and the association-segregate. Ecol. 16:514–519.

———. 1935a. The vegetation of Pine Mountain, Kentucky. Amer. Midl. Nat. 16:517–565.

———. 1936. Forests of the Illinoian till plain of southwestern Ohio. Ecol. Monog. 6:89–149.

———. 1940. An ecological transect of Black Mountain, Kentucky. Ecol. Monog. 10:193–241.

———. 1942. Forests of the Cumberland Mountains. Ecol. Monog. 12:413–447.

———. 1947. Development of the deciduous forests of eastern North America. Ecol. Monog. 17:211–219.

———. 1950. Deciduous forests of Eastern North America. Blakiston–McGraw-Hill. New York.

CLEMENTS, F. E. 1916. Plant succession. Carnegie Inst. Wash. Pub. 242.

———. 1928. Plant succession and indicators. H. W. Wilson. New York.

COOPER, W. S. 1926. The fundamentals of vegetational change. Ecol. 7:391–413.

COWLES, H. C. 1899. The ecological relations of the vegetation on the sand dunes of Lake Michigan. Bot. Gaz. 27:95–117, 167–202, 281–308, 361–391.

———. 1901. The physiographic ecology of Chicago and vicinity. Bot. Gaz. 31:73–108, 145–182.

———. 1911. The causes of vegetative cycles. Bot. Gaz. 51:161–183.

GLEASON, H. A. 1926. The individualistic concept of the plant association. Bull. Torrey Bot. Club 53:7–26.

NICHOLS, G. E. 1923. A working basis for the ecological classification of plant communities. Ecol. 4:11–23, 154–179.

OOSTING, H. J. 1948. The study of plant communities. W. H. Freeman. San Francisco.

THORNTHWAITE, C. W. 1941. Atlas of climatic types in the United States, 1900–1939. U.S. Dept. Agric. Soil Cons. Service Misc. Pub. 421.

WHITTAKER, R. H. 1951. A criticism of the plant association and climatic climax concepts. Northwest Science 25:17–31.

———. 1953. A consideration of climax theory: the climax as a population and pattern. Ecol. Monog. 23:41–78.

———. 1956. Vegetation of the Great Smoky Mountains. Ecol. Monog. 26:1–80.

RECENT EVOLUTION OF ECOLOGICAL CONCEPTS IN RELATION TO THE EASTERN FORESTS OF NORTH AMERICA

R. H. Whittaker

17

RECENT EVOLUTION OF ECOLOGICAL CONCEPTS IN RELATION TO THE EASTERN FORESTS OF NORTH AMERICA[1]

R. H. Whittaker

INTRODUCTION: THE SYSTEM OF CLEMENTS. The eastern forests have always been the real homeland of American ecology. It is here that the largest number of American ecologists have lived and worked, the most extensive ecological research has been carried out. From study of the eastern forests have developed many of the concepts which ecologists have applied to interpretation of other areas. Results of ecological research in other areas have also influenced interpretation of the eastern forests; but the commerce of ecological ideas has been more one of export from than of import into the eastern forests. Of all the exports from the ecology of the eastern United States, the most widely influential was the system of vegetation interpretation designed by Clements. Discussion of changing views of ecological concepts, especially in relation to the eastern forests and Clements' system of interpretation, is the object of this paper.

The major unit of vegetation in Clements' system is the plant *formation.* The formation is a great regional unit of vegetation characterized by its dominant growth form, as the eastern forests are characterized by deciduous trees, the prairie where Clements did his early work by grasses. Every formation is a product of climate and is controlled and delimited by climate and climate alone; the formation occurs in a natural area of essential climatic unity and

[1] Based on the conclusion to a symposium on "Approaches to Interpretation of the Eastern Forests of North America," including also the cited papers of Braun (1956, published in this volume), Wang (1956), Raup (1956), and Bray (1956b), at the meetings of the American Institute of Biological Sciences, Storrs, Conn., August 28, 1956.

expresses the climate and its unity (Weaver and Clements, 1938, p. 89). The visible unity of the formation is due to its dominant species, for all these are of the same growth form and certain species and genera of dominants range widely through the formation and link together its various associations (Weaver and Clements, 1938, pp. 91, 489). The formation is not an abstraction or a mere unit of classification; it is a definite and concrete organic entity, covering a definite area marked by a climatic climax (Clements, 1928, p. 128). The formation is, in fact, a *complex organism,* and as such it arises, grows, matures, reproduces, and dies (Clements, 1928, p. 125).

The formation is the *climax* in its climatic region; the terms climax and formation are in fact synonymous (Weaver and Clements, 1938, pp. 91, 478). The climax is the final, mature, stable, self-maintaining, and self-reproducing state of vegetational development in a climatic unit. The climax formation is the adult organism, the fully developed community, of which all other communities are but stages of development (Clements, 1928, p. 126). Since climate alone determines the climax formation, there is but one true, or climatic, climax in a climatic region. Communities differing from the climax because of distinctive soils or other habitat characteristics are developmental. Relatively stabilized vegetation other than the climatic climax may occur in a region, however, because vegetation is held indefinitely in stages preceding the climax by factors other than climate (subclimaxes and serclimaxes), because local soils and topography offer conditions favorable to the climaxes of other regions (preclimax and postclimax), or because disturbance causes modification or replacement of the true climax (disclimax) (Weaver and Clements, 1938, pp. 81–86).

The formation, like other organisms, has an evolutionary history and arises from the modification of earlier formations; it has a *phylogeny.* The grouping of formations on different continents whose descent from a common ancestor is indicated by possession of similar dominant growth forms is a panclimax (Clements, 1936; Clements and Shelford, 1939, p. 243). Every climax formation consists of two or more subdivisions known as *associations,* each marked by one or more dominant species peculiar to it (Weaver and Clements, 1938, p. 93). Like the formation, the association is a regional vegetation unit; it is the climatic climax of a subclimate within the general climate of the formation. In the eastern forest formation, climatic influences have resulted in a threefold differentiation—into the maple-beech association in the northern, the oak-chestnut association in the eastern, and the oak-hickory association in the western part of the formation. Of these, the maple-beech is the typical association of the formation, because of the greater environmental requirements and smaller number of its dominants and its closer similarity to the European member of the panclimax (Weaver and Clements, 1938, p. 510). Each association is similar throughout its extent in physiognomy or outward appearance, in its ecological structure, and in general floristic composition (Weaver and Clements, 1938, p. 94). Differentiation of the association in relation to local habitats, however, requires the recognition of other units subordinate to it—the consociation, fasciation, lociation, etc.—and a parallel hierarchy of units exists for successional communities (Clements, 1936).

Each climax is the direct expression of its climate; the climate is the cause, the climax the effect, which, in turn, reacts upon the climate (Weaver and Clements, 1938, p. 479). Since the habitat is cause and the community effect, it is inevitable that the unit of the vegetative covering should correspond to the unit of the earth's surface, the habitat (Clements, 1905, p. 202; 1928, p. 119). During succession *dominant species* occupy the habitat, modify the habitat in ways unfavorable to themselves, and are replaced by other dominant species. Every succession ends in a climax when the occupation by and reactions of a dominant are such as to exclude the invasion of another dominant (Weaver and Clements, 1938, p. 237). The climax dominants are the species best adjusted to habitat and able to take possession of the habitat and hold it against all comers. The dominant receives the full impact of the environment and determines conditions of life for other species in the community both by effects on environment (reactions) and by relations to other species (coactions) (Clements and Shelford, 1939, p. 239). The dominant species characterize the community, express its environment, and indicate the actual or probable presence of other, associated species (Clements, 1928, p. 253). The dominant is the real basis of indicator study, so commanding is its role in the processes of vegetation (Clements, 1928, p. 236). Dominance, in fact, is one of the master keys to the understanding of vegetation, as successional process is another.

Such was Clements' system in essential features. Details which would too much lengthen the present paper have been omitted, but the abbreviated account makes clear one of the system's most significant characteristics—its coherence. Each of the preceding statements of the system has necessary logical interconnections with most of the other statements. Each major concept is defined, and seems explained through, other concepts; some of the concepts (climatic unit, formation, and climax; dominance, reaction, and community unit) are as strictly interdependent as the terms of an equation. Almost all there is about vegetation seems accounted for by the system and its complications (the various special climaxes, the subordinate vegetation units, etc.) introduced to accommodate it to the complexities of vegetation. It is a system which, in its orderliness, seems to imply the orderliness of vegetation. Clements' own role in creating this coherent and orderly system went far beyond the bringing together of inductive discoveries about vegetation from Warming, Cowles, Moss, and other predecessors and his own work. Influenced by the ideal of the deductive system represented in philosophy, mathematics, and physics, Clements fashioned from limited evidence and liberal assumption a coherent, and apparently inclusive, deductive system of vegetation interpretation.

The structure of the system began to appear early in the century (Clements, 1905), was developed in its major features in the monograph on plant succession (Clements, 1916, 1928) and textbook of plant ecology (Weaver and Clements, 1929, 1938), stated in definitive form in an essay on the climax (Clements, 1936, 1949), and soon thereafter applied to the study of natural communities through the plant-animal formation, or biome (Clements and Shelford, 1939). In its course of about half a century it has accumulated increasing criticism; although the most influential product of American ecology, its influence has declined in the last decades. In considering the meaning of this decline, the author has no desire to minimize the great contribution which Clements made or to criticize Clements or his system for the uncritical manner in which it was applied by some of his followers. The object is not to show that Clements was mistaken, or that time and experience have been unkind to his system, as they have been to others which were brilliant in their times. It is to show, with this system as background, how the climate of interpretation in a field of science has changed and seems to be changing today.

CHANGING VIEWS OF ECOLOGICAL CONCEPTS. One of Clements' central concepts is that of formation. Vegetation of the eastern United States seemed to support his view of formations as regional and climatic units, each distinct from others and unified within itself by a single dominant growth form. Three great formations are easily recognized in temperate eastern North America—the eastern deciduous broad-leaf forest, the grassland or prairie, and the northern evergreen needle-leaf forest, or taiga. These seem indeed distinct, natural, climatically determined units; there is only the problem of the mixed broad-leaf–needle-leaf forests of the Lake States, to be treated as either a transition between the first two or an additional formation. Experiences elsewhere in the world have not supported the view that formations are determined only by climate, that climatic regions and vegetation formations must always be identical. The tropical grasslands, or savannas, surely among the world's great plant formations, are thought to owe their existence to factors of both climate and soil (Beard, 1953). The formation is not simply a unit reflecting climate alone, but a grouping of communities which are of similar physiognomy and express a similar set of ecological conditions, of which climate is only one (Beadle and Costin, 1952). It seemed simple enough to recognize three major formations of temperate eastern North America by one character only—dominance of a single growth form. But in a broader view, many formations must be defined by mixtures of growth forms, and many others must be distinguished by differences of environments, not of growth-form dominance. Clements and others have been impressed by the discontinuity between grassland and forest in eastern North America;

but in a world view there is extensive continuity and intergradation among formations, as is especially evident in the tropics (Beard, 1955). This continuity is not easily reconciled with the view of the formation as a distinct entity comparable to an organism. All these observations contribute to a changed view of the formation as a vegetation unit. The formation seems not a distinct, concrete vegetation unit determined solely by climate, but an abstract grouping of communities of similar physiognomy and environmental relations, a grouping dependent in the end on man's choice of what constitutes sufficient similarity of physiognomy and environment.

The nature of the association has been more bitterly contested and is a problem many ecologists do not feel has been resolved. The essence of the concept lies in the association, i.e., occurrence together, of species populations to form community units which may be recognized by species composition. Since associations are thought to be "natural units," very often compared with species as units of classification, it has been natural to assume that they are distinct in the sense of being separated from one another by well-defined boundaries. The basis of the association concept was challenged by Ramensky (1924) with two propositions: (1) The principle of species individuality—each species responds uniquely to external factors and enters the community as an independent member; there are no two species which relate themselves to environments and communities in quite the same way. (2) The principle of community continuity—composition of the plant cover changes continuously in space; sharp boundaries between communities are special circumstances requiring special explanation. Ideas closely related to Ramensky's were expressed about the same time by Gleason (1926) and Lenoble (1926, 1928); they are most familiar to Americans as the former's "individualistic concept of the plant association" (Gleason, 1926, 1939).

If the association is a "natural unit" of species populations, its unity is presumably expressed in species distributions. Most, or at least some, of the associated species should have closely similar distributions. Recent studies in the eastern forests (Whittaker, 1951, 1952, 1956; Curtis and McIntosh, 1951; Brown and Curtis, 1952) have shown that, in fact, no two species are distributed alike, that species show the heterogeneity of distribution and lack of organization into distinct groups of associates which Ramensky's first principle asserts. Relations to other species must, to be sure, affect their distributions; species distributions cannot be "independent" in the sense that they are unaffected by competition and other interrelations. These interrelations, however, do not result in the organization of species into definite groups of associates; and species distributions are "individualistic" in the sense that each species is distributed according to its own way of relating to the range of total environmental circumstances, including effects of and interrelations with other species, which it encounters. Both continuity and discontinuity of vegetation have been very widely observed; but the same recent studies of the eastern forests support Ramensky's view that vegetational continuity is the more general, discontinuity the more special, condition (Whittaker, 1956). As discussed by Bray (1956b), these studies and some other current work (Goodall, 1954; Poore, 1955; Hale, 1955; Culberson, 1955; Bray, 1956a) are concerned less with associations than with ways of dealing effectively with vegetational continua.

The association thus ceases to be the clearly defined natural unit assumed by Clements. Two further difficulties are suggested in relation to Clements' associations. The first of these involves the dominance concept. The dominant trees of a forest strongly affect conditions of life for other species of the forest, but different tree species may affect the forest environment similarly. It is not the species of the dominant but the total environment in the forest which determines whether subordinate plant species can or cannot occur there. The principle of species individuality applies to relations of subordinate species to dominant species and total environmental complexes influenced by dominants. Dominants do not control the distribution of other species and organize them into well-defined groups of associates in the manner, or to the extent, assumed by Clements (Whittaker, 1956). Furthermore, as has been observed by European critics (Lippmaa, 1931; Braun-Blanquet, 1951, p. 560; Ellenberg, 1954), the dominant species are often very wide-ranging ones, occurring in most varied environments with most varied associates; the overemphasis of dominants by Clements and others is sometimes a rather crude and superficial approach to relations of whole communities to environments. Dominance cannot provide the simple, comfortable solution to many problems, the master key for the recognition and classification of communities, indication of community environments, and understanding of community function, assumed by Clements.

The other difficulty is in the treatment of associations as regional units. It is easily shown that geographically, as locally, no two species of an association have the same distribution (e.g., Billings, 1949). In consequence, the association as a particular combination of species is a localized phenomenon (Lippmaa, 1933; Bourne, 1934; Cain, 1947; Knapp, 1948; Ellenberg, 1954); and associations may be geographically, as locally, continuous with one another (Walter and Walter, 1953). Species are distributed in relation to total environmental complexes, not to climate alone; and species do not fit into sharply defined groups corresponding to climatic regions and regional associations. Approaches to generalization about geographic relations of species in communities are being developed in Europe (Böcher, 1938; Meusel, 1939, 1943; Walter, 1954b), but these provide no simple answers to the definition of associations and determination of their boundaries. The major concepts of such geographic treatment—species groups based on distributional relations, natural areas defined by species distributions, distributional centers—have no very close relation to the association concept or to Clements' regional vegetation units.

One direction in the modernization of Clements' association concept has been followed by Braun (1935, 1938, 1947, 1950, 1956) in her work on the eastern forests. Braun's associations are, in manner of definition, derived from Clements'; each is a large-scale, relatively heterogeneous unit which characterizes a geographic region as its climax. Clements' three-part division was shown by Braun to be unrealistic; and Braun recognizes six regional associations (not including the Lake Forest, the evergreen forest of the Gulf Coastal Plain, and the transitional Western Mesophytic Region). Of importance at least equal to this change of view is the immense complexity of detail in these forests described by Braun (1950). Because of this complexity, Braun's six associations are by no means all that an ecological splitter might choose to recognize. The association is a more or less arbitrary unit (Braun, 1947, 1950, p. 525); Braun's associations are no more than Clements' the unique and inescapable solution to division of the eastern forests. Braun's division is not simply true where Clements' was false; it is a better pattern of abstraction from the eastern forests because more realistic, more frankly cognizant of the underlying complexity in seeking that simplification from complexity which is necessary to any abstraction. Thus in Braun's work, retention of some of the essential form of Clementsian ecology is combined with a fundamental change in perspective.

Very different directions have been followed in Europe. In Braun-Blanquet's (1921, 1932, 1951) system, the one most widely applied by phytosociologists, communities are classified by character species—species, often rather obscure ones, which are centered in or largely confined to a given community type and which, hence, are more truly characteristic of it than wide-ranging dominants. An association is in general a vegetation unit of the lowest rank which can be defined by character species. For most areas, the approach through dominants is thought suitable only for preliminary and superficial work; but another vegetation unit, the sociation, is used in areas such as Scandinavia, where communities contain relatively few species. This unit is one of small scale, defined by dominants of the various strata, not of the uppermost stratum alone; and it is unrelated to Clements' regional associations. For neither the sociation nor the association is climax stability or regional prevalence required, as for the association of Clements. Although regional vegetation types are recognized there, the American association as such is used by no one in Europe. In contrast to Clements' view of formations and associations as "concrete" entities, most European authors emphasize that the association is an abstraction—the association first comes into being at the phytosociologist's desk (Klapp, 1949, p. 10). Increasingly detailed knowledge of the vegetation of Europe has led to increasing recognition of the limitations inherent in even this most successful definition of plant associations. Most communities are actually intermediate to associations (Klapp et al., 1954; Ellenberg, 1954); and (because of species individuality) the number of "good" character species becomes progressively smaller as knowledge of their distributional relations increases (Ellenberg, 1954).

In spite of its role in the work of Braun, the Clementsian association seems a concept of declining significance. Many American ecologists do not use it in current work, and it has little use outside this country. So far as the term *association* has an accepted international meaning, it is the meaning given it by Braun-Blanquet, recommended by the Botanical Congress of 1935 (Du Rietz, 1936), and accepted by much the largest number of students of vegetation around the world. The diversity of vegetation units, of which the formation, association, American association, and sociation are only a few, has a deeper significance in relation to Clements' system, however. It implies that there are many possible approaches to classification of vegetation, that nothing forces ecologists to choose one or another of these, and that none results in units really comparable to species. An association is not a concrete natural community; it is an abstraction from the unlimited complexity and intergradation of communities, a class produced by an ecologist's choice of a class concept or definition (Whittaker, 1956).

Clements' treatment of climax and succession seemed to imply three other assumptions about the orderliness of vegetation (although the difficulties were known to Clements and allowed for in his system): (1) Succession is an orderly growth process, leading from varied beginnings to the same maturity or climax. (2) The climax is determined only by climate; consequently climaxes and climatic regions must correspond. (3) Vegetation consists of climaxes and their successional communities, and the climax and successional conditions are clearly distinguishable. It is consistent with the "individuality" of species that they enter successions in the most varied ways; there appears also to be a strong element of chance, of accidents of dispersal and timing, affecting the composition and sequence of successional communities. The complex interrelations among these which result led Cooper (1926) to compare succession with a braided stream. Successional processes give an impression of relative irregularity and disorderliness in detail, together with a degree of orderliness in general pattern and trend (Whittaker, 1953).

The assertion that there is only one true, climatic climax in a climatic region—the monoclimax theory—has been one of the most frequently criticized parts of Clements' system. As the system was usually applied, only the regionally prevailing vegetation, that undisturbed vegetation occupying the largest part of the land surface, was regarded as truly climax. For undisturbed vegetation, climax stability, regional prevalence, and maximum mesophytism seemed to be identified with one another. Observations have been accumulating from all over the world which suggest that this identification is untenable, that many communities which are not regionally prevalent are as much stabilized, as much "climax," as the "climatic climax." Many ecologists have been led to a polyclimax view which defines climaxes by an essentially stabilized, or self-maintaining, condition, regardless of prevalence, and considers that a number of climaxes may exist in a given area. The differences of monoclimax and polyclimax approaches are in part semantic ones (Cain, 1947); for authors of the two viewpoints use different terms for the same observations, or the same terms for somewhat different conceptions. The essential difference seems one of relative emphasis. Monoclimax authors emphasize the essential unity of climax vegetation within a region, with allowance also for stabilized communities other than the climatic climax. Polyclimax authors emphasize the inherent complexity of climax vegetation, with allowance also for prevailing climax communities which characterize the vegetation of a region and express its relation to climate. This difference of emphasis is one of great importance, for it affects the manner in which evidence on natural communities is selected, treated, and interpreted; in a rather subtle and often unconscious way it may color and condition the ecologist's whole perspective in the interpretation of vegetation.

Braun's (1950) approach to the eastern forests follows Clements in recognition of climatic climaxes as regional units; other stabilized communities are subclimaxes which "only theoretically could be replaced by the climax." One region of the eastern forests, the Western Mesophytic Region, is characterized not by a monoclimax but by a mosaic of unlike climaxes and subclimaxes. In the present author's view, Braun's approach is primarily monoclimax both in its expression (Braun, 1950, pp. 12–13) and in the manner in which the description throughout the book is influenced by this conception. It is, however, quite different from the rather doctrinaire climate-climax-formation identification of Clements. Monoclimax conceptions far removed from Clements have recently been expressed also by Dansereau (1954) and Walter (1954a). Polyclimax, or climax-pattern, conceptions asserting that the climax state is determined by the environments of individual communities, not regional climate, have been stated by Schmithüsen (1950) and Whittaker (1953); the latter identifies the climax as a community steady state and substitutes "prevailing climax" for "climatic climax." The varied current interpretations of "climax" by the author and others suggest that none of these can claim exclusive truth or univocal determination by properties of vegetation.

Raup (1956) has discussed the role of hurricane winds in patchwise disturbance of New England forests at time intervals that are short in relation to the life cycles of trees. The eastern forests, like the prairies (Malin, 1956), are probably much less stable than the climax ideal has suggested to many ecologists. The more closely vegetational dynamics are observed, the less clear-cut becomes the distinction between climax and successional communities (Whittaker, 1953). Vegetation does not really consist of climaxes and successions leading toward them. In a long-range perspective, the vegetation of the earth's surface is in incessant flux; what we observe in the field are not simply successions and climaxes, but only different kinds and degrees of vegetational stability and instability, different kinds and rates of population change. Vegetation change does not consist of successions toward climaxes in quite the sense ecologists have implied. Rather than this, from the diversity of communities we observe, some can be arranged in meaningful sequences of temporal development; and thus we bring into order and comprehension some part, but never more than part, of the flux of populations in natural communities. Climaxes do not simply exist in nature; rather than this, ecologists must define the terms of relative vegetational stability by which their climaxes are to be recognized—or, in a sense, created.

The flux of populations and "braided" relation of communities bear also on community phylogeny. Common ancestry for the deciduous-forest formations of eastern North America, Europe, and eastern Asia is implied in Clements' concept of panclimax; and maple-beech was regarded as the "typical" association of the American deciduous forest. Braun's (1950) interpretation of eastern forest history leads to a very different view of the eastern-forest associations. The Mixed Mesophytic Association of the Appalachian plateaux is seen as the central, the oldest, and the most complex association of the Deciduous Forest Formation. From the mixed mesophytic, or its ancestral progenitor, the mixed Tertiary forest, all other climaxes of the deciduous forest have arisen (Braun, 1950, p. 39). Braun's treatment, too, suggests a "phylogeny" for these forest associations, though in a sense less literal than Clements'. A somewhat different, nonphylogenetic interpretation of eastern forest history has been discussed by Wang (1956).

As the relations of species in space and successional time are, in a limited sense, "free" so that they may combine and recombine in most varied ways, so, we may believe, are their relations in evolutionary time (Mason, 1947). Species of one association do not simply evolve together into a new association; the species may change their distributional relations in time, entering in varied ways into new "associations" with other species. Interrelations of communities in evolutionary time are consequently intricately reticulate. It cannot really be said that two modern communities are descended from one Tertiary one in the same sense as two modern species populations may be descended from one Tertiary one. Rather than a formal phylogeny, or descent in a strict sense, historical interrelation of community types must be based on judgments of floristic relation, of relative floristic continuity of a past and a present community vs. relative floristic divergence and dilution.

The historic data on which such judgments must be based are fragmentary, and to some extent ambiguous or indeterminate. The student of vegetational evolution must select from the available data that which is most relevant, provide the interpretation of its relevance, and fashion the interpreted data into a meaningful pattern according to his general perspective and chosen ecological concepts. Some conclusions of historians of man are pertinent, although the vegetational historian may be spared concern with influences of culture, personality, choice, and volition—except those affecting the historian. History is an imaginative reconstruction of the past which is scientific in its determinations but approaches the artistic in its formulation; and history is more genuinely scientific in spirit as it takes into account the inescapable limitations on its objectivity and rigor (Muller, 1954, p. 35). It is no criticism of Clements', Braun's, and Wang's interpretations of eastern forest history to observe that these are human conceptual reconstructions which cannot be simply inherent in, or uniquely determined by, the information which they bring into a pattern of interpretation. It is relevant to Clements' system, however, that vegetational evolution is not a phylogeny.

Natural communities are not organisms, except in Whitehead's sense in which "organism" is equivalent to "system." Their manner of function and organization, their interrelation and classification, their development and maturity, and their evolution present problems which are distinct from, and significantly different in character from, those of individual biological organisms. In Clements' system the complex organism became the central unifying theme, the background concept from which the meanings of other concepts were to be understood. The organismic analogy has been accepted by some authors (Phillips, 1934–1935; Tansley, 1920, 1935; Allee et al., 1949), rejected by many others (Gleason, 1917, 1926; Gams, 1918; Meusel, 1940; Schmid, 1942; Ellenberg, 1950, 1954); but in current writing it is not the central concept of vegetational understanding—it is seldom referred to. The treatment of environment as cause and community as effect has been criticized as a fundamental weakness of Clements' system (Egler, 1951); probably Clements would not have ventured the parallel statement: environment is the cause, and the plant is the result. The cause-and-effect view seems poorly suited to the manner in which the community and environment are interrelated; a "transactional" approach to the functional system formed by community and environment may be more appropriate (Whittaker, 1954). A central concept different from Clements' complex organism and its cause-and-effect relation to environment, and different from the kind of synthesis of plant-animal ecology attempted by Clements and Shelford (1939), appears in contemporary ecology. It is the concept of the functional whole formed by community and environment—the ecosystem.

CONCLUSION: A CHANGING OUTLOOK. Scarcely a major feature of Clements' system as an intellectual structure remains intact, in this author's view. To observe this alone, however, may be to underestimate both the lasting significance of Clements' contribution and the real significance of the change that has occurred. Taken all together, the changes in ecological conceptions discussed amount to a general re-orientation of viewpoint in a field of science.

The changes have usually been regarded piecemeal, as particular changes affecting particular concepts; but this conclusion will attempt a more general interpretation of their meaning.

One over-all feature of the change is the dissolution of a coherent, well-ordered, deductive system for the interpretation of vegetation, an attempt at deductive whole-knowledge of vegetation, and its replacement by less coherent and less interdependent, inductive part-knowledges of different vegetational problems. Implied in this is a change in the view of natural communities themselves. These seem no longer to form an area of clear-cut, well-ordered, simply defined, and neatly interdependent phenomena to which a deductive system like Clements' may be appropriate. Although the deductive system is an ideal of science, students of natural communities find themselves authors of an inductive science, with problems more akin to those of the social sciences than those of physics and geometry.

The lack of clear-cut orderliness in natural communities is a necessary consequence of the multiplicity of factors and complexity of interrelations with which the ecologist must deal. This complexity is as much a fundamental circumstance of the study of natural communities as of the study of man's communities. A basis of lack of simple orderliness may be found also in Ramensky's principle of species individuality. Species populations are distributed "individualistically," and they are not organized in terms of man's ecological concepts and classifications. Major problems of community classification, of succession and climax, and community evolution stem directly from this individuality of species distributions. Many ecologists view the implications of species individuality in a more conservative light than that of this paper; but to the author species individuality is one major theme of current changes in ecological concepts. A further aspect of the lack of simple orderliness is in substantial effects of chance, of largely unpredictable factors of dispersal and population interaction, on species distributions and communities (Palmgren, 1929; Egler, 1942; Whittaker, 1953). Species individuality, multiplicity of factors, and effects of chance all contribute to giving most statements about natural communities a quality of probability, not necessity, partial correlation, not strict interdependence, inductive generalization for which limitations and exceptions are granted, not exact prediction. And if one turns to Braun's description of the eastern forests with these things in mind, a view of these forests quite different from Clements' results. One is struck first by their immense, their almost overwhelming, complexity of pattern. If one seeks to view this complexity in perspective, in terms of species populations in space and successional and evolutionary change and without the intervention of man's ecological abstractions, then the view of the forest is not one of clear and orderly associations, successions, and phylogeny. It is one of a veritable shimmer of populations in space and time.

If this population shimmer is not simply and clearly orderly, it is also not devoid of order. It is neither well ordered as it seemed in Clements' system nor simply disorderly; the fundamental character of natural communities with which ecologists must deal might be described as *loosely ordered complexity*. In this, it is the task of the ecologist not to discover simple order inherent in his material, but to find such means of effective abstraction and generalization as the loosely ordered condition permits. In the author's view, it is the gradual coming to terms with this loosely ordered complexity, as it affected one ecological concept after another, that has characterized the changing climate of interpretation in synecology. The negative aspect of this change is a loss of the kind of uncritical faith in ecological concepts and systems which once prevailed. Ecological concepts, man's means of interpreting natural communities, were projected back into natural communities and seen as part of nature itself. They were hypostatized or reified and were directly identified with the natural communities from which they were abstractions. Not only were plant associations part of the order of nature (Conard, 1939; cf. Clements, 1916, 1928; Tansley, 1920; Du Rietz, 1921, 1929; Alechin, 1925); plant communities *were* formations and associations, vegetation *was* succession and the climax. More recent experiences have led to increasing wariness of such identification, and increasing awareness of the complexities which bedevil all ecological generalizations, the limitations affecting all ecological concepts. The positive aspect of the change has been the building of inductive knowledge and a more realistic understanding of the function of ecological concepts. Older concepts of continuing usefulness, like the climax and association, are seen in new lights, while newer concepts, like the ecosystem and continuum, share with them in the understanding of vegetation.

The changing view of the role of ecological concepts may be summarized in a few points:

1. All are necessarily abstractions; they are essentially human creations serving to order, interrelate, and interpret some of the information about natural communities available to us. None can be thought inherent in vegetation in the sense earlier authors assumed; none can be thought to represent the real, whole, ultimate truth about natural communities.

2. There is nothing in the nature of vegetation which compels us to adopt one or another of these concepts as the primary basis of vegetation interpretation in general—as Americans have often adopted succession and the school of Braun-Blanquet associations. Nothing in the nature of vegetation forces us to choose a particular way of defining a given concept. The various approaches and concepts have only different degrees of general usefulness and different appropriateness to particular circumstances and purposes.

3. There is in all these concepts a dependence on choice and assumption, an element of subjectivity and artistry, which can never be wholly escaped, even though it is an objective of science to minimize its influence. The best means of controlling subjectivity is through a recognition, as frank and clear-

sighted as possible, of its presence and its underlying sources and necessity.

4. All these concepts and approaches have a partial character; they bring into order and comprehension only a small fraction of the information available to us. No one of them, however exhaustively applied, nor all of them together, can ever bring into a pattern of understanding all the available information about natural communities.

5. Finally, these concepts are to some extent interdependent, since the definition of one may influence another; but they are not interdependent in any simple, direct, and necessary way. There is no necessary correspondence between climax stability and regional prevalence of vegetation, between climax regions and floristic natural areas, between limits of dominance and of community types, between associations in the international and associations in the American sense. Rather than assuming that such necessary correspondence exists, it may be best to pursue each approach on its own terms for its own merits, with a minimum of assumption derived from other approaches, and to see then what degree of correspondence may or may not appear, in what way the results of one approach may or may not illuminate the results of another.

Experiences with the eastern forests and other natural communities since the time of Clements suggest a change of attitude from unquestioning faith in ecological concepts as part of nature and unquestioning assurance in the sufficiency of one's own system of interpretation toward greater modesty before the complexities of natural communities and the limitations of ecologists' understanding. They suggest an attitude of tolerance and open-mindedness, and eclecticism in practice, with regard to the varied possible concepts and approaches to natural communities, no one of which is uniquely determined by nature, each of which may contribute to understanding in a complementary relation to others. They suggest a greater self-consciousness in the use of ecological concepts, a continuing awareness of the role of the man in the interaction of ecologists and natural communities by which ecological understanding grows, of the extent to which the function of the scientist is not simple discovery, but the creation of means of understanding.

SUMMARY

1. Study of the eastern forests of North America has been the source of many ecological concepts, some of which became part of the widely influential system of Clements. Such concepts as formation and association, succession and climax, dominance, vegetational phylogeny, and the complex organism were synthesized by Clements into an orderly, coherent, deductive system of vegetation interpretation.

2. Further experiences in the eastern forests and elsewhere have led to changing views of these concepts:

a. The formation and association seem no longer distinct, clearly bounded entities comparable to organisms or species. With recognition of the significance of vegetational continuity and the principle of species individuality, these have come to be regarded as man-made classes of natural communities.

b. Succession seems a less orderly process than in Clements' view; only a part of the incessant flux of populations in natural communities can be understood in terms of succession. Relative stability of vegetation is not determined by climate alone, and varied approaches to the definition of "climax" are possible.

c. Dominant species do not control the distribution of other species and characterize communities and their environments in the way assumed by Clements.

d. Since plant species are free to change their distributional relations to one another through evolutionary time, evolutionary relations of communities are reticulate. Natural communities do not evolve by phylogenetic descent from past communities in the same sense as organisms.

e. The concept of the community as a complex organism, central to Clements' system, has been largely abandoned as inappropriate or unproductive. Current interpretations emphasize the functional system formed by community and environment, the ecosystem.

3. These changes in individual concepts, taken together, amount to a fundamental re-orientation of the field. The deductive system of Clements has broken down into the more detailed, more complex, and less coherent understanding of an inductive science. Because of multiplicity of ecological factors, effects of chance, and individuality of species distributions, natural communities are not an area of orderly, clear-cut, exactly predictable phenomena to which the system of Clements might be appropriate. A fundamental characteristic of natural communities affecting all ecological concepts is the condition of loosely ordered complexity. Ecological concepts cannot be thought inherent in, or uniquely determined by, vegetation; they are the means of human abstraction by which some of the diverse information about natural communities can be brought into comprehensible forms.

LITERATURE CITED

ALECHIN, W. W. 1925. Ist die Pflanzenassoziation eine Abstraktion oder eine Realität? Bot. Jahrb. 60 (Beibl. 135):17–25.

ALLEE, W. C., A. E. EMERSON, O. PARK, T. PARK, AND K. P. SCHMIDT. 1949. Principles of animal ecology. 837 pp. Saunders. Philadelphia.

BEADLE, N. C. W., AND A. B. COSTIN. 1952. Ecological classification and nomenclature. With a note on pasture classification by C. W. E. Moore. Proc. Linn. Soc. N.S. Wales 77:61–82.

BEARD, J. S. 1953. The savanna vegetation of northern tropical America. Ecol. Monog. 23:149–215.

Beard, J. S. 1955. The classification of tropical American vegetation-types. Ecology 36:89–100.

Billings, W. D. 1949. The shadscale vegetation zone of Nevada and eastern California in relation to climate and soils. Amer. Midland Nat. 42:87–109.

Böcher, T. W. 1938. Biological distributional types in the flora of Greenland. A study on the flora and plant-geography of South Greenland and East Greenland between Cape Farewell and Scoresby Sound. (Danish summ.) Meddel. om Grønland 106(2):1–339.

Bourne, R. 1934. Some ecological conceptions. Empire Forestry Jour. 13:15–30.

Braun, E. Lucy. 1935. The undifferentiated deciduous forest climax and the association-segregate. Ecology 16:514–519.

———. 1938. Deciduous forest climaxes. Ecology 19:515–522.

———. 1947. Development of the deciduous forests of eastern North America. Ecol. Monog. 17:211–219.

———. 1950. Deciduous forests of eastern North America. 596 pp. Blakiston–McGraw-Hill. New York.

———. 1956. The development of association and climax concepts; their use in interpretation of the deciduous forest. Paper given at meetings of Amer. Inst. Biol. Sci., Storrs, Conn., Aug. 28, 1956.

Braun-Blanquet, J. 1921. Prinzipien einer Systematik der Pflanzengesellschaften auf floristischer Grundlage. Jahrb. St. Gall. Naturw. Gesell. 57(2):305–351.

———. 1932. Plant sociology, the study of plant communities (Transl. by G. D. Fuller and H. S. Conard). 439 pp. McGraw-Hill. New York.

———. 1951. Pflanzensoziologie. Grundzüge der Vegetationskunde. 2 Aufl. 631 pp. Springer. Vienna.

Bray, J. R. 1956a. A study of mutual occurrence of plant species. Ecology 37:21–28.

———. 1956b. The use of ordination techniques in interpreting the eastern forest. Paper given at meetings of Amer. Inst. Biol. Sci. Storrs, Conn., Aug. 28, 1956.

Brown, R. T., and J. T. Curtis. 1952. The upland conifer-hardwood forests of northern Wisconsin. Ecol. Monog. 22:217–234.

Cain, S. A. 1947. Characteristics of natural areas and factors in their development. Ecol. Monog. 17:185–200.

Clements, F. E. 1905. Research methods in ecology. 334 pp. Univ. Publ. Co. Lincoln, Nebr.

———. 1916. Plant succession, an analysis of the development of vegetation. Carnegie Inst. Wash. (D.C.) Publ. 242:1–512.

———. 1928. Plant succession and indicators: A definitive edition of plant succession and plant indicators. 453 pp. H. W. Wilson. New York.

———. 1936. Nature and structure of the climax. Jour. Ecol. 24:252–284.

———. 1949. Dynamics of vegetation. Selections from the writings of F. E. Clements, compiled and edited by B. W. Allred and E. S. Clements. 296 pp. H. W. Wilson. New York.

——— and V. E. Shelford. 1939. Bio-ecology. 425 pp. Wiley. New York.

Conard, H. S. 1939. Plant associations on land. Amer. Midland Nat. 21:1–27.

Cooper, W. S. 1926. The fundamentals of vegetational change. Ecology 7:391–413.

CULBERSON, W. L. 1955. The corticolous communities of lichens and bryophytes in the upland forests of northern Wisconsin. Ecol. Monog. 25:215–231.

CURTIS, J. T., AND R. P. MCINTOSH. 1951. An upland forest continuum in the prairie-forest border region of Wisconsin. Ecology 32:476–496.

DANSEREAU, P. 1954. Climax vegetation and the regional shift of controls. Ecology 35:575–579.

DU RIETZ, G. E. 1921. Zur methodologischen Grundlage der modernen Pflanzensoziologie. 267 pp. Holzhausen. Vienna.

———. 1929. The fundamental units of vegetation. Proc. Internatl. Congr. Plant Sci., Ithaca, 1926, 1:623–627.

———. 1936. Classification and nomenclature of vegetation units 1930–1935. Svensk Bot. Tidskr. 30:580–589.

EGLER, F. E. 1942. Vegetation as an object of study. Philos. of Sci. 9:245–260.

———. 1951. A commentary on American plant ecology, based on the textbooks of 1947–1949. Ecology 32:673–695.

ELLENBERG, H. 1950. Landwirtschaftliche Pflanzensoziologie. I. Unkrautgemeinschaften als Zeiger für Klima und Boden. 141 pp. Eugen Ulmer. Stuttgart.

———. 1954. Zur Entwicklung der Vegetationssystematik in Mitteleuropa. Angew. Pflanzensoz. [Wien], Festschr. Aichinger 1:134–143.

GAMS, H. 1918. Prinzipienfragen der Vegetationsforschung. Ein Beitrag zur Begriffsklärung und Methodik der Biocoenologie. Vrtljschr. Naturf. Gesell. in Zürich 63:293–493.

GLEASON, H. A. 1917. The structure and development of the plant association. Bull. Torrey Bot. Club 44:463–481.

———. 1926. The individualistic concept of the plant association. Bull. Torrey Bot. Club 53:7–26.

———. 1939. The individualistic concept of the plant association. Amer. Midland Nat. 21:92–110.

GOODALL, D. W. 1954. Vegetational classification and vegetational continua. (Germ. summ.) Angew. Pflanzensoz. [Wien], Festschr. Aichinger 1:168–182.

HALE, M. E., JR. 1955. Phytosociology of corticolous cryptogams in the upland forests of southern Wisconsin. Ecology 36:45–63.

KLAPP, E. 1949. Landwirtschaftliche Anwendungen der Pflanzensoziologie. 56 pp. Eugen Ulmer. Stuttgart.

——— ET AL. 1954. Die Grünlandvegetation des Eifelkreises Daun und ihre Beziehung zu den Bodengesellschaften. Angew. Pflanzensoz. [Wien], Festschr. Aichinger 2:1106–1144.

KNAPP, R. 1948. Einführung in die Pflanzensoziologie. I. Arbeitsmethoden der Pflanzensoziologie und Eigenschaften der Pflanzengesellschaften. 100 pp. Eugen Ulmer. Stuttgart.

LENOBLE, F. 1926. À propos des associations végétales. Bull. Soc. Bot. France 73: 873–893.

———. 1928. Associations végétales et espèces. Arch. Bot. [Caen] 2(Bull. Mens. 1):1–14.

LIPPMAA, T. 1931. Pflanzensoziologische Betrachtungen. Acta Inst. et Horti Bot. Univ. Tartu 2(3/4):1–32.

LIPPMAA, T. 1933. Aperçu général sur la végétation autochtone du Lautaret (Hautes-Alpes) avec des remarques critiques sur quelques notions phytosociologiques. (Estonian summ.) Acta Inst. et Horti Bot. Univ. Tartu 3(3):1–108.

MALIN, J. C. 1956. The grassland of North America: Its occupance and the challenge of continuous reappraisals. *In* Man's role in changing the face of the Earth, ed. by W. L. THOMAS, pp. 350–366. Univ. Chicago Press. Chicago.

MASON, H. L. 1947. Evolution of certain floristic associations in western North America. Ecol. Monog. 17:201–210.

MEUSEL, H. 1939. Die Vegetationsverhältnisse der Gipsberge im Kyffhäuser und im südlichen Harzvorland. Ein Beitrag zur Steppenheidefrage. Hercynia 2: 1–372.

———. 1940. Die Grasheiden Mitteleuropas. Versuch einer vergleichend-pflanzengeographischen Gliederung. Bot. Arch. [Leipzig] 41:357–418, 419–519.

———. 1943. Vergleichende Arealkunde. Berlin. (Fide Walter 1954b.)

MULLER, H. J. 1952. The uses of the past; profiles of former societies. 394 pp. Oxford Univ. Press. New York. 1954. 384 pp. Mentor. New York.

PALMGREN, A. 1929. Chance as an element in plant geography. Proc. Internatl. Congr. Plant Sci., Ithaca, 1926, 1:591–602.

PHILLIPS, J. 1934–35. Succession, development, the climax, and the complex organism; an analysis of concepts. Parts I–III. Jour. Ecol. 22:554–571; 23:210–246, 488–508.

POORE, M. E. D. 1955. The use of phytosociological methods in ecological investigations. III. Practical application. Jour. Ecol. 43:606–651.

RAMENSKY, L. G. 1924. Die Grundgesetzmässigkeiten im Aufbau der Vegetationsdecke. (Russian) Wjestn. opytn. djela Woronesch. 37 pp. (Bot. Centbl., N.F. 7:453–455, 1926.)

RAUP, H. M. 1956. Vegetational adjustment to the instability of the site. Paper given at meetings of Amer. Inst. Biol. Sci., Storrs, Conn., Aug. 28, 1956.

SCHMID, E. 1942. Über einige Grundbegriffe der Biocoenologie. Ber. Geobot. Forschungsinst. Rübel, Zürich 1941:12–26.

SCHMITHÜSEN, J. 1950. Das Klimaxproblem, vom Standpunkt der Landschaftsforschung aus betrachtet. Mitt. der Florist.-Soziol. Arbeitsgemeinsch. [Stolzenau/Weser] N.F. 2:176–182.

TANSLEY, A. G. 1920. The classification of vegetation and the concept of development. Jour. Ecol. 8:118–149.

———. 1935. The use and abuse of vegetational concepts and terms. Ecology 16:284–307.

WALTER, H. 1954a. Klimax und zonale Vegetation. Angew. Pflanzensoz. [Wien], Festschr. Aichinger 1:144–150.

———. 1954b. Einführung in die Phytologie. III. Grundlagen der Pflanzenverbreitung. II. Teil: Arealkunde. 246 pp. Eugen Ulmer. Stuttgart.

——— AND E. WALTER. 1953. Einige allgemeine Ergebnisse unserer Forschungsreise nach Südwestafrika 1952/53: Das Gesetz der relativen Standortskonstanz; das Wesen der Pflanzengemeinschaften. Ber. Deutsch. Bot. Gesell. 66:228–236.

WANG CHI-WU. 1956. A paleoecological interpretation of the eastern forests of North America. Paper given at meetings of Amer. Inst. Biol. Sci., Storrs, Conn., Aug. 28, 1956.

WEAVER, J. E., AND F. E. CLEMENTS. Plant ecology. (1st ed., 1929) 2d ed., 520 pp. McGRAW-HILL. New York. 1938.

WHITTAKER, R. H. 1951. A criticism of the plant association and climatic climax concepts. Northwest Sci. 25:17–31.

———. 1952. A study of summer foliage insect communities in the Great Smoky Mountains. Ecol. Monog. 22:1–44.

———. 1953. A consideration of climax theory: The climax as a population and pattern. Ecol. Monog. 23:41–78.

———. 1954. Plant populations and the basis of plant indication. (Germ. summ.) Angew. Pflanzensoz. [Wien], Festschr. Aichinger 1:183–206.

———. 1956. Vegetation of the Great Smoky Mountains. Ecol. Monog. 26:1–80.

H. A. GLEASON
"INDIVIDUALISTIC ECOLOGIST"
1882-1975

Robert P. McIntosh

BULLETIN OF THE TORREY BOTANICAL CLUB
VOL. 102, NO. 5, pp. 253–273 SEPTEMBER–OCTOBER 1975

H. A. Gleason—"Individualistic Ecologist" 1882–1975: His contributions to ecological theory

Robert P. McIntosh
Department of Biology,
University of Notre Dame,
Notre Dame, Indiana 46556

In 1959 Dr. Henry Allan Gleason was cited by the Ecological Society of America as "Eminent Ecologist," although his last clearly ecological publication had been twenty years earlier (Cain 1959). In 1953 Dr. Gleason had been labeled "Distinguished Ecologist" (Gleason 1953), although his major botanical work had always been in plant taxonomy. Stanley Cain's (1959) tabulation of Dr. Gleason's publications as of 1959, at age seventy-seven while he was still actively writing and publishing, suggests that it is, quantitatively at least, incorrect to claim Gleason as an ecologist. From the beginning of his long professional career in the late 1890s up to 1958 less than 15% of Dr. Gleason's more than 200 publications were ecological; even during his major period of ecological work, 1907–1920, less than 50% of his publications were primarily concerned with ecology. If ecology can't claim Dr. Gleason as an ecologist on quantitative grounds, it will have to turn to qualitative reasons for acclaiming him as one of its most distinguished figures. Here the reasons are indisputable, for some of his relatively few ecological publications are certainly among the most significant in the pioneer years of ecology in America.

Gleason's professional life spanned very nearly the whole of the rise of ecological science in the United States. Moreover, during his early career he worked in the focal area of the genesis of self-conscious ecology, the American mid-west. Ecology surfaced, rather abruptly, as a formal discipline in the United States in the 1890s and early 1900s particularly at the universities of Chicago, Nebraska, Illinois and Wisconsin (Egerton, in press; McIntosh, in press). Here, as Gleason wrote, there appeared many of the scientists who were not instructed by ecologists, but who themselves form the pantheon of American ecology. Few would dispute Gleason's list of the pioneers of plant ecology in America, "Pound and Clements, Cowles, Gleason, Harshberger . . . Harper, Transeau, Shantz," although others (e.g., Beal, MacMillan, Shreve, Gates) might be added. The pioneers of American limnology, S. A. Forbes and E. A. Birge, and of animal ecology, C. C. Adams and V. Shelford, were also active at this time in the same area; and Gleason, early in his career, was associated with Forbes at the Illinois State Laboratory of Natural History, one of the seminal sites of ecology in the United States.

Gleason took his B.S. and M.A. at the University of Illinois in 1901 and 1904, respectively. Although he went to Columbia University for the Ph.D. (1906), he promptly returned to the University of Illinois until 1910. He then went to the University of Michigan (1910–1919) before leaving for the New York Botanical Garden where he remained until retirement. Thus, his most active period of ecological work was while he was associated with the region, the institutions and personalities significantly involved in laying the groundwork of ecology in America. It is striking that though Dr. Gleason's ecological work was quantitatively the smaller portion of his extensive publications, it is qualitatively so significant an aspect of ecology. Stanley Cain (1959) commented that Gleason "has always been a man of ideas." In an era of prolific accumulation of facts and of extended descriptions of vegetation but limited development of principles and concepts, Dr. Gleason's lasting contributions were a few ideas which represented new insights into important aspects of the young science of ecology. One idea, commonly given the eponym "Gleasonian," is his "Individualistic Concept of the Plant Association," which is a viable and even expanding tenet of current ecological thought.

There is no doubt an "ecology of ecologists," to borrow Paul Sears' (1956) apt phrase, and H. A. Gleason's ecological views were conditioned by his geographic location and by working at a singularly exciting time in biology. Living in the prairie-forest border region and on the glacial border in Illinois, Gleason was in a salient area to note the relationship of prairie and forest and the evidences of migration and successional change in these communities. Working at a time when the seminal views of S. A. Forbes on ecology and quantitative methods, H. C. Cowles on plant succession and F. E. Clements on just about everything in ecology (competition, experimentation, quantitation, paleo-ecology and, particularly, succession, climax and the nature of the plant association) were being developed, Gleason had unparalleled stimulus to check these ideas against his own observations. Growing up professionally with a newly-christened embryonic aspect of biological science which was gradually developing its own structure out of an amorphous body of biological natural history, hydrobiology, physiology, morphology and evolutionary thought (Gleason 1919), he had the advantage of not being excessively bound by an undue accretion of intellectual restrictions. These began to form rapidly in ecology, and one of Gleason's major contributions to ecology was that he strove to keep the conceptual mold from hardening prematurely.

The prairie-forest border. Gleason (1953) attributed his early interest in the relation between prairie and forest to the influence of his father, Henry Milton Gleason, who had homesteaded on the unbroken prairie, and Professor T. J. Burrill. His B.S. thesis at the University of Illinois (1901) was "The Flora of the Prairies." His earliest published ecological work (Hart and Gleason 1907, Gleason 1909a, b, 1910) was concerned primarily with prairie and prairie-forest border problems. Even then (Gleason 1910) he lamented ". . . the prairies of Illinois were connected into cornfields long before the development of ecology or phytogeography in America. . . ." Gleason studied the plant associations of sand areas of river valleys in Illinois which, like Cowles' studies of Lake Michigan dunes and Clements' early work on Nebraska sand hills, focused his attention on problems of the dynamics of vegetational change (Hart and Gleason 1907, Gleason 1909a, 1910). In these studies he accumulated the background and initiated some of the divergent views which were later to make him, as he phrased it, "an ecological outlaw" or "a good man gone wrong" (Gleason 1953). In his early ecological work, Gleason identified key problems of the prairies of Illinois, described the associations of plants on the riverine sand dunes, and particularly, addressed the biogeographical and ecological problems posed (Gleason 1909a, 1910, 1912, 1913).

According to Dansereau (1957) and Benninghoff (1964), Gleason was the first to identify the "Prairie Peninsula," which was subsequently delineated and named by E. N. Transeau. Gleason (1909a), himself, attributed the recognition of the eastern arm of the prairie province to Pound and Clements; but according to Benninghoff, he "laid down the outlines of post-glacial history of vegetation in the middle west on a purely floristic basis." Gleason's (1953) own version of his work was that it "occurred to me that we . . . could look back and deduce the past of the vegetation." What he proceeded to do was link the processes of migration and succession into an interpretation of the post-glacial history of the midwest which "ascribed the origin of this eastward extension of grassland vegetation to an earlier period of greater dryness," (Benninghoff 1964). This ascription of a post-glacial period of warmer, drier climate permitting western vegetation to advance eastward was later, in the 1930s, developed as xerothermic theory; and Gleason's interpretation was confined by pollen analysis (P. B. Sears, personal communication; Cain 1959). Gleason (1953) commented, rather ruefully, that in 1911 he and a graduate student (Fred Loew) at Michigan had the idea of studying fossils in peat bogs. Unfortunately, they only looked for macro-fossils missing the thousands of microscopic fossil pollen grains, which were not exploited until around 1916 in Europe and later in America.

Gleason, in his studies of the prairies of Illinois, noted the importance of fire in relation to prairie and forest and recognized that fire was a potent cause of succession. He wrote that forest was "everywhere pushing out upon the prairie"

(Gleason 1909a) and that prairie fires determined forest distribution, forests generally being in protected areas growing on the leeward (eastern) sides of streams or sloughs, or on bluffs or rugged valleys, where fire was less prevalent (Gleason 1912). The relation of prairie to forest on the prairie-forest border has become part of the conventional wisdom of ecologists. Gleason's own early work was based on extensive observation and what he described as "biological ratiocination" or shrewd deductions beginning with his observations, but extending to conclusions concerning conditions "not directly perceptible." (Gleason 1910). He was more aware than most of the first ecologists of the nature of the evidence and the pitfalls of the subjective inferences of ecologists in linking observation (e.g., zonation) to causes and results of ecological processes (Gleason 1927).

Succession. The concept of succession was one of the first phenomena to attract the attention of ecologists. Beginning with the classic work of H. C. Cowles (1901) on the sand dunes of Lake Michigan, it engaged the attention of most of the early plant ecologists in the United States. Notable among these was F. E. Clements, who very early formalized his famous view of the plant association as an organism. A key element of Clements' concept of vegetation was that succession was always progressive to a single climax association under the control of the regional climate. Gleason, along with W. S. Cooper and others, dissented from the rigid Clementsian concepts of succession. Gleason (1910) wrote, ". . . it is impossible to state whether there is one definite climax association in each province; it seems probable that there are several such associations each characteristic of a limited portion." Thus, he clearly came out against the monoclimax concept proposed by Clements and endorsed a much less rigid view. Gleason (1927) asserted, ". . . succession is an extraordinarily mobile phenomenon, whose processes are not to be stated as fixed laws, but only as general principles of exceedingly broad nature, and whose results need not, and frequently do not, ensue in any definitely predictable way."

Most of Gleason's ideas on succession were developed early in his career during his studies on the prairie-forest border. As a consequence, he was much influenced by problems of scale of time as well as of space. Thus, he incorporated into his concept of succession the migration of the four vegetational provinces he saw as shifting in their distribution in Illinois. He commented, "Even now a biotic migration is in progress, which is probably the direct continuation of early postglacial movements, and is doubtless as rapid and far-reaching in its effects as any of the past. . . . The actual steps in the migration are due to a series of successions, by which associations of the prairie or of the coniferous forest are replaced by others, with similar environmental demands, from the deciduous forest," (Gleason 1910). He distinguished between intraprovince succession and that involving components of two or three provinces (e.g., prairie, deciduous and coniferous forests) and attempted to portray "one phase of this great vegetational movement and of the consequent struggle for supremacy which is still being waged," (Gleason 1910).

Gleason wrote that as early as 1908 he became convinced that succession could be retrogressive and that the Clementsian concept of succession, as an irreversible trend leading to the climax, was untenable. He, of course, allowed that succession was influenced by climatic change, while Clements presumed stable climatic conditions. Gleason was not alone in rejecting Clements' views, for he comments that his ideas on succession harmonized with those of W. S. Cooper (Gleason 1927). Gleason expanded and codified his views of succession relating these to his concept of the plant association (Gleason 1926a, 1927). By his own account (Gleason 1953), at one point he encountered some difficulty in publishing his views in the botanical literature and required the endorsement of H. C. Cowles and E. N. Transeau before his "Vegetational History of the Middle West" was published (Gleason 1923). Gleason (1927), acutely aware of the influence of geographical location and scale of time on successional concepts, commented, "In the center of an association, away from any lateral successions which may be in progress at its margin, we have no past or future time phases for comparison, and we see only the fluctuations in structure from year to year." Gleason's view of succession between vegetational provinces had its counterpart in Clements' concept of the clisere; and he differed from Clements in including interformational sequences as successional, whereas Clements regarded succession as proceeding to a climax determined by stable climatic conditions. Certainly in middle America, the concept of stable climate and long-term trends to climax has proven a wraith, and the interpretation of great variability espoused by Gleason and Cooper is the more characteristic view of succession in modern ecology.

It is tempting to speculate how Clements' views may have been formed by his location, early in his career, near the center of the grassland (Nebraska), while Gleason, located on the prairie-forest border (Illinois), arrived at diametrically opposed views. These were substantially retained by both men, in spite of subsequent extensive experience in diverse areas. Is there an imprinting phenomenon in ecologists?

Gleason contributed little in the way of detailed studies of succession, but his consideration of succession effectively resisted too rigid a formalization, and his early ecological instincts appear sound, even conventional, by today's hindsight. Vegetational seres are not fixed developmental patterns following strictly ordered series: retrogressive succession does occur even within a vegetational province, and rates of succession vary between phases of rapid change and relative stability. Gleason (1910) also stated the now familiar attributes of plants in the early stages of succession, high seed production, mobile seeds and better adaptation to aberrant environment, recently enshrined as "*r*-selection."

Quantitative methods. The development of sampling methods and primitive statistical analysis of samples was substantially simultaneous with the origin of ecology. Ernst Haeckel, who coined and defined the word ecology in 1866, afterwards engaged in a dispute over the validity of statistical methods in ecology (Lussenhop 1974). Quantitative sampling methods using small areas or quadrats were introduced in a few of the earliest studies of American vegetation (Pound and Clements 1898), and in Europe. Raunkiaer (1934) is usually credited with introducing, as early as 1909, frequency or valence analysis based on the determination of the presence or absence of species in a number of quadrats.

In his first published ecological study (Hart and Gleason 1907), actually done in 1903, Gleason used quadrats and counted or estimated the number of individual plants of each species, recognizing eight density classes. Although his subsequent publications did not use the quadrat method extensively, Gleason obviously thought about sampling and statistical considerations and used them in his teaching (Gleason 1920, 1953). It is not clear how much, or in what way, he may have been influenced by his association with S. A. Forbes at the Illinois State Laboratory of Natural History. Forbes was an extremely productive and imaginative ecologist who explored the possibilities of sampling and statistical analysis, one such study (1907) published in the same journal issue as Gleason's first ecological publication. Methods developed in the study of marine populations were also tried on freshwater plankton in the Illinois River as early as 1897 (Lussenhop 1974), so quantitative methods were in the air during Gleason's early work in Illinois. Whatever the source of ideas or the case for priority, H. A. Gleason was clearly one of the earliest and most insightful proponents of the use of quantitative methods in terrestrial ecology. Goodall's (1962) extensive bibliography of statistical plant sociology shows only Jaccard and Raunkaier, in Europe, publishing studies on use of quantitative techniques earlier than Gleason; and four of only eleven articles he cites in the United States' literature in the 1920s were Gleason's. Gleason's interest in quantifying was a distinctive personal trait. For some years he was a member of a group which met regularly for bridge on Sunday afternoons. During the afternoon they would recess, have a few drinks, and then play again. Gleason kept a cumulative record of the scores and found that he and another non-drinker had greatly improved scores during the second session (A. Cronquist, personal communication).

Gleason (1920) wrote that he began an intensive study of the relation between frequency index and random distribution in 1911, but his first paper on the subject did not appear until 1920. Gleason was cognizant of the problem of heterogeneity, of the effects of quadrat size and species number, and of the necessity of locating quadrats randomly or, at least, objectively, in contrast to the common practice of sampling with a single quadrat strategically located to represent the presumably homogeneous association. He also recognized that the use of contiguous quadrats could bias the results and that no effective increase in accuracy was gained by increasing the number of quadrats beyond a certain number. Gleason was acutely aware of the problem of distribution of plants and noted that the quadrat method depends on the uniformity of the association (Gleason 1917, 1920). He recognized that uniformity was not perfect, although he said that the distribution of organisms tended toward randomness in older and longer established stands of vegetation. This was borne out in subsequent work (Whitford 1949, Curtis and McIntosh 1950, Greig-Smith 1964). Gleason (1920) wrote, "If plants were distributed absolutely at random . . . that is if the association were absolutely uniform throughout, separate quadrats would never be necessary." Strikingly, although Gleason recognized that plants were not necessarily distributed at random but were sometimes clumped, he assumed that plants were characteristically distributed randomly producing a uniform association. He wrote that the correlation of frequency and density would be exact ". . . if all plants were distributed through the extent of an association precisely in accordance with the laws of probability and chance. This sort of distribution is in fact characteristic of vegetation," (Gleason 1929). Although Gleason was correct in asserting there is a tendency for many species to approach randomness in older stands, random distribution is not characteristic of plants or of organisms generally. As late as 1952 Goodall wrote, "Unfortunately, the persistence of this view still bedevils much of the current literature on quantitative ecology," and much later, the same untrue assumption is still common in much ecological theory and modeling. The problem of heterogeneity or pattern persists in current sampling work and in ecological theory; and that Gleason did not resolve it, in what was one of the earliest efforts to do so in terrestrial studies, should not diminish his insights into the problem. Goodall (1952) stated that Gleason, in 1920, was, "The first man to study quantitatively the distribution of individual plants. . . ." Gleason demonstrated that, if a plant species was randomly distributed, the proportion of sample quadrats in which it would occur could be determined by the equation $1-[1-(1/q)]^n$ where n individuals had been found in q quadrats. He recognized that a departure from the expected values of this relation, generally a lesser value, indicated a nonrandom distribution and, thus, was the first to devise a measure of the aggregation of species using the mathematics of probability.

Much of the early work in plant ecology was preoccupied with description of vegetation. Gleason urged the advantages of the quadrat method as a necessary adjunct to photography and verbal description but not to their exclusion. He went further, recognizing that having described an area of vegetation the next order of business is to compare it with other areas similarly described. Jaccard, in 1902, had developed a coefficient of similarity to compare two communities based on presence-absence data. Gleason (1910), in one of his earliest ecological articles, commented that the weakness in this was that a rare species is given as much weight in the comparison as a common one. He suggested the desirability of weighting each species by the area it occupied (cover) to improve the coefficient but said, "Unfortunately no practical method for this has been devised." A decade later Gleason (1920) was the first ecologist to suggest weighting a similarity coefficient by a quantitative measure, the frequency index of the species. The use of similarity coefficients in diverse ways, with and without weighting, has subsequently been extensively pursued as a basis for classification and ordination of communities and testing homogeneity of vegetation (Goodall 1973). Gleason's initial effort was another example of his keen grasp of the promising lead, if not a final success.

Gleason's facility in identifying problems of key ecological interest was further demonstrated in his work with species-area curves. It had long been familiar to ecologists that the number of species increased with the area studied, and Arrhenius (1921) produced a mathematical equation relating species and area. Gleason (1922) showed that Arrhenius' equation estimated excessively large numbers of species for large areas, and he amended Arrhenius' equation by recording the number of species in samples of increasing size. Plotting the cumulative number of species against the logarithm of the area, he found a nearly straight line expressed by the equation $X = a + b \log y$ (Goodall 1952). He tested this empirical observation using it to predict the number of species in larger areas of aspen and beech-maple forest and showed that the estimated number of species closely approximated the observed. Species-area relations were pursued extensively in subsequent ecological studies; and the recent interest in the subject of diversity, its expression mathematically and its importance to ecological theory, is a continuation of this early exploration.

Goodall (1952) and Greig-Smith (1964) both credit Gleason (1925) with being the first plant ecologist to use a quantitative measure of interspecific association. Again, it is not clear how much he may have been influenced by his earlier association with S. A. Forbes, who had been interested in measures of interspecific association as early as 1907. Gleason was basically concerned with the general phenomenon of distribution of plants, the relation of species to area and the problem of rare species. In effect, he was one of the earliest ecologists seriously to pursue the use of mathematics in problems which have become central concerns of modern ecology under the rubrics of homogeneity, pattern and diversity. He asserted (1925) that the questions of the nature of plant distribution and association can be tested by statistical methods. In his pioneer studies of interspecific association he assumed random distribution and assessed the associations between species using the elementary mathematics of probability. He compared actual counts of species co-occurrence with the expected, although he did not use a test of significance. Gleason concluded that the species in aspen and beech-maple forests were distributed independently, reinforcing the idea, derived from his previous studies of distribution of individual plants, that plants were randomly distributed. This also suggested to him that the environment was locally homogeneous with respect to the plants.

One of the earliest attempts to assert a mathematical relation in ecology was Raunkaier's famous reversed "J"-shaped curve of frequency distribution which came to be called "Raunkiaer's Law." This model showed a decreasing number of species in successive 20% frequency-class intervals but an increase in the 80–100% class and was widely interpreted as being a fundamental community characteristic indicating homogeneity. Gleason (1929) clearly pointed out the error in the "Law" which, however, persisted in the ecological literature for another twenty-five years in spite of numerous subsequent demonstrations of its inadequacy (McIntosh 1962).

In Gleason's small flurry of four papers on quantitative ecology (1920–1929), he accurately touched on many of the key issues of subsequent interest in quantitative ecology. However, his enthusiasm for the use of frequency as a rapid means of determining density somehow persuaded him that plants are distributed at random over a small area (Gleason 1925, 1929, 1936). This is not usually the case, and it is something of an enigma that Gleason, who recognized that species often formed clumps due to vegetative reproduction and that the degree of clumping changed during succession, rested so much of his later thinking on the idea that species were uniformly, by which he meant randomly, distributed in an association. However, that he was in error in some of his interpretations does not vitiate the fact that he took first steps into areas which have proved extremely productive and, moreover, are, in modern guise, still highly significant and active areas of ecology e.g., pattern, interspecific association, species diversity, and species-area relationships. It may be regretted that he did not pursue this part of his ecological work, for clearly he had a sure instinct for significant problems in ecology requiring quantitative analysis and more mathematical facility than most of his contemporaries. His own assessment was, "The general value of quantitative methods has been so thoroughly shown . . . that we may predict a still greater development in the quarter century before us," (Gleason 1936). Certainly his prediction has been amply substantiated.

The individualistic concept. Gleason's most significant and most lasting contribution to ecology was his "individualistic concept." It persists in the current research literature and recent textbooks of ecology as one of the basic tenets of modern ecology, although it earned him little credit when he propounded it. According to Gleason (1953), following its first full-dress statement and subsequent discussion at the International Botanical Congress in 1926, he was anathema to ecologists; and he added, ". . . for ten years or thereabout, I was an ecological outlaw, sometimes referred to as 'a good man gone wrong.' " This view of his personal standing may have been somewhat exaggerated, and Gleason's then heretical idea is now widely recognized as part of the conventional wisdom of ecology. Collier et al. (1973) wrote that the individualistic concept "constitutes one of the most influential and widely accepted views at the present time." According to Poole (1974), "Forty years later most of Gleason's views became acceptable," and Krebs (1972) commented, "The present view of the nature of the community lies closer to the individualistic view than to Clements' superorganismic interpretation."

The specific stimulus for Gleason's earliest (1917) formal exposition of his views on the nature of the plant association was, in fact, F. E. Clements' description of the plant community as an organism in the first American book on ecology (Clements 1905) and in subsequent writings. Gleason (1926b) recorded his response to Clements' "hard-and-fast crystallization of ecological phenomena into fixed and inviolable laws." According to Clements' very influential view of the community, the "association" was an organic entity. He held the rather extreme position that the successional development of a community is comparable to the development of an individual organism. Not all ecologists, in these early days of ecology, subscribed to Clements' supraorganismic concept of the plant community; but the most widely held concept of the community was that of a group of organisms constituting a well-defined, integrated, ecological entity. This concept was not limited to plant ecologists but pervaded ecology at large (Dice 1952, Bodenheimer 1958, Mills 1969). The organismic analogy had been used by Thienemann, one of the pioneers of limnology, "Every life community forms together with the environment which it fills, a unity, and often a unity so closed in itself that it must be called an organism of the highest order," (*vide* Bodenheimer 1958). Animal ecologists, although less concerned about formalizations of community than plant ecologists, also viewed the community as an organized and even evolving entity. Allee et al. (1949) defined the community as ". . . a natural assemblage of organisms which together with its habitat has achieved a survival level such that it is relatively independent of adjacent assemblages of equal rank and given radiant energy, it is self-sustaining." Alfred Emerson, one of the authors of this distinguished book on animal ecology, was an explicit proponent of the community as evolving an organismic unity (Emerson 1960). Bodenheimer (1953) wrote that the highly integrated supraorganismic concept of the community was stressed in nearly every textbook of ecology and, ". . . backed by established authority, is generally regarded if not as a fact, then at least as a scientific hypothesis not less firmly founded than the theory of transformation." Gleason's individualistic concept was, therefore, contrary to a view widely held, from the beginning of ecology as a science and throughout his career as part-time ecologist, of communities as relatively organized and discrete entities which were traditionally classified into ecological units on various grounds. The major differences between most ecologists, particularly plant ecologists, were on how natural groups of organisms should be classified, but there was near unanimity on the idea that organisms formed organized aggregations subject to being classified as natural groups (Whittaker 1962).

Gleason's own field observations started him, early in his career, toward a quite different concept of the dynamics of vegetation, and the major themes of his individualistic concept are clearly seen in one of the earliest of his ecological studies (Gleason 1910). He commented that this study was based almost exclusively on the observational method, which, he asserted, has led and will lead in the future to some of the most important results in ecology. His later more explicit expressions of the individualistic concept are anticipated in several respects. He wrote, "*. . . the presence of a particular flora in the sand is due partly to the selection from the surrounding associations of various species with certain physiological requirements,*" (italics mine). He noted the difficulty for vegetation study created by extensive human disturbance, but at this stage he accepted the view that, ". . . experience in natural conditions justifies the statement that associations are definite organized units and that all vegetation is composed of them, either mature and fully differentiated or in process of organization." He warned, however, "... it is as difficult to formulate a satisfactory definition of an association as of a species and as unnecessary." In a second statement anticipating his later ideas he wrote, "*No two areas of vegetation are exactly similar, either in species, the relative number of individuals of each or their spatial arrangement, and the smaller the areas to be compared the greater proportionately are the differences between them,*" (italics mine). He also raised the often repeated query, "how great a variation may occur in the structure of the vegetation without the identity of the association being changed." A third statement suggesting his later views was, "*The more widely the different areas of an association are separated the greater are the floral discrepancies,*" (italics mine). Gleason traced the genesis of his thought, in a footnote in a later formal expression of the individualistic concept, to these early writings (Gleason 1926a).

Gleason first brought together his maturing ideas on the plant association in 1917 (in the Bulletin of the Torrey Botanical Club), and initially he used the term "individualistic concept" in the more encompassing phrase "the individualistic concept of ecology" rather than the more restricted and more familiar "The individualistic concept of the plant association." He wrote that it is almost axiomatic that vegetation is the resultant of its component individuals; and he asserted that, "... no two species make identical environmental demands." In the same year Joseph Grinnell (1917), usually credited with being the founder of the animal niche concept, wrote, "It is, of course, axiomatic that no two species regularly established in a single fauna have precisely the same niche relationships."

Gleason's first emphasis was the individualistic nature of the species and of the habitat. The community in any area, he said, was controlled by two factors, "the nature of the surrounding population, determining the species of immigrants ... and the environment, selecting the adapted species." "In the same limited region, that is, with the same surrounding population, areas of similar environment, whether continuous or detached, are therefore occupied by similar assemblages of species," which he called a plant association (Gleason 1917). He emphasized that the plant populations modified and came to control the environment and that the species of one association may be excluded from another by the effect of the organisms of that association in modifying the environment. Adjacent associations may then meet in a narrow transition zone even though the change in physical environment is gradual. Gleason observed that removal of one association would allow the expansion of the other, suggesting that the control of the environment by organisms was, in fact, limiting the spread of an association. He particularly emphasized the tendency toward uniformity of distribution of species within the association which, he said, increased with age of the association. Gleason, in his quantitative studies, and in his early considerations of the community concept, regarded the association as uniform and the species in it uniformly distributed, by which he meant randomly distributed, a position which he modified in his later articles.

In later years, Gleason's critics (Becking 1957, Daubenmire 1966, Harper 1967) sometimes suggested that he held that species which occupied an area had little effect on each other. Although he emphasized selection by the environment among species, Gleason did recognize that, "... as soon as the ground is occupied competition restricts it [the species] to its proper proportion." He also noted the capability of initial occupants which occupied the ground closely to exclude other species.

Gleason summarized his view of the association bringing together the effect of surrounding populations which supply migrants, historical accidents, time for migration, and selection by the local environmental conditions from the available migrants producing areas of differing species and proportionate composition, which differences increased with distance. Gleason (1917) succinctly stated his concept of the community:

> Whether any two areas, either contiguous or separated, represent the same association, detached examples of the same one, consocies or different associations and how much variation of structure may be allowed within an associa-

tion without affecting its identity, are both purely academic questions, since the association represents merely the coincidence of certain plant individuals and is not an organic entity of itself. While the similarity of vegetation in two detached areas may be striking, it is only an expression of similar environmental conditions and similar surrounding plant populations. If they are for convenience described under the same name, this treatment is in no wise comparable to the inclusion of several plant individuals in one species.

In general, Gleason asserted, environment varied gradually in space, and change in vegetation often was gradual and progressive as was the change in environment. He commented wryly, "This condition has in fact led some students to complain that it was difficult or impossible to distinguish associations in the vegetation of the western states. This would be a sad state of affairs if all vegetation were composed of definite organic entities, but is quite to be expected when vegetation depends upon environmental selection of favored individuals," (Gleason 1917). Gleason, in effect, threw down the gauntlet to the proponents of the organismic view of community, but apparently very few noticed, although Tansley (1920) denied Gleason's concept.

Nearly a decade later Gleason (1926a) reasserted his avowedly heretical ideas as "The Individualistic Concept of the Plant Association." He noted the "chronic inability" of ecologists to agree on the nature and classification of plant associations and diagnosed their failure as due to "attacking the problem from the wrong angle." This absence of consensus persists today, and Daubenmire (1966) referred to a "spectrum of concepts, terms and methods so broad as to discourage the novice and confuse even the specialist." It is this persistent confusion which suggests the failure of the traditional community concept as a theoretical framework for ecology. Gleason fortified, by 1926, with his work in sampling and statistical ecology, modified his earlier assertions about the homogeneity of the plant association. Although attesting to the reality of the individual stand or plant association in the concrete instance as having measurable extent and a high degree of structural uniformity making different parts look reasonably alike, he noted the possibilities for disagreement concerning the association. He wrote (1926a), "More careful examination of these areas, especially when conducted by some statistical method will show that the uniformity is only a matter of degree, and that two sample quadrats with precisely the same structure can scarcely be discovered." He stated the principal point, ". . . precise structural uniformity of vegetation does not exist, and . . . we have no general agreement of opinion as to how much variation may be permitted within the scope of a single association . . . we are treading on dangerous ground when we define the association as an area of uniform vegetation." He also noted the lack of concordance between recognized associations and environment, which complicated the problem, and the failure of many plants to adhere to a particular association, i.e., to exhibit "fidelity," as described by traditional phytosociologists. All this led, he said, to a community "so heterogeneous as to lead observers to conflicting ideas as to its associational identity." It was in this version of the individualistic concept that Gleason introduced his familiar example of the transition of forests from the upper to the lower Mississippi to illustrate his idea. He was, more than most of his contemporaries, impressed by the heterogeneity and variation of vegetation both in space and time and posed his ultimate question, "Are we not justified in coming to the general conclusion, far removed from the prevailing opinion, that an association is not an organism, scarcely even a vegetational unit, but merely a *coincidence?*" (italics his). He answered it himself reiterating his ideas of dispersion of seeds from the surrounding populations, the individualistic physiological properties of plant species which do not have identical optimal environments, and the selection of species by a particular complex of environmental conditions modified by the organisms. Each area, he said, is a resultant of a unique mixture of migrants, environment and historical sequence, and there are no grounds for recognizing one as normal or typical. Gleason wrote, "Neither are we given any method for the classification of associations into broader groups."

As in the 1917 version of his concept, Gleason related it to succession. If environment changed progressively in one direction, the vegetation also changed but not, he added, in a regular and fixed sequence. Succession is also individualistic. He summarized, "The sole conclusion we can draw from all the foregoing considerations is that the vegetation of an area is merely the resultant of two factors, the fluctuating and fortuitous immigration of plants and an equally fluctuating and variable environment." Gleason, however, recognized similarity of vegetation in an area, due to similar causes over the whole area, which led to the recognition of associations by plant ecologists. Under the individualistic concept, he said, the fundamental idea is "the visible expression, through the juxtaposition of individuals, of the same or different species and either with or without mutual influence, of the result of causes in continuous operation." He noted that similar juxtaposition of plants is simply due to the similarity in contributing causes. He specifically addressed a question that many critics of the individualistic concept and its intellectual descendants, the continuum concept and gradient analysis, have said refutes the individualistic concept and proves the reality of distinct communities. Gleason wrote, "Where one or both of the primary causes changes abruptly, sharply delimited areas of vegetation ensue." The observation of discontinuity on the ground is entirely compatible with the individualistic concept. Gleason reiterated his beliefs that rigid definition of the association is impossible and that logical classification of associations into larger groups or successional series has not (and probably will not) be achieved.

Gleason's paper was published in the spring (1926a). That August it was discussed by George Nichols (1926) at the International Congress of Plant Sciences. In Gleason's (1953) recollection, twenty-seven years later, "George Nichols pulverized my theory. Worse than that he ridiculed it." The taxonomists in the audience assured Gleason he was correct, but not one of the ecologists believed his ideas or would even argue the matter, he said. The record does not show what went on in the halls of the Congress, but Nichols' paper was not a particularly venomous attack upon Gleason's concept and does not justify Dr. Gleason's recollection of Nichols "pulverizing" his theory. Nichols accepted most of Gleason's factual observations, conceded numerous discrepancies between fact and theory, but asserted that the facts ". . . would seem to be equally capable of a very different interpretation from the one which he (Gleason) has given them." This is a familiar aspect of scientific work, and is one of the interesting facets of science which historians and philosophers of science deal with. In his rebuttal to Nichols, Gleason also noted their points of agreement and acknowledged that Nichols had stated his (Gleason's) views with a degree of accuracy and fairness which is seldom found in a critique.

Nichols drew an extensive comparison between the difficulties of recognizing and classifying species and the systematic study of plant communities. To Nichols the perennial disagreements cited by Gleason as grounds for doubting the possibility of classifying communities were no worse than similar problems faced by plant systematists, and he saw no more reason to refuse to recognize a vegetational unit than a species because of the existence of intermediates and differences of opinion. Nichols (1926) summarized his position, "The point at issue is simply this: that in the systematic study of plant communities, intergradation and absence of sharp lines of demarcation are to be expected even more than in the systematic study of plants. The existence of this condition no more invalidates the concept of the association as an entity or unit than does the corresponding discrepancy in taxonomic botany invalidate the generally accepted concept of the species." He also attacked Gleason's interpretation of the complexities of the tropical forest as buttressing the individualistic concept saying that tropical forest came closer to realizing the uniform climatic climax than temperate forest. Three or four decades later the status of tropical forest was still in doubt (Greig-Smith 1952, Richards 1963, Williams et al. 1969).

According to Nichols, ecologists agreed in defining the community as exhibiting "essential uniformity" in composition and structure, "*provided each be permitted to place his own construction upon what is meant by essential uniformity,*" (italics mine). Nichols asserted that classification is an essential of all science in bringing order out of chaos, a position echoed by Daubenmire forty years later (1966). He concluded with the concession that, ". . . we are dealing with merging phenomena," and quoted W. F. Ganong, ". . . the only possible procedure is to select the extreme marked types of the groups and giving these careful study and description to describe the intermediate kinds according to their positions between the types," (cf. Poore 1964). Anyone who has read the extensive discussion in the last two decades of the pros and cons of the possibility of, or techniques for, classifying vegetation will very likely reread this discussion with a feeling of *déjà vu*.

Gleason (1926b) objected to Nichols' analogy between association and species on the grounds that species reproduce and have genetic continuity. Moreover, he said that the species question was so chaotic, that no ecological principles "should be supported on so weak a crutch." Both Gleason and Nichols commented that they were not so far apart in their views; and Nichols said, "Practically everyone will agree with him (Gleason) on his three fundamental theses," as enunciated in the 1917 paper. Given so much agreement, one wonders at the failure of Gleason and Nichols or the positions they represent to arrive at some common ground then or since.

The individualistic concept, as expressed by Dr. Gleason in the 1917 and 1926a versions, had little impact on the general tone of ecological thought about the community. Weaver and Clements' (1929) *Plant Ecology*, the most influential textbook of plant ecology, ignored it; and the other early standard ecology texts and references which were beginning to codify the facts and principles of ecology in the 1920s and 1930s, similarly, paid it no attention. Chapman (1931) quoted Gleason's views on succession, as opposed to those of Clements, but did not refer to Gleason's consideration of the community. It must have remained buried in the consciousness of some ecologists, for over a decade later the individualistic concept made its third and most formal appearance, in a symposium, Plant and Animal Communities (Just 1939), which Allee (1939) described as "the first ambitious attempt to arrange a stock taking of ecology." Gleason described his 1939 article as "merely a restatement of the subject in different terms; it is in no way different in principles or conclusions from the first presentation in 1926," (Gleason 1939). He allowed that uniformity, area, boundary and duration are essentials of a plant community, but that uniformity, without which the other essentials become poor criteria, is insufficient to recognize a community as an organized unit. Moreover, he reiterated that differences in a series of communities cumulate with distance "*. . . so that the ends of the series may be strikingly different, although connected by imperceptible or apparently negligible intermediates,*" (italics mine). Gleason, in his several presentations of the individualistic concept, argued from a limited number of theses that he regarded as first principles. In the discussion following the 1939 paper, two of the speakers (T. Lippma and H. S. Conard) accepted Gleason's theses and logic, much as Nichols had agreed with most of his 1926 paper, but could not accept his conclusions. Conard commented that the traditional association had been the most fertile and productive concept in ecology. He added, "It is therefore so useful that whether logical or not, I am for it." A. E. Emerson, a proponent of the organismic view of the community, also opposed Gleason's conclusions commenting in the discussion, "I do not think the only reality is the individual and that the associations are not real integrated units."

The individualistic concept—retrospectus, conspectus, prospectus. Dr. Gleason published his last strictly ecological article in 1939, and to the great loss of ecology, he devoted the remainder of his long career in botany to plant taxonomy (Maguire 1975). He is, perhaps the only person cited in bibliographies of ecological articles as author of a major ecological concept and as the source for the plant nomenclature used. The somewhat belated recognition, in 1953 and 1959, of his contributions to ecology, which was mentioned at the beginning of this article, could be considered as an effort to make amends for an earlier lack of recognition of Dr. Gleason's contributions to ecology. These, particularly the individualistic concept, are of notable interest in the history of ecology, are at the core of certain aspects of current ecology in a period of enormous change and it seems likely, will continue to be recognized in future developments of ecology.

It is a striking fact that almost none of the major ecological texts or references published in the United States prior to ca. 1949 cited any of the three articles in which Gleason put forth his individualistic concept. In fact, the only standard, pre-

1950, plant ecology text which cited Gleason's concept was W. B. McDougall's (1927) *Plant Ecology,* which hardly mentioned it, however, and adopted the Clementsian organismic view completely. Thus, the individualistic concept, though clearly presented no less than three times over two decades, had little impact on ecology when first published. Nichols (1926) attacked it; the famous British ecologist, Sir Arthur Tansley, denied it (1920); but most ecologists simply ignored it, a far worse fate. As Egler (1951) stated, "The weight of the opposition against this presumptuous iconoclast was so powerful that it is only in the last few years that his ideas have been gaining credence."

It is not clear from the available evidence why the individualistic concept, when first put forth by Gleason, had so little impact. It appeared in well-known journals, and certainly the 1926 and 1939 articles, and the discussion of them, had substantial visibility. It is not clear that there was so closed an establishment in that day that Gleason's divergent views could not get a hearing. He did assert (Gleason 1953) that his paper on the history of vegetation in the midwest faced difficulties, of an unstated sort, getting published; but he did not mention similar difficulties with his several papers on the individualistic concept, although he commented that this idea made him persona non grata in ecological circles. His ideas were not so different from those of some other ecologists who did not subscribe to the full range of the reigning Clementsian position. Not all ecologists, as Egler (1951) puts it, "were won by its dogma and ritual." Nevertheless, there were powerful intellectual forces which limited any substantial acceptance of the individualistic concept even among non-Clementsian ecologists; and, no doubt, underground currents not appearing in the published records militated against it. Winning converts to a new scientific concept, particularly one which departs radically from the conventional scientific thought of an era, is not easy; nor does science necessarily advance by application of cool, clear-headed logic to a set of established facts as the discussion of the individualistic concept by Gleason, Nichols, Emerson and Conard suggests (Kuhn 1970, Brush 1974). Gleason's concept languished some twenty to thirty years, depending on which dates one takes as its premier presentation and subsequent reappearance.

The individualistic concept was clearly given a new birth in the American literature in Ecological Monographs, 1947. Three distinguished botanists (two in the same symposium), Drs. F. E. Egler, S. A. Cain and H. L. Mason, strongly supported Gleason's views (McIntosh 1967). Gleason (1953) mistakenly indicated that Mason had arrived at the idea independently, but Mason (1953) graciously asked to be included among those who had been influenced by Gleasons thought. Cain described his own views as "a 20th century affirmation of Gleason's individualistic *hypothesis,*" (italics mine) suggesting, by implication, that Gleason's concept, so long unrecognized, was a product of an earlier century. All, apparently, agreed with Egler who adopted "wholeheartedly and without exception the individualistic concept of the association as developed by Gleason"; and Egler added that he considered "these all-but-forgotten papers (of Gleason's) as being of top significance in the entire development of American vegetational thought." In 1949 W. D. Billings stated that vegetational boundaries in shadscale are arbitrary but useful due to the individualistic boundaries of the dominant species of shadscale (McIntosh 1967). Also in 1949, the major American textbook of animal ecology (Allee et al. 1949) at least cited Gleason's 1926 and 1939 papers on the individualistic hypothesis, although its views of the community and succession were not markedly influenced by his ideas. Obviously, the recognition of a scientific idea does not develop in a continuous manner, for a survey of the major references in plant and general ecology shows that while nearly all pre-1949 books did not cite Gleason's papers on the individualistic hypothesis, nearly all post-1949 books cite one or more of them. Odum's (1953) important textbook on general ecology does not cite Gleason's concept in the second edition but it appears in the third. Most of the recent spate of general ecology textbooks in the 1970s not only cite Gleason's concept but give it, or its intellectual derivatives, rather extensive, and usually favorable, coverage as the quotations earlier in this article suggest.

In the early 1950s there appeared, in

the United States, the leading edge of an extended source of new support for and documentation of Gleason's individualistic concept. This work appeared, simultaneously and independently, in the work of John Curtis and his students as the "vegetational continuum" and in that of Robert Whittaker as "gradient analysis" (McIntosh 1967, Whittaker 1967). The conception of the samples or stands of a community as forming a continuously varying series on an environmental gradient, which was extensively documented in the work of Curtis, Whittaker, and many others, was clearly anticipated by Gleason (1939), who referred to strikingly different ends of a vegetational series "connected by imperceptible or apparently negligible intermediates" and a "vegetation which varies constantly in time and continuously in space." Gleason's largely subjective or intuitive ideas on the community were supported by a variety of studies many of which were based on quantitative indices of association and coefficients of similarity both of which he had pioneered. Curtis and McIntosh (1951) asserted that their vegetational continuum supported Gleason's concept, and Curtis (1959) flatly stated, "The entire evidence of the P.E.L. [Plant Ecology Laboratory] study in Wisconsin can be taken as conclusive proof of Gleason's individualistic *hypothesis* of community organization. . . ." (italics mine). Conclusive proof is a difficult and elusive thing, and subsequent developments in continuum studies, gradient analysis and ordination have not achieved the complete consensus that conclusive proof would appear to require. Gleason's individualistic concept has, however, in the last twenty-five years, and fortunately in the latter part of its author's lifetime, been brought out of the dust bin, dusted off and today occupies a respected position in plant and general ecology. It has not had an entirely triumphal march to a scientific throne to receive a crown of gold and jewels signifying its "truth," but it is clearly, now, one of the major elements of ecological thought being tested in the current hurly-burly of ecological modelling and theory construction, often without recognition of its origins in Gleason's prescient thought.

Gleason had entitled his idea the individualistic concept and wrote it, specifically, as opposed to the widely heralded organismic "theory" of community, particularly as propounded by F. E. Clements (Gleason 1917, Egler 1951). Subsequently others came to calling it the individualistic hypothesis (Curtis 1959, Cain and Castro 1959, Whittkaer 1975); and, in a review of the continuum concept (McIntosh 1967), I, too, mistakenly attributed this use to Gleason, an error which shocked some ecologists (Langford and Buell 1969). Ecologists have been relatively slow in developing an effective body of theory and reticent in applying the approved appelation "hypothesis" and "theory" to their ideas. The hypothetic-deductive ideal of science has not easily been met in ecology, and, as Macfadyen (1975) notes, few ecologists pay much attention to philosophical or historical aspects of science. Recent zestful excursions into ecological theory have led to wider, and sometimes promiscuous, use of these terms even as applied to rather casual thoughts. In any event, if niche theory is accorded a place in the current lexicon of ecologists, it does not seem inappropriate to apply the term hypothesis to Gleason's more tentative concept which has been as extensively tested in recent decades as any ecological idea.

The resurgence of Gleason's individualistic hypothesis, nee concept, is strongly linked with the development in plant ecology of the widespread use of quantitative methods for analyzing vegetation (McIntosh 1967, 1974). This is particularly appropriate in bringing together two areas of Gleason's major contirbutions to ecology. It is neatly exemplified in Poore's (1964) study of integration in the plant community juxtaposing the still unsettled organismic-individualistic controversy, species-area curves and coefficients of community, to all of which Gleason made distinctive contributions. It has been widely asserted that Whittaker's gradient analysis and Curtis' continuum have lineal intellectual descent from Gleason's individualistic concept (Ponyatovskaya 1961, McIntosh 1967, Whittaker 1967). Gleason (1953) himself commented, "Now Curtis and his colleagues at Wisconsin are getting much better supporting evidence and will undoubtedly revise and extend my ideas in many ways but without changing their essential nature," (cf. Egler in Dansereau, 1968). It is clear that Gleason's concept represented one of two sometimes polarized views in ecology which, according to Ponyatovskaya (1961), arise "from two different conceptions of the basic object of investigation." It is a tribute to Gleason's intellectual capacity and the courage of his convictions that he created and defended the individualistic concept through decades of neglect while an almost dichotomously opposed view held sway. One assessment of the qualities which moved Gleason is given by A. S. Watt (1964), who described him as "the empiricist, the mathematically-minded analyst, systematist and plant geographer, to whom vegetation is a flux with every bit unique, impossible of prediction and even of classification." It is symbolic of the change in recognition of Gleason's contribution to ecology that this evaluation appeared in the Jubilee Symposium of the British Ecological Society, in an article by Watt attempting to integrate Gleason's concept with the organismic view of Clements.

Some have suggested that the conceptual dichotomy over which plant ecologists have argued, almost as long as ecology as a discipline has existed, is a "non-existent problem" (Anderson 1965) or that the apparent antithesis persists because of failure to state the hypothesis clearly and to devise adequate tests (Goodall 1963). An extreme, and characteristically tart, view is that of Egler (1962) who says that the controversies and "misunderstandings of each other are on a par with racial prejudices." In spite of manifest difficulty in resolving all of the issues to everyone's satisfaction, it does seem clear that Gleason, in his statement of the individualistic concept, demonstrated his nose for the key intellectual concern in ecology. It need not be argued that he was entirely correct or that his idea would not require revision and extension. These it is certainly getting, but Gleason's contribution must be seen not simply as one of historical interest but very likely as one of the key concepts of modern and, perhaps, future ecological thought.

The individualistic concept is not at present widely appealed to in unmodified form any more than Darwinian evolution is unmodified by findings and events since the publication of the *Origin of Species*. There are questions concerning techniques, logic, and even philosophy which critics of the individualistic concept and its intellectual heirs still raise (Becking 1957, Daubenmire 1966, Langford and Buell 1969, Morrison and Yarranton 1974). The extended literature on the subject has been considered by McIntosh (1958, 1960, 1967, 1974), Curtis (1959), Ponyatovskaya (1961), Goodall (1963), Whittaker (1962, 1967), and Dansereau (1968), and the recent Handbook of Vegetation Science, Vol. 5, Ordination and Classification of Communities edited by R. H. Whittaker (1973b) provides much additional up-to-date material (see articles 2–5, 8, 17). Much of the discussion concerning Gleason's concept and its successors, the continuum and gradient analysis, pits them against the classical organismic view of the community which few, if any, modern ecologists subscribe to. Goodall (1963) and others point out that there is no necessary antithesis between a continuum and classification, and Gleason and recent proponents of his individualistic concept, such as Curtis and Whittaker, agree that classification is possible but arbitrary (McIntosh 1967). Gleason, in a semi-popular book, "The Natural Geography of Plants," described major vegetational units, but reasserted the elements of his individualistic concept conceived over a half-century earlier (Gleason and Cronquist 1964).

It is not perhaps surprising to find in the extended discussion of the nature of community organization, methods of analysis and community classification, that Gleason's original concept is sometimes misinterpreted. Various straw men have been set up, in place of Gleason's actual ideas, and resolutely beaten to a pulp. I have reviewed the development and substance of Gleason's several versions of the individualistic concept earlier, but it seems desirable to emphasize certain aspects in view of many criticisms directed at it explicitly, or often, implicitly. It should not be necessary to comment in relation to Gleason's concept that, "stands of trees are not chance combinations of trees without relation to their environment." It is not proper to say, as some have, that Gleason regarded the association as purely a random phenomenon bringing together all possible combinations of species. He certainly attributed great importance to the chance arrival of immigrants from the surrounding populations, but the association was a function of the selection among these by the environment of the site as modified by the organisms entering it, and he recognized that the time or sequence of arrival was also a crucial factor. J. T. Curtis (1959), in asserting that his classical work on the vegetation of Wisconsin was in full support of the "individualistic hypothesis," wrote, "It must not be assumed, however, that the vegetation of Wisconsin is a chaotic mixture of communities, each composed of a random assortment of species, each independently adapted to a particular set of external environmental factors. Rather there is a certain pattern to the vegetation, with more or less similar groups of species recurring from place to place." Curtis attributed this pattern to the potentiality for dominance of a limited number of species in the total flora. These few dominants are the species which modify the environment, producing a series of microenvironments in which most of the remaining species must grow. This certainly follows the main thrust of Gleason's thought.

Gleason's expression of the individualistic concept has sometimes been criticized as ignoring the interactions between organisms as a force in molding an integrated community (Daubenmire 1966, 1968, Harper 1967). Gleason, clearly and explicitly, described the plant as adapting to the environment as modified by the effects of other organisms. He recognized that competition, as well as control of the physical environment, could play a role in inhibiting spread of an association, and also that competition did operate to restrict a species to its proportion in the community once the ground was fully occupied (Gleason 1917). It is not a fair interpretation of Gleason's thought to assert as Daubenmire (1968) did, "No organism lives in a biologic vacuum, as implied by the 'individualistic' concept." Gleason, it is true, did not explore or develop the role of interactions among species comprising the community, but this aspect has certainly been pursued in recent work supporting his concept (Whittaker 1956, 1967).

Gleason really stated two individualistic concepts:

1. The individualistic species—each species responding in its own way to each environmental variable, including other species, which affect it.
2. The individualistic association—this is a consequence of the first coupled with his idea of the variation in the environment.

Gleason (1926b) commented, ". . . there is no physiological evidence that plants are segregated by their environmental demands into groups which are nearly or quite exclusive." The concept of the individualistic species is perhaps the key to Gleason's thought. He wrote, ". . . an exact physiological analysis of the various species in a single association would certainly show that their optimum environments are not precisely identical. . . ." (Gleason 1926a). This problem of the segregation of species distributions on ordination axes (Whittaker 1967, 1973a), or their separation in the formalization of the niche by G. E. Hutchinson (1957) and numerous subsequent studies of niche separation among animal species (Terborgh 1970, James 1971), comprise the crux of modern ecological theory and community analysis. "Species do not form well-defined groups of associates with similar distributions, clearly separate from other such groups, but are distributed according to the principle of species individuality." (Whittaker 1973a). Whittaker (1967) and Whittaker and Woodwell (1972) emphasized the role of the influence of other species, especially competition, in leading to the individualistic paterns of species distribution and the failure of species to evolve toward formation of organized groups.

A particularly intriguing aspect of Gleason's individualistic concept is that it is an excellent example, in modern science, of three aspects of the history of science which are primary concerns of historians and philosophers of science in their efforts to understand the nature of science (Stauffer 1957, Kuhn 1959, White 1968). Most of their work has been concerned with earlier periods, e.g., the sixteenth to nineteenth centuries, and essentially with the physical sciences. Studies of the biological sciences, particularly relatively contemporary aspects of biology such as ecology, are limited (Egerton 1976, McIntosh 1976).

The individualistic concept is a remarkable example of the multiple and independent discovery with which the history of science is studded (Ihde 1948, Kuhn 1959). The multiple discovery of the individualistic concept has frequently been noted (Ponyatovskaya 1961, McIntosh 1967,

Whittaker 1973a, Sobolev and Utekhin 1973, Aleksandrova 1973). It has been attributed to Gleason (ca. 1917) in the United States, to Ramensky (ca. 1910, according to Sobolev and Utekhin) in Russia, to Negri (ca. 1914) in Italy, to Lenoble (ca. 1926) in France, as well as to others (Ponyatovskaya 1961). Perhaps, the most dramatic parallel is between Gleason and Ramensky. Ramensky, like Gleason, was a pioneer in the use of quantitative methods and statistical concepts in ecology. Ramensky's views were opposed to those of the great Russian phytosociologist, Sukachev, whose views of the community as a relatively distinct entity paralleled those of F. E. Clements in dominating the Russian ecology of the day. Like Gleason's ideas, Ramensky's were not mentioned in the prominent Russian textbooks of the 1950s and were not well known among Russian botanists (Ponyatovskaya 1961). Many of the quantitative approaches developed in the United States in the 1950s, and generally considered as buttressing or extending the ideas of Gleason and Ramensky, similarly had multiple and independent discovery in Poland and Russia (Curtis 1959, McIntosh 1967, Aleksandrova 1973, Sobolev and Utekhin 1973).

In each of the several instances where the individualistic concept was presented, it had, at first, little impact on ecology. The reactions varied from simply ignoring it to overt criticism or, as Gleason (1953) put it, ridicule. There is a clear suggestion that in Russia, France and in the United States the individualistic concept was opposed to a much more widely accepted and influential view of the community "as composed of quite definite, well-delimited units. . . ." (Ponyatovskaya 1961). Resistance to new or unconventional ideas is as familiar in science as in other human activities (Barber 1961, Kuhn 1970).

Certainly one of the striking aspects of Gleason's individualistic concept is that it suggests so drastic a change in point-of-view that it may be looked on as representing a "revolution" or a new "paradigm" in the sense of Kuhn (1970) or the even more encompassing "thema" of Holton (1975). Since most of the allusions in these articles are to physical science, it is not easy to draw parallels to a largely non-experimental biological science which developed as a discrete discipline primarily in the twentieth century. The terms revolution or revolutionary have been frequently applied to ecology in recent years, but not usually in the restricted meaning intended by Kuhn. It is, as Macfadyen (1975) stated, a matter of interest to ecologists that Kuhn contends that at an early stage in its development a body of science lacks widely-accepted coherent traditions or "paradigms."

Kuhn noted that effective research in a scientific area can hardly begin before there is some consensus on certain questions such as:

1. What are the fundamental entities of which the universe is concerned?
2. How do they interact?
3. What are legitimate questions about such entities?
4. What techniques can be used in seeking solutions?

It may be urged by the cynical that ecology or, at least phytosociology, bogged down on the first question. However, one of the earliest distinctive concepts in ecology was the community, and later its corollary concept, succession, appeared. There arose very early, possibly out of the long tradition of order and design in nature, a concept of the community as an organized entity (McIntosh 1960). This was developed in the United States notably by F. E. Clements, one of the most prolific and influential ecologists of the early decades of the twentieth century. His concept of the community as an organism rapidly became dominant in plant ecology in the United States (Egler 1951) and, although less explicit, it characterized most of the animal ecology texts and references before 1950. Community concepts and classifications in Europe were split into a number of schools (Whittaker 1962) which differed on details but converged on the idea that well-defined communities existed which could be identified and classified. Within this framework, ecology made progress in description of the geographical distribution of organisms, their autecology, and, particularly on the botanical side, their aggregation into communities or associations.

Whether this state of ecology constitutes a paradigm, *sensu* Kuhn, is not entirely clear. It did attract an extensive following, and allowed ecologists to fit multifarious new facts into the Clementsian framework, and descriptions of communities and diagrams of seres flourished. A paradigm, according to Kuhn, provides a common framework within which most people in a field operate collecting facts and filling in the holes within the constraints created by the paradigm. It allows the formation of a body of cognoscenti who share substantially the same rules and standards in their scientific work. This leads to a "normal" science, which is what is expounded in textbooks, and which Egler (1951) attacked in his article on American plant ecology, based on the textbooks of 1947–1949. These, he said, left him with an "inescapable sense of insufficiency," because they followed dogmatic theory. Kuhn also asserted that a paradigm implies a new and more rigid definition of the field —those who don't adopt it are left out. This certainly was the fate of Gleason, whose individualistic concept, as a major departure from the Clementsian theory of the climax association and its rigid developmental seres, had little impact and left him out of ecology. Gleason, apparently, felt that he was an outsider and, effectively gave up publishing on ecological subjects following the last version of the individualistic concept in 1939; he had already abandoned active field work in ecology years before that.

Four additional aspects of a paradigm are, at least, suggestive:

1. The existence of a paradigm makes possible awareness of an anomaly or something which contradicts the expectations created by the paradigm. Such recognition may have dramatic results: Darwin commented that it was like confessing to a murder when he finally concluded that species were not immutable.

2. Awareness of anomaly is followed by an extended exploration of the anomaly requiring an adjustment of the paradigm.

3. There is resistance to the recognition of the observations generating the anomaly, because a well-developed paradigm is a useful base for scientific work and is not surrendered easily.

4. A paradigm is not abandoned because it conflicts with nature; it is abandoned in favor of a new paradigm which replaces it.

Gleason's individualistic concept emerged in 1917 and 1926 specifically in response to Clements' views of the community as an organism which were then being formalized and becoming widely accepted, in whole or in part, as an important part of the framework of ecology. There were some qualms about the extreme application of Clements' organismic analogy, and Tansley proposed the less rigid quasi-organism. Nevertheless, the Clementsian position was clearly the dominant view of both animal and plant communities prior to 1950. In the 1940s a number of established plant ecologists and biogeographers, using established methods of ecological study, stated their dissatisfaction with the reigning views on community and resurrected Gleason's concept. One of the symptoms of crisis leading to a search for and acceptance of new ideas is lack of clarity and proliferation of controversy about the current paradigm. Certainly there had been developing in the 1940s widespread recognition that Clements' monoclimax association theory and its corollary, progressive succession, were not satisfactory explanations of many observations; and lack of clarity of concept and terminology was rampant leading to much ineffectual disputation. All these suggested that the time was ripe for change.

The reappearance of Gleason's concept in 1947 followed closely, in the 1950s, by extensive support for it based on extensive use of quantitative data and new analytic techniques, was certainly a dramatic development, if not a revolution in the sense of Kuhn. The rise, almost simultaneously, of the "new ecology" with its emphasis on functional ecology and introduction of new approaches to the study of ecosystems diverted attention from the old controversies concerning community analysis and classification. In fact, concern with "mere" description and classification was widely considered passé, if not downright unscientific, in an era of modelling and systems analysis (McIntosh 1974, 1976). Nevertheless, Gleason's concept reappeared in the ecological research literature on both plants and animals and emerged in the numerous new ecology textbooks and extensive research literature published in the late 1960s and early 1970s. It was recognized both as Gleason's individualistic concept and often via what Egler (Dansereau 1968) described as its "bastard" intellectual offspring—the continuum concept and gradient analysis.

The recent incorporation of Gleason's individualistic concept (hypothesis) as part of the essence of ecology is more than a casual change in point-of-view, and creates intellectual problems which are not unique to ecology. Lovejoy (1936) commented, "There are not many differences in mental habit more significant than that between the habit of thinking in discrete, well-defined class concepts and that of thinking in terms of continuity. . . ." Holton (1975) describes as an example of antithetical themata, which lead to controversy and marked advance, the thema of atomism or discreteness as opposed to the thema of continuity. Such differences in mental habit or thema, whatever their source, certainly lead to greatly differing conceptual models, even though the facts available are common to both. One of the fundamental problems of ecology is dealing with phenomena that, as Gleason put it, vary continuously in time and space. The individualistic species and the way in which groups of species partition the available habitat and resources constitute the crux of current niche theory; and, according to Gleason's concept, the individualistic association is a product of the interactions of the individualistic environment and the individualistic species with a modicum of chance or, in current parlance "stochastic events" thrown in.

It is clearly recognized by historians and sociologists of science that science is not a product of cooly-objective intellectual automata but is the creation of imaginative, intelligent, often passionate, sometimes obtuse human beings. Recent interest in the history of science stresses the relation of a scientific view to its own time not solely as to whether it has led to a permanent contribution to current science. Perhaps the best measure of Gleason's threefold contributions to ecology in his interpretations of the prairie-forest border, his insightful ventures into quantitative ecology, and his major conceptual contribution, the individualistic concept, is not whether he was right or wrong, or truly first, but in the clarity of his thinking and expression. Francis Bacon put it, "Truth emerges more readily from error than confusion." Gleason questioned the conventional wisdom of early ecology, provided a lucid and forceful statement of an alternative view, and, unfortunately for ecology, devoted most of his energy and capacity to taxonomy. Even as part-time ecologist, he truly deserves the tributes belatedly offered by the Ecological Society of America—Distinguished Ecologist and Eminent Ecologist.

Acknowledgements. I wish first to acknowledge my debt to Dr. John T. Curtis who, three decades ago, introduced me to the individualistic concept as a possible alternative to then conventional ways of thinking about vegetation. I have spent the rest of my professional life as an ecologist standing on the intellectual shoulders of H. A. Gleason and John Curtis hoping to see farther but, perhaps, only seeing longer. I am, also, most appreciative of the invitation by Dr. Bard to write of Dr. Gleason's contributions to ecology in the journal which published the first two versions of his individualistic concept, for, by coincidence, I am currently editor of the *American Midland Naturalist,* which published the third version.

In the preparation of my article, I have had the benefit of comments from W. S. Cooper, a most distinguished contemporary of Dr. Gleason's in the early decades of ecology, and from Paul B. Sears and Frank E. Egler, whose interpretations of the state of ecology, where it was and where it is going, enlightened me over the years. I also wish to thank my colleagues at the University of Notre Dame, Philip Sloan and Michael J. Crowe, for sharing their knowledge and interest in the history of science.

Literature Cited

ALEKSANDROVA, V. D. 1973. Russian approaches to classification. *In* Ordination and Classification of Communities. R. H. Whittaker, ed. Pp. 493–528. Dr. W. Junk b.v. The Hague.

ALLEE, W. C. 1939. An ecological audit. Ecology 20:418–421.

————, A. E. EMERSON, O. PARK, T. PARK, and K. P. SCHMIDT. 1949. Principles of Animal Ecology. W. B. Saunders Co. Philadelphia. 837 p.

ANDERSON, D. J. 1965. Classification and ordination in vegetation science. Controversy over a non-existent problem? J. Ecol. 53: 521–526.

ARRHENIUS, O. 1921. Species and area. J. Ecol. 9:95–99.

BARBER, B. 1961. Resistance by scientists to scientific discovery. Science. 134:596–602.

BECKING, R. 1957. The Zurich-Montpellier School of phytosociology. Bot. Rev. 23:411–488.

BENNINGHOFF, W. S. 1964. The prairie peninsula as a filter barrier to postglacial plant mi-

gration. Proc. Ind. Acad. Sci. 72:116–124.
BILLINGS, W. D. 1949. The shadscale vegetation zone of Nevada and eastern California in relation to climate and soils. Am. Midl. Nat. 42:87–109.
BODENHEIMER, F. S. 1953. The concept of biotic organization in synecology. Bull. Res. Coun. Israel 3:114–121.
——————. 1958. Animal Ecology Today. Uitgererij, Dr. W. Junk, Den Haag. 276 p.
BRUSH, S. 1974. Should the history of science be rated X? Science 183:1164–1172.
CAIN, S. A. 1959. Henry Allen Gleason—Eminent ecologist. Bull. Ecol. Soc. Am. 40:105–110.
——————, and G. M. DE CASTRO. 1959. Manual of Vegetation Analysis. Harper and Row, New York. 325 p.
CHAPMAN, R. N. 1931. Animal Ecology. McGraw-Hill Book Company, Inc. 463 p.
CLEMENTS, F. E. 1905. Research Methods in Ecology. Univ. Pub. Co. Lincoln, Neb. 334 p.
COLLIER, B. D., G. W. COX, A. W. JOHNSON, and P. C. MILLER. 1973. Dynamic Ecology. Prentice-Hall, Inc. Englewood Cliffs, N.J. 563 p.
COWLES, H. C. 1901. The physiographic ecology of Chicago and vicinity. Bot. Gaz. 31:73–108, 145–182.
CURTIS, J. T. 1959. The Vegetation of Wisconsin. Univ. of Wisconsin Press. Madison. 657 p.
——————, and R. P. MCINTOSH. 1950. The interrelations of certain analytic and synthetic phytosociological characters. Ecology 31:434–455.
——————, and R. P. MCINTOSH. 1951. An upland continuum in the prairie-forest border region of Wisconsin. Ecology 32:476–496.
DANSEREAU, P. 1957. Biogeography. An Ecological Perspective. Ronald Press, Co. New York. 394 p.
——————. (ed.) 1968. The continuum concept of vegetation: responses. Bot. Rev. 34: 253–332.
DAUBENMIRE, R. 1966. Vegetation: identification of typal communities. Science 151:291–298.
——————. 1968. Plant Communities. Harper and Row, Publishers. New York. 300 p.
DICE, L. R. 1952. Natural Communities. Univ. Michigan Press. Ann Arbor. 547 p.
EGERTON, F. 1976. Ecological studies and observations in America before 1900. *In* Evolution of Issues, Ideas and Events in America: 1776–1796. Benjamin J. Taylor, ed. Univ. Oklahoma Press, Norman. In press.
EGLER, F. E. 1951. A commentary on American plant ecology based on the textbooks of 1947–1949. Ecology 32:673–695.
——————. 1962. On American problems in the communication of biologic knowledge to the public. Dodonaea 30:264–304.
EMERSON, A. E. 1960. The evolution of adaptation in population systems. *In* Evolution after Darwin. Vol. I. S. Tax, ed. Pp. 307–348. Univ. Chicago Press. Chicago.
FORBES, S. A. 1907. On the local distribution of certain Illinois fishes: an essay in statistical ecology. Bull. Ill. State Lab. Nat. Hist. 7:273–303.
GLEASON, H. A. 1901. The flora of the prairies. B. S. Thesis. Univ. of Illinois. Unpublished.
——————. 1909a. Some unsolved problems of the prairies. Bull. Torrey Bot. Club 36: 265–271.
——————. 1909b. The vegetational history of a river dune. Trans. Ill. Acad. Sci. 2:19–26.
——————. 1910. The vegetation of the inland sand deposits of Illinois. Bull. Ill. State Lab. Nat. Hist. 9:21–174.
——————. 1912. An isolated prairie grove and its phytogeographical significance. Bot. Gaz. 53:38–49.
——————. 1913. The relation of forest distribution to prairie fires in the middle west. Torreya 13:173–181.
——————. 1917. The structure and development of the plant association. Bull. Torrey Bot. Club 44:463–481.
——————. 1919. What is ecology? Torreya 19:89–91.
——————. 1920. Some applications of the quadrat method. Bull. Torrey Bot. Club 47:21–33.
——————. 1922. On the relation between species and area. Ecology 3:158–162.
——————. 1923. The vegetational history of the Middle West. Ann. Assoc. Am. Geog. 12:39–85.
——————. 1925. Species and area. Ecology 6:66–74
——————. 1926a. The individualistic concept of the plant association. Bull. Torrey Bot. Club 53:1–20.
——————. 1926b. Plant associations and their classification: A reply to Dr. Nichols. Proc. International Congress of Plant Sciences. Ithaca, New York 1926.
——————. 1927. Further views on the succession concept. Ecology 8:299–326.
——————. 1929. The significance of Raunkiaer's law of frequency. Ecology 10:406–408.
——————. 1936. Twenty-five years of ecology, 1910–1935. Mem. Brooklyn Bot. Gard. 4: 41–49.
——————. 1939. The individualistic concept of the plant association. Am. Midl. Nat. 21:92–110.
——————. 1953. Autobiographical letter. Bull. Ecol. Soc. Am. 34:40–42.
——————, and A. CRONQUIST. 1964. The Natural Geography of Plants. Columbia Univ. Press. New York 420 p.
GOODALL, D. W. 1952. Quantitative aspects of plant distribution. Biol. Rev. 27:194–245.

———. 1962. Bibliography of statistical plant sociology. Exerpta Botanica, Sectio B. Bd. 4:254–322.

———. 1963. The continuum and the individualistic association. Vegetatio 11:297–316.

———. 1973. Sample similarity and species correlation. *In* Ordination and Classification of Communities. R. H. Whittaker, ed. Pp. 105–156. Dr. W. Junk b.v. The Hague.

GREIG-SMITH, P. 1952. Ecological observations on degraded and secondary forest in Trinidad, British West Indies. II. Structure of communities. J. Ecol. 40:316–330.

———. 1964. Quantitative Plant Ecology. Butterworths. London. 256 p.

GRINNELL, J. 1917. The niche relationships of the California thrasher. Auk 34:427–433.

HARPER, J. J. 1967. A Darwinian approach to plant ecology. J. Ecol. 55:247–270.

HART, C. A., and H. A. GLEASON. 1907. On the biology of the sand areas of Illinois. Bull. Ill. State Lab. Nat. Hist. 7:137–272.

HOLTON, G. 1975. On the role of themata in scientific thought. Science 188:328–338.

HUTCHINSON, G. E. 1957. Concluding remarks. Cold Spring Harbor. Symp. Quant. Biol. 22:415–427.

IHDE, A. J. 1948. The inevitability of scientific discovery. Sci. Monthly 67:427–429.

JAMES, F. C. 1971. Ordinations of habitat relationships among breeding birds. Wilson Bull. 3:215–236.

JUST, T. (ed.) 1939. Plant and Animal Communities. Am. Midl. Nat. 21:1–255.

KREBS, C. J. 1972. Ecology. Harper and Row, Publ. New York. 694 p.

KUHN, T. S. 1959. Energy Conservation as an example of simultaneous discovery. *In* Critical Problems in the History of Science. M. Clagett, ed. Pp. 321–356. Univ. of Wisconsin Press. Madison.

———. 1970. The Structure of Scientific Revolutions. 2nd ed. Univ. Chicago Press. Chicago. 210 p.

LANGFORD, A. N., and M. F. BUELL. 1969. Integration, identity and stability in the plant association. Adv. Ecol. Res. 6:84–135.

LOVEJOY, A. O. 1936. The Great Chain of Being. Harvard Univ. Press. Cambridge, Mass. 382 p.

LUSSENHOP, J. 1974. Victor Hensen and the development of sampling methods in ecology. J. Hist. Biol. 7:319–337.

MACFADYEN, A. 1975. Some thoughts on the behaviour of ecologists. J. Anim. Ecol. 44:351–363.

MAGUIRE, B. 1975. Henry Allen Gleason, Jan. 2, 1882–April 21, 1975. Bull. Torrey Bot. Club 102:

MASON, H. L. 1953. Communications. Bull. Ecol. Soc. 34:95–96.

MCDOUGALL, W. B. 1927. Plant Ecology. Lea and Febiger. Philadelphia. 326 p.

MCINTOSH, R. P. 1958. Plant communities. Science 128:115–120.

———. 1960. Natural order and communities. The Biologist 42:55–62.

———. 1962. Raunkiaer's "Law of frequency." Ecology 43:533–535.

———. 1967. The continuum concept of vegetation. Bot. Rev. 33:130–187.

———. 1974. Plant ecology 1947–1972. Ann. Mo. Bot. Gard. 61:132–165.

———. 1976. A history of ecology in America 1900–1976. *In* Evolution of Issues, Ideas and Events in America: 1776–1976. Benjamin J. Taylor, ed. Univ. Oklahoma Press. Norman. In press.

MILLS, E. L. 1969. The community concept in marine zoology, with comments on continua and instability in some marine communities: A review. J. Fish. Res. Bd. Can. 26:1415–1428.

MORRISON, R. G., and G. A. YARRANTON 1974. Vegetational heterogeneity during a primary sand dune succession. Can. J. Bot. 52:397–410.

NICHOLS, G. 1926. Plant associations and their classification. Proc. International Congress of Plant Sciences. Ithaca, New York. 1926.

ODUM, E. P. 1953. Fundamentals of Ecology. W. B. Saunders Co. Philadelphia. 2nd ed., 1958. 3rd ed., 1971. 574 p.

PONYATOVSKAYA, V. M. On two trends in phytocoenology. (Translated by J. Major). Vegetatio 10:373–385.

POOLE, R. W. 1974. An Introduction to Quantitative Ecology. McGraw-Hill, Inc. 532 p.

POORE, M. E. D. 1964. Integration in the plant community. J. Ecol. 52(Supplement): 213–226.

POUND, R., and F. E. CLEMENTS. 1898. A method of determining the abundance of secondary species. Minn. Bot. Stud. 2:19–24.

RAUNKAIER, C. 1934. The Life Forms of Plants and Statistical Plant Geography. Clarendon Press. Oxford. 632 p.

RICHARDS, P. W. 1963. What the tropics can contribute to ecology. J. Ecol. 51:231–241.

SEARS, P. B. 1956. Some notes on the ecology of ecologists. Sci. Monthly 83:22–27.

SOBOLEV, L. N., and V. D. UTEKHIN. 1973. Russian (Ramensky) approaches to community. *In* Ordination and Classification of Communities. R. H. Whittaker, ed. Pp. 75–104. Dr. W. Junk b.v. The Hague.

STAUFFER, R. C. 1957. Haeckel, Darwin and ecology. Quart. Rev. Biol. 32:138–144.

TANSLEY, A. G. 1920. The classification of vegetation and the concept of development. J. Ecol. 8:118–149.

TERBORGH, J. 1970. Distribution on environmental gradients: Theory and a preliminary interpretation of distributional patterns in the avifauna of the cordillera Vilcabamba, Peru. Ecology 52:24–40.

WATT, A. S. 1964. The community and the individual. J. Ecol. 52 (Supplement):203–211.

WEAVER, J. E., and F. E. CLEMENTS. 1929. Plant Ecology. McGraw-Hill Book Co. New York. 2nd ed., 1938. 601 p.

WHITE, L., JR. 1968. Dynamo and Virgin Reconsidered. Mass. Inst. Tech. Press. Cambridge, Mass. 186 p.

WHITFORD, P. B. 1949. Distribution of woodland plants in relation to succession and clonal growth. Ecology 30:199–208.

WHITTAKER, R. H. 1956. Vegetation of the Great Smoky Mountains. Ecol. Monogr. 26:1–80.

——————. 1962. Classification of communities. Bot. Rev. 28:1–239.

——————. 1967. Gradient analysis of vegetation. Biol. Rev. 42:207–264.

——————. 1973a. Direct gradient analysis. *In* Ordination and Classification of Communities. R. H. Whittaker, ed. Pp. 33–52. Dr. W. Junk b.v. The Hague.

——————. (ed.) 1973b. Ordination and Classification of Communities. Dr. W. Junk b.v. The Hague. 737 p.

——————. 1975. Communities and Ecosystems. Macmillan Publ. Co. Inc. New York. 2nd ed. 385 p.

——————, and G. M. WOODWELL. 1972. Evolution of natural communities. *In* Ecosystem Structure and Function. J. A. Wiens, ed. Pp. 137–159. Oregon State Univ. Press. Corvallis.

WILLIAMS, W. T., G. N. LANCE, L. J. WEBB, J. G. TRACEY, and M. B. DALE. 1969. Studies in the numerical analysis of complex rain-forest communities. III. The analysis of successional data. J. Ecol. 57:515–535.

THE ECOLOGICAL SOCIETY OF AMERICA
HISTORICAL DATA
AND SOME PRELIMINARY ANALYSES

Robert L. Burgess

THE ECOLOGICAL SOCIETY OF AMERICA

Historical Data and Some Preliminary Analyses

Robert L. Burgess
Environmental Sciences Division

OAK RIDGE NATIONAL LABORATORY
Oak Ridge, Tennessee 37830
operated by
UNION CARBIDE CORPORATION
for the
ENERGY RESEARCH AND DEVELOPMENT ADMINISTRATION
Contract No. 7405-eng-26

CONTENTS

THE ECOLOGICAL SOCIETY OF AMERICA

Historical Data and Some Preliminary Analyses[1,2]

Robert L. Burgess

Environmental Sciences Division
Oak Ridge National Laboratory[3]
Oak Ridge, TN 37830

INTRODUCTION

Since its inauspicious beginnings in 1914, the Ecological Society of America, like most professional scientific groups, has generated an interesting and complex history. In spite of periodic attempts at record keeping and archiving, however, the Society has not made a concerted and continuous effort at documenting its meetings, activities, deliberations, and accomplishments. As a Society, there has been no processing of the temporal changes in its role that accompany both societal growth and the technological, cultural, and populational shifts in the internal structure of a great nation. From time to time, several papers have appeared on aspects of the history of ecology (Brewer 1960, Cowles, 1904, McIntosh 1976, Reed 1905). These have not, however, dealt with the development of an organized professional group, its founders, its builders, its objectives, or its accomplishments. In the United States, much of the general scientific history reflects, and is a reflection of, the history of a professional society. Consequently, much interpretation drawn from the writings of both historians and ecologists (Egerton 1976, McIntosh 1974, 1976) has been used to highlight this preliminary inquiry into the developmental history of the Ecological Society of America. Various scientific groups have documented their development with essentially this same technique (Abbott 1958, Kathren and Tarr 1974, Laude *et al.* 1962, Reese 1976, Sullivan 1976), natural resource research and management history has been recently published (Doig 1976, Price 1976), and numerous papers have appeared (in English) on aspects of the history of ecology and related fields (Allee *et al.* 1949, Brewer 1960, Egerton 1976, Egler 1951, Gleason 1936, McIntosh 1974, 1975, 1976, Odum 1968, Raup 1942, Roche 1976, Rübel 1927, Sears 1969, Tansley 1947). In addition, at least two major foreign language works exist on the history and development of ecology (DuRietz 1921, Trass 1976). All of these are useful, not only for their comprehensiveness, but for the approaches used.

In an attempt to delve into Society history, as must be true of most similar endeavors, one is met immediately with either a lack of data, or data in rather dispersed form. The *Bulletin of the Ecological Society of America*, begun in 1917, contains a wealth of information, albeit scattered and discontinuous, on the mechanics of the Society during its 62 years of operation. Lists of officers, and reports of meetings and committees appeared in most early issues. In addition, *Ecology*, the Society's major journal carried business proceedings and reports from 1926 through 1946. These sources were scrupulously searched, primarily for material that could be tabulated, in order to place any future efforts toward a comprehensive history of the ESA on a firm foundation.

---

[1] Research supported by the Eastern Deciduous Forest Biome, US/IBP, funded by the National Science Foundation under Interagency Agreement AG 199, DEB 76-00761 with the Energy Research and Development Administration – Oak Ridge National Laboratory.

[2] Contribution No. 284, Eastern Deciduous Forest Biome, US/IBP, and Publication No. 1037, Environmental Sciences Division, Oak Ridge National Laboratory.

[3] Operated by Union Carbide Corporation for the Energy Research and Development Administration.

The data and analyses which follow, gleaned solely from the above sources, cover membership, officers, meetings, committees, publications, and a suite of Society activities that have resulted in major impacts on the ecological setting in the United States. I stress the preliminary nature of these analyses, but believe them to be appropriate and utilitarian ventures toward a more definitive history of the Ecological Society of America.

THE BEGINNINGS

The initial move toward an Ecological Society appears to be a letter from Robert H. Wolcott, Professor of Zoology at the University of Nebraska, to Victor E. Shelford, then at the University of Chicago, dated March 27, 1914 (Shelford 1938). We do not know the extent of the influence of Tansley's founding of the British Ecological Society the year before. Wolcott suggested a society composed of both botanists and zoologists, strongly oriented toward field work (rather than formal meetings and presentation of papers), but limited geographically to the upper Mississippi valley. He specifically suggested limits of Kansas and North Dakota on the west (but equivocated about including Colorado), western Ohio (Sandusky Biological Station) to the east, and Missouri to the south. Subsequent correspondence confirmed Wolcott's concept of a regional, rather than national, organization.

Following the Shelford-Wolcott exchanges, Henry Chandler Cowles organized a meeting of both animal and plant ecologists on December 30, 1914 in the lobby of the Hotel Walton in Philadelphia. Present were C. C. Adams, H. H. Bartlett, F. H. Blodgett, W. L. Bray, C. T. Brues, W. A. Cannon, Cowles, A. P. Dachnowski-Stokes, R. F. Griggs, J. W. Harshberger, A. F. Hill, O. E. Jennings, D. T. MacDougal, Z. P. Metcalf, G. E. Nichols, R. C. Osburn, A. S. Pearse, H. L. Shantz, Shelford, Forrest Shreve, Norman Taylor, and Wolcott (and perhaps a few others) (Shelford 1938). Of this group, all but Bartlett and Brues became charter members of ESA, and nine later served the Society as president.

The 1914 meeting appointed Harshberger to chair an organizing committee and prepare for another meeting of interested individuals at the American Association for the Advancement of Science (AAAS) sessions in Columbus, Ohio the next December. That took place at the Hotel Hartman on December 28, 1915, with about 50 in attendance. Chairman Harshberger also had about 50 letters in favor of a society from those unable to attend the Columbus meeting. The group voted to form the Ecological Society of America, adopted a brief constitution, elected officers, and set the next meeting in New York, again with AAAS. Shelford (1938) notes that W. C. Allee and F. E. Clements objected to the formation of "just another society" but both were charter members of ESA. Allee served as president in 1929, but until his death in 1945, Frederick Edward Clements, perhaps the single most influential personage in early American ecology, never held an elective office in the Society.

Present issues of ESA journals indicate that the Society was incorporated under the laws of Wisconsin in 1915. Either the new Society acted quickly after the December 28 meeting, incorporation procedures were already implemented and needed only the December ratification vote, or the date is wrong. The Proceedings of the 1944 meeting [*Ecology* 26(2): 216–234, 1945] state that 1927 articles of incorporation were archived at the University of Cincinnati, and Article 2 of the Bylaws states rather unequivocally that "The Society . . . was incorporated . . . in the State of Wisconsin, December 20, 1927."

THE MEMBERSHIP

The 22 people in attendance in Philadelphia at the close of 1914 had fertilized the germ of a professional society. A year later, 286 had become charter members (Moore 1920a). Biographical sketches of this original group were published in a 1917 "Handbook of the Ecological Society of America," a group surely representative of the nation's ecological community at that time, but probably missing a number of then present and soon-to-be ecologists.

By 1921, the year after the Society began its own journal, membership was up to 458. Another 140 were added in the next two years, and in spite of dues of only $4.00, the financial success of the young association seemed assured. During the twenties, growth slowed, and following the crash of 1929, Society membership declined during the great depression, down to 546 in 1934 (from 645 in 1928), but was back up to 680 in 1937. Stability followed through World War II, a period of little production in colleges and universities, and of course, some casualty losses. The growth curve (Fig. 1) starts to climb during the 1950's, reaching 2000 by early 1960. Acceleration continued in the 60's: 3000 by 1966 and 4000 by 1970. Doubling time during this period ranges between nine and 13 years, depending on the slope of the curve. By 1973, 5000 members were recorded, and the 1976 total stood at 5890. The 6000 mark should be reached sometime during 1977.

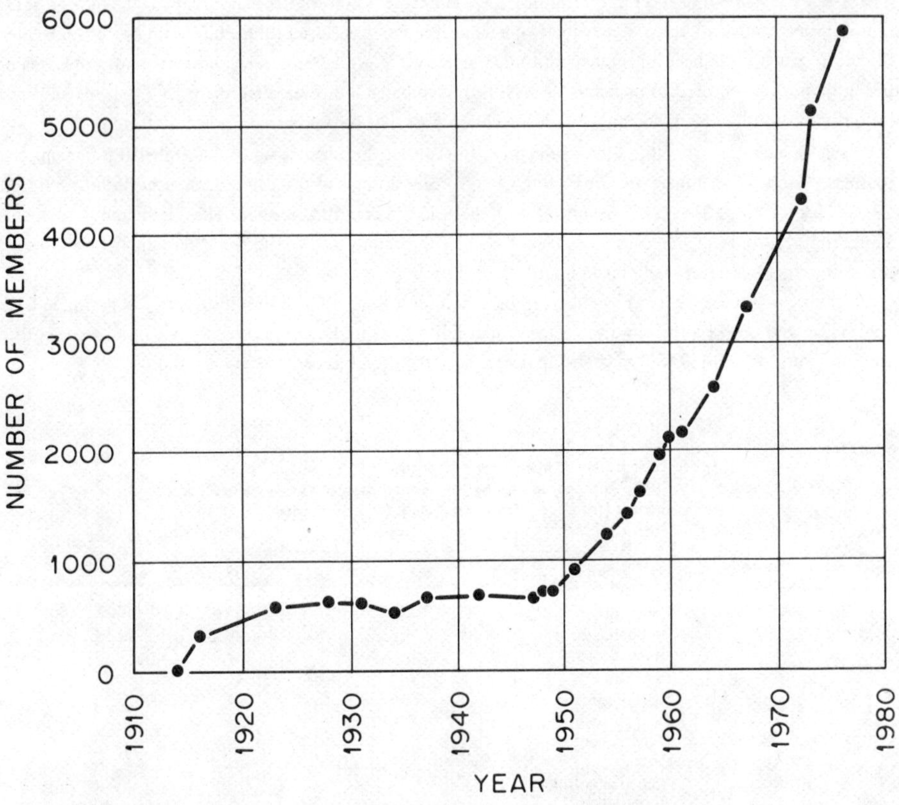

Fig. 1. Growth in membership in the Ecological Society of America from its start in 1914 through 1976, the latest year for which data are available.

Much of the recent growth appears to be a complex result of several factors. During the 1960's, the post-World War II "baby-boom" was hitting the colleges. Many students, simply as a function of the expanding population, found their way into ecological studies, environmental careers, and ultimately, the ESA. Secondly, Earth Day 1970, spawning the "environmental decade," spurred an unprecedented interest in, and concern for, the world around us. While ESA gained some members from the "bandwagon," it came nowhere near the 100,000 new members enjoyed by organizations like the Sierra Club and the National Audubon Society. Finally, beginning with the National Environmental Policy Act (NEPA) in late 1969, a new breed of scientist arose – the environmental consultant. Applied expertise for hire did two things. First, it created jobs – a demand for ecological knowledge that rapidly permeated all levels of government. Secondly, it attracted numbers of other kinds of environmental scientists – geologists, chemists, meteorologists – to a society that had been (and still was) predominantly biological. This does not imply that in the previous 50 years ESA did not have such members. Rather, following NEPA, percentages of these kinds of members increased, along with significant numbers of interested laymen.

Table 1 gives a classification of the 307 members (284 charter members plus 23 elected to membership at the 1916 annual meeting) according to disciplinary interest. Fifty-seven percent classed themselves as either plant or animal ecologists, but even then, both applied fields and non-biological environmental scientists were represented. Geographical distribution (Table 2) shows the major seats of the new Society. Illinois was a hotbed of activity, but the Federal government was close behind. Interestingly, 59 years later (Table 3) growth has been phenomenal in most areas, while Washington, D.C. has apparently only maintained its population of ESA members. It must be remembered, however, that many ecologists live or work (or both) in suburban Maryland and Virginia, and current totals for these states reflect this fact.

Table 4, adapted from the 1976 Directory, shows that the main reason for membership remains the journals. "Active" members receive *Ecology*, and "Sustaining" also receive *Ecological Monographs*. In 1976, ESA still had 31 "Life" members, a category long since discontinued. The 71 "Family" memberships, for additional members where at least one in the family is "Active" or "Sustaining," is a recent innovation that has never really taken hold.

Eighteen of the 284 charter members (6 percent) were women. Of this group, only E. Lucy Braun later held office, but both Edith Schwartz Clements and Edith Bellamy Shreve are well known both for scientific work in their own right as well as for being the wives of very famous ecologists.

Table 1. Membership in the Ecological Society of America in 1917, grouped according to disciplinary areas of major interest [from the "Handbook," Bull. Ecol. Soc. Amer. 1(3), 1917]

Plant ecology	88
Animal ecology	86
Forestry	43
Entomology	39
Marine ecology	14
Agriculture	12
Plant physiology	7
Plant pathology	4
Climatology	4
Geology	4
Animal parasitology	3
Soil physics	3
	307

Table 2. Ranked geographical distribution of members, of the Ecological Society of America, 1917 [from the "Handbook," Bull. Ecol. Soc. Amer. 1(3), 1917]

Illinois	32	Montana	4
District of Columbia	30	New Jersey	4
New York	30	British Columbia	3
California	20	Indiana	3
Massachusetts	14	Nebraska	3
Minnesota	14	Quebec	3
Michigan	10	Texas	3
Pennsylvania	10	Hawaii	2
Colorado	9	North Dakota	2
Ohio	9	Philippine Islands	2
Ontario	9	South Carolina	2
Wisconsin	9	Vermont	2
Maryland	8	Alberta	1
Iowa	7	British Guiana	1
Kansas	7	Canal Zone	1
New Mexico	7	Florida	1
Arizona	6	Louisiana	1
Connecticut	6	Maine	1
Missouri	6	North Carolina	1
Washington	6	New Hampshire	1
Oregon	5	Sweden	1
Utah	5	Tennessee	1
Idaho	4	Wyoming	1
Total	307		

Table 3. Ranked geographical distribution of members of the Ecological Society of America, 1976 [from the "Directory," Bull. Ecol. Soc. Amer. 57(3a), 1976]

California	655	Iowa	55
New York	348	Montana	49
Illinois	230	Oklahoma	46
Michigan	213	New Hampshire	44
Pennsylvania	190	Alaska	42
Ohio	167	Rhode Island	40
Massachusetts	166	United Kingdom	39
Texas	160	Idaho	38
Florida	158	Hawaii	36
Wisconsin	155	Mississippi	36
Maryland	153	Asia	35
Washington	152	Louisiana	35
North Carolina	150	North Dakota	35
Colorado	148	Quebec	35
New Jersey	130	Alabama	33
Virginia	126	Washington, DC	33
Tennessee	123	Wyoming	31
Oregon	114	Kentucky	29
Minnesota	112	Maine	29
Connecticut	105	Nevada	28
Arizona	101	Delaware	26
Georgia	100	Africa	23
Ontario	99	Vermont	21
Utah	87	Puerto Rico, Guam, Virgin Islands	20
Indiana	79	Saskatchewan	20
Kansas	70	South Dakota	20
Australia	66	Oceania	19
Europe	65	Manitoba	18
New Mexico	62	Arkansas	17
Missouri	61	West Virginia	16
British Columbia	60	Nebraska	15
Alberta	59	Nova Scotia	15
South Carolina	59		
Latin America	59		

Table 4. Numbers, dues and classes of membership in the Ecological Society of America, 1976. All members receive the *Bulletin of the Ecological Society of America;* Active and Student Active members receive *Ecology* in addition; Sustaining and Student Sustaining members receive all three ESA journals, including *Ecological Monographs.* A Family member resides in the same household as a member of another class. Emeritus membership is available after 30 years of continuous Active or Sustaining membership, and journals are available at cost. [from the "Directory," Bull. Ecol. Soc. Amer. 57(3a), 1976]

Class	Current dues	Number
Life		31
Emeritus		39
Associate	$ 7.00	456
Active	$25.00	2639
Student active	$20.00	589
Sustaining	$35.00	1537
Student sustaining	$30.00	400
Family	$ 3.00	71
Total		5762

THE OFFICERS

Since the 1915 organizational meeting, ESA has always had a president and a vice-president. The other offices have experienced some changes, albeit rather minor ones, and numerous new offices and some reorganization has occurred, mostly in the last 25 years.

The presidency, throughout most of ESA history, has been a prestige position. It has been called sort of "an eminent ecologist award" (Simkins 1971). The evidence for this lies in the prohibition, following establishment of an Eminent Ecologist citation by the Society in 1953, of presidents or past presidents of ESA from consideration. Within the last 10 years this policy has changed, but it did, in fact, permeate the election process for several decades. A second policy, also recently allowed to die in peace, was a stipulation that the office of president should alternate between a botanist and a zoologist – a policy that, on reflection, was self-defeating of the original aims of the Society – to bring *ecologists* together.

Nevertheless, the Ecological Society of America has consistently been led by good ecologists. Their success or failure as presidents has probably been as much a function of the scientific and political climate of the times and the generally conservative tenor of a professional scientific association as it was the philosophy, nature, or perseverance of the presidents. The list (Table 5) reads much like a "who's who" in American ecology, from Shelford and Cowles, Adams, Transeau and Juday, through Nichols, Vorhies, and Emerson, to Cain, Blair, Odum, and Stearns. Each president has had, at least on paper, some semblance of a platform (Coker 1938, Dreyer 1945). The early ones stressed growth, consolidation, a journal. Later came a preoccupation with things like research, preservation of research areas, ecological programs – many aimed, however, at an individual or institutional level. With few exceptions, the Society has not undertaken major projects that have involved the Society as a unit. More recently, platforms have stressed fiscal necessity, social responsibility, and problem responsiveness (Hollander 1976, Nelkin 1976, Simkins 1971). A broad view again indicates that many of the above were (and are) a reflection of the scientific, political, and socio-economic conditions of the period.

Table 5. Officers of the Ecological Society of America, 1916-1976

Year	President	Vice-President
1916	Victor E. Shelford	William Morton Wheeler
1917	Ellsworth Huntington	John W. Harshberger
1918	Henry Chandler Cowles	R. E. Coker
1919	Barrington Moore	T. L. Hankinson
1920	Barrington Moore	George Elwood Nichols
1921	Stephen A. Forbes	Edgar Nelson Transeau
1922	Forrest Shreve	H. E. Crampton
1923	Charles C. Adams	Gustav Adolph Pearson
1924	Edgar Nelson Transeau	W. C. Albee
1925	A. S. Pearse	John Ernst Weaver
1926	John W. Harshberger	R. C. Osburn
1927	Chancey Juday	William Skinner Cooper
1928	Homer Leroy Shantz	R. N. Chapman
1929	W. C. Allee	Walter P. Taylor
1930	John Ernst Weaver	G. P. Burns
1931	A. O. Weese	Francis Ramaley
1932	George Elwood Nichols	Joseph Grinnell
1933	E. B. Powers	Herbert C. Hanson
1934	George D. Fuller	Paul S. Welch
1935	Walter P. Taylor	Emma Lucy Braun
1936	William Skinner Cooper	J. G. Needham
1937	R. E. Coker	H. DeForest
1938	Herbert C. Hanson	Lee R. Dice
1939	Charles T. Vorhies	C. F. Korstian
1940	Francis Ramaley	Orlando Park
1941	Alfred E. Emerson	B. C. Tharp
1942	C. F. Korstian	C. E. Zo Bell
1943	Orlando Park	Paul B. Sears
1944	Robert F. Griggs	Alfred C. Redfield
1945	Alfred C. Redfield	John M. Aikman
1946	John M. Aikman	Aldo Leopold
1947	Aldo Leopold	Paul B. Sears
1948	Paul Bigelow Sears	William A. Dreyer
1949	Z. P. Metcalf	Charles E. Olmstead
1950	Emma Lucy Braun	R. V. Truitt
1951	S. Charles Kendeigh	Fred W. Albertson
1952	Frank C. Gates	David E. Davis
1953	Lee R. Dice	Stanley Adair Cain
1954	John Ernst Potzger	Samuel Eddy
1955	William J. Hamilton, Jr.	Murray Fife Buell
1956	Henry J. Oosting	W. Frank Blair
1957	William A. Dreyer	William T. Penfound
1958	Stanley Adair Cain	Frank Preston
1959	Thomas Park	Aaron J. Sharp
1960	Charles E. Olmstead	W. Dwight Billings
1961	Arthur D. Hasler	Edward S. Deevey, Jr.
1962	Murray Fife Buell	Bostwick H. Ketchun
1963	W. Frank Blair	Lora Mangum Shields
1964	John F. Reed	LaMont C. Cole
1965	Eugene Pleasant Odum	John E. Cantlon
1966	Bostwick H. Ketchum	Robert B. Platt
1967	Rexford Daubenmire	George M. Woodwell
1968	LaMont C. Cole	George E. Sprugel, Jr.
1969	John E. Cantlon	Pierre Dansereau
1970	Edward S. Deevey, Jr.	Paul G. Pearson
1971	Frank Herbert Bormann	Robert H. Whittaker
1972	Stanley Irving Auerbach	Forest W. Stearns
1973	Robert B. Platt	Frank B. Golley
1974	Frederick E. Smith	Charles R. Goldman
1975	Richard S. Miller	Arthur S. Cooper
1976	Forest W. Stearns	Gordon Orians

Years	Secretary-Treasurer
1916-1919	Forrest Shreve
1920-1930	A. O. Weese
1931	Alfred E. Emerson
1932-1933	Raymond Kienholtz
1934-1935	Arthur G. Vestal
1936-1937	Orlando Park

Years	Secretary
1938	Orlando Park
1939-1941	William J. Hamilton, Jr.
1942-1947	William A. Dreyer
1948-1950	William A. Castle
1951-1953	Murray Fife Buell
1934-1957	John F. Reed
1958-1961	John E. Cantlon
1962-1964	Paul G. Pearson
1965-1969	Stanley Irving Auerbach
1970	William A. Niering
1971-1976	J. Frank McCormick

Years	Treasurer
1938-1940	Stanley Adair Cain
1941-1943	Royal E. Shanks
1944-1949	Henry J. Oosting
1950	William T. Penfound
1951-1954	Frederick H. Test
1955-1957	Alexander C. Hodson
1958	Jack S. Dendy
1959-1962	Kirby L. Hays
1963-1965	Ralph W. Kelting
1966-1969	William Clark Ashby
1970-1971	Shelby D. Gerking
1972-1975	Forest W. Stearns
1976	Paul G. Pearson

Only one president has succeeded himself, Barrington Moore in 1919 and 1920. Moore was the first editor of *Ecology*, and the continuance through reelection appears to be related to a new journal and the attendant complications. And only a single woman, E. Lucy Braun of Cincinnati, has been elected president – 26 years ago.

The distribution of presidents (Fig. 2) reflects distribution of the membership, the meetings, and to some extent, the centers of ecology in the United States. Illinois has been home to 11 presidents, two from Northwestern, two from the University of Illinois, and seven from the University of Chicago. New York and Wisconsin have furnished five presidents each, while four have come from Ohio, Massachusetts, North Carolina, and Arizona. Identical to the distribution of annual meetings (though without substantive correlation), 14 ESA presidents have come from west of the Mississippi, while 47 were eastern. Perhaps more importantly, only four of the 14 western presidents have served since 1940 (Gates in Kansas, Blair in Texas, Reed in Colorado, and Daubenmire in Washington) and the four from Arizona (Shreve, Shantz, Taylor, and Vorhies) all had served by 1939. California has not yet had a president, even though over 10 percent (655 of 5762) of the ESA membership was located in California in 1976.

The vice-presidency, similar to the situation in many organizations, has been an ill-defined role. The office has not been used as a replacement for the president, as no ESA president has either resigned or died in office. No vice-president has ever been re-elected, but Paul B. Sears served in both 1943 and 1947, before moving up to the presidency in 1948. Of the 61 vice-presidents, 30 (49 percent) have subsequently been president, with an average intervening time span of 6.75 years. This ranges from 19 years for Robert E. Coker (VP in 1918, president in 1937) to one year for Redfield, Aikman, Leopold, and Sears (second term). This last sequence, from 1945–1948, indicates an initial attempt, although unspecified, at the

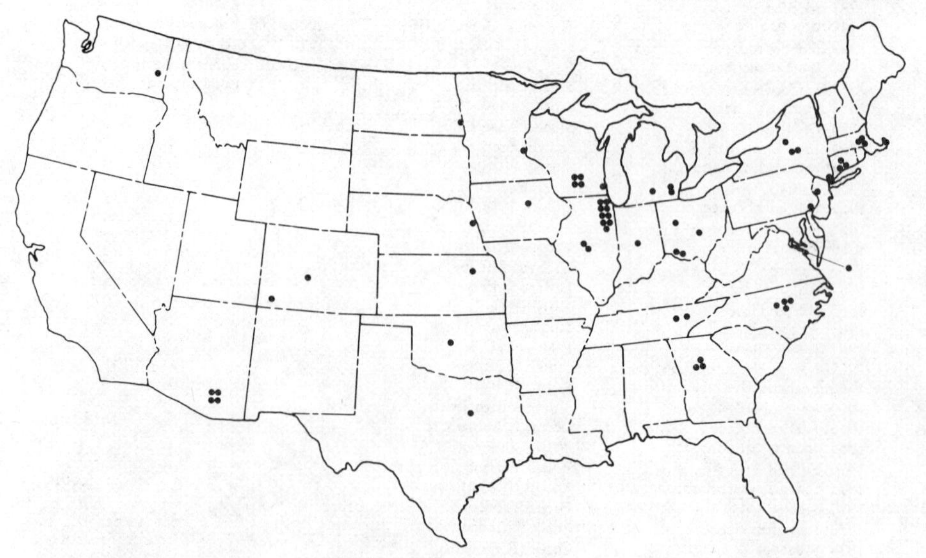

Fig. 2. Geographical locations of presidents of the Ecological Society of America, 1915–1976, as of the time of their election.

concept of a "president-elect." The necessary constitutional changes were finally implemented in 1951, so that ESA now annually elects a president-elect, rather than a president. Two women have served as vice-president, E. Lucy Braun in 1935 (later president in 1950), and Lora Mangum Shields of New Mexico in 1963.

For the first 22 years of its existence, ESA had a combination secretary-treasurer (Table 5). Recognizing that a degree of continuity was essential to the administration of a young society, these officers were elected for more than one year. Forrest Shreve (1916–1919) and A. O. Weese (1920–1930) carried the Society through its formative (and crucial) years. Shelford (1938) states, "The society was guided and its policies given continuity for the first 16 years of its existence through the unbroken services of these two men whose enthusiasm kept life and progress in the organization while maintaining an unusual harmony because of their kindly and cordial personalities."

In 1938, the offices were divided, with Orlando Park choosing to remain as secretary and Stanley Cain elected treasurer. This arrangement has persisted to the present, with three- to five-year terms, usually staggered, being the general rule.

Both major journals have always had a business manager, starting with the Brooklyn Botanic Garden for *Ecology* in 1920. In 1931, with the beginning of *Ecological Monographs*, Duke University Press furnished a business manager for both journals. The need for this type of professionalism continued to grow, and in 1952, ESA arranged for a business manager for the Society, the first being Henry J. Oosting. Duties were many, but included overseeing the business (i.e., fiscal) operations of the Society – journals, income, expenses, trusts, and general cash flow. He worked closely both with the treasurer and the business manager for the journals, an individual still provided by Duke University Press. In 1970, the importance of a business manager was recognized by placing virtually all financial responsibilities with that office. The treasurer became more of a planning position and figurehead, a situation which continues, but without any movement to abolish the office.

ESA Council was organized in 1946 as the true governing body of the Society, following the failure of an amendment for a "Board of Governors" in 1941. In addition to the elected officers, it included representatives to various organizations, and chairmen of sections and standing committees. With the addition of an elective Board of Editors in 1970, Council swelled to 30 members, dominated, of course, by the editors. Some members thought this good, as they believed that the prime function of ESA lay with its journals. Others disapproved of the dominance, and sought ways to reduce the size of the Council. At present, constitutional amendments are pending which would maintain Board of Editors representation, but curtail the actual number as members of Council.

With a Council of considerable size, it rapidly became unwieldy and unresponsive to decisions that had to be made and actions that needed implementation. Consequently, an Executive Committee, empowered to act on behalf of Council (which in turn was empowered to act on behalf of ESA) was formally designated. The Executive Committee consists of the president, immediate past-president, vice-president, secretary, treasurer, and business manager. As far back as 1935, the term "executive committee" had been used, and references to it appear throughout the meeting summaries, minutes, and proceedings. It appears to have been synonymous with the "officers."

THE MEETINGS

The meetings of the Ecological Society of America fall readily into two classes, the annual meetings, and secondary or ancillary meetings. With few exceptions, the official annual meeting has been held in conjunction with a larger scientific body. Until the early 1950's these were with the AAAS that almost always met the week between Christmas and New Year. Since then, meetings have been with the American Institute of Biological Sciences (AIBS), on college campuses (in contrast to downtown hotels), and in summer, usually in August. Annual meetings include elections (or election results), business sessions, reports of officers and committees, presentation of awards, a banquet and presidential address, presentation of symposia and contributed papers, and field trips.

The distribution of annual meetings (Fig. 3) shows a distinct favoritism for the northeastern quadrant of the United States. Ohio has hosted seven meetings, Pennsylvania six, and Massachusetts five. Of 61 meetings, only 14 have been west of the Mississippi River, only six west of the 100th meridian (all since 1957, and three in the 1970's). The annual meetings, therefore, show strong correlation with the total population, the distribution of colleges and universities, and undoubtedly with the location of a membership. They do not correlate with reigning presidents or other items internal to ESA. The explanation lies in the continued programming of the official annual meeting with a larger, umbrella-type, organization. Thus, meeting locations are beyond the jurisdiction of the Society, its officers, or its membership. This is true despite the fact that ESA has a representative to AIBS who participates in the setting of meeting locations.

In contrast, the geographic distribution of secondary meetings (official meetings other than the "annual meeting") (Fig. 4) shows an almost complete reversal of pattern. Forty-one have been held west of the Mississippi, and only 15 to the east. Thirty-nine were held west of the 100th meridian, and 23 in California. In addition, two meetings have been held in Vancouver, and one each in Toronto and Montreal. Annual meetings have never been held outside the conterminous United States. The possible reasons for this distribution are both numerous and cloudy.

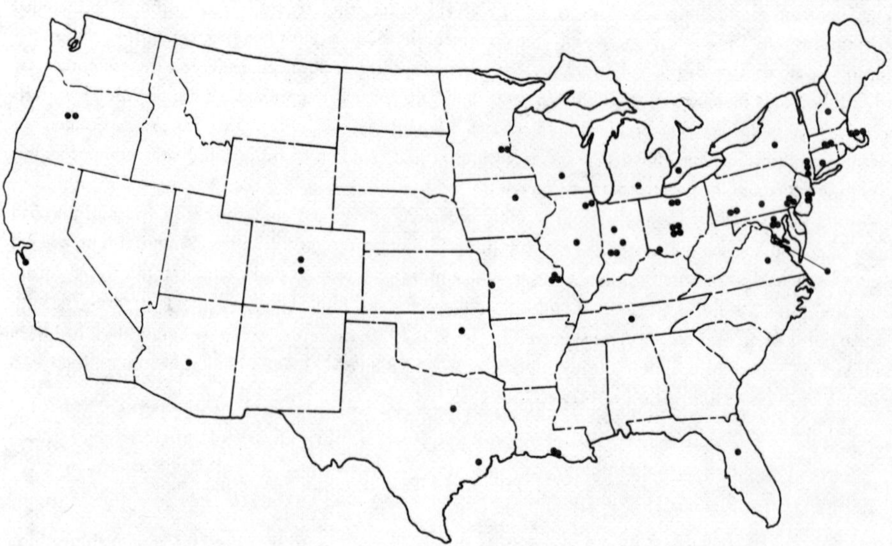

Fig. 3. Geographical distribution of annual meetings of the Ecological Society of America from 1914 through 1976.

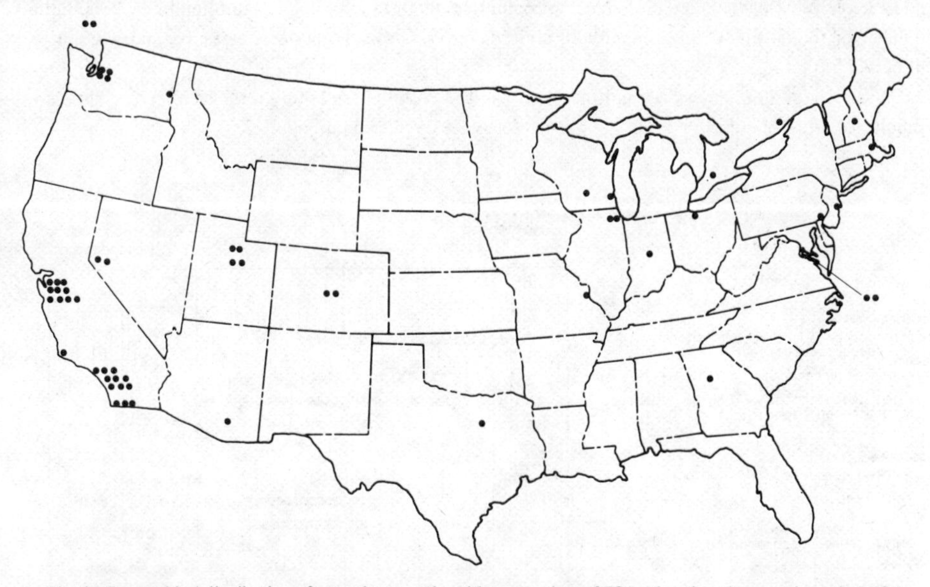

Fig. 4. Geographical distribution of secondary meetings (those meetings of ESA other than the annual meeting) of the Ecological Society of America from 1914 through 1976.

First, many of the California meetings were in conjunction with the Pacific Division of AAAS, and the Society often had a "western" meeting on this pretext. More important, perhaps, were the regularly scheduled field trips. With the preponderance of ecologists in the east and midwest, the opportunity for new and exciting ecological experiences was undoubtedly a stimulus. Unfortunately, we have little information on either the number or the distribution of those in attendance. Finally, the west coast was building an ESA population that must have found it difficult, in the pre-1950 period, to finance long train or bus trips to the annual meetings. Consequently, these secondary meetings in the western states provided opportunities for papers, discussions, acquaintanceships, and scientific exchange, in addition to the ever-present field trips. No report of any meeting has yet come to light that did not offer at least one field trip. The ESA, 61 years after its founding, is still a field oriented society.

Meeting cost was recognized in the early days. One of the prescribed duties of the secretary was to negotiate with the railroads for a reduced rate for members attending the annual meeting. Evidently this correspondence was usually successful, and the most common rate was three-fourths of the normal round-trip fare. It must be remembered that in these days before research grants, professional meetings were something that the scientist attended *at his own expense*. They were a business cost, and many of the Society stalwarts attended regularly, with perhaps an occasional token sum contributed by their home institution. Today, of course, there is little of this sacrifice, and most of the attendees that come unsupported are the job-seekers.

THE COMMITTEES

The various committees established by ESA through its first 61 years constitute a very difficult subject. This issue is also clouded by many that I consider to be "housekeeping" committees, and for the moment, these have been ignored. Examples in this group include membership, program, finance, and publication committees. Most of these are now standing committees, defined by either constitution or by-laws. But a number of special, or *ad hoc* committees, also fall into this societal business category, and will not be discussed further.

A select group of committees and their life span and evolutionary history is shown in Fig. 5. These are subdivided into five groups, subjectively chosen, solely for discussion purposes.

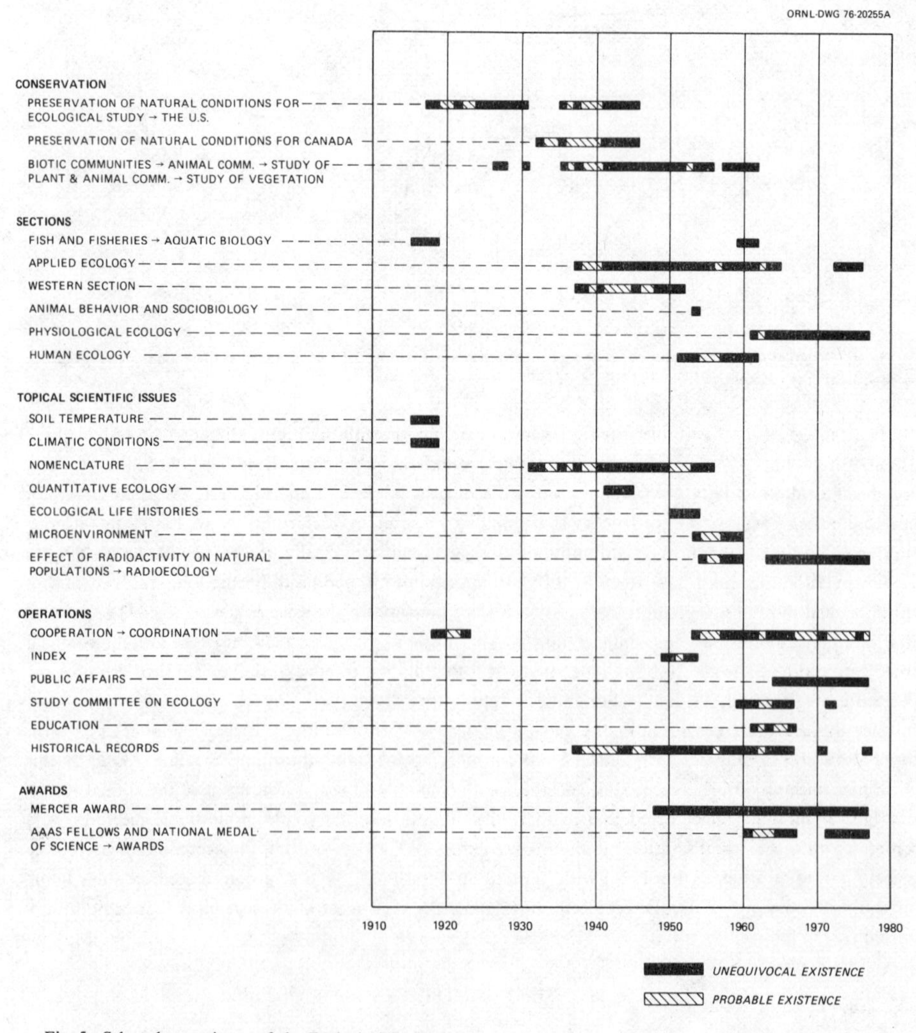

Fig. 5. Selected committees of the Ecological Society of America, grouped by general categories, and indicating time spans, longevity, continuity, and interrelationships.

Conservation

The first group covers those committees addressing a combination of preservation motives, and the objects of study of the science of ecology. ESA has been involved in "natural area" preservation right from the beginning (Harshberger 1918, Moore 1920, Shelford 1943), and while not actively involved as a Society at present, its members lend a great deal of strength to the natural area and wilderness preservation movement. A committee on "Preservation of Natural Conditions for Ecological Study" was established in 1917. The name was changed to "Preservation of Natural Conditions in the United States" in 1932, the same year that spawned a similar committee for Canada. No evidence of their existence after 1946 has been found, but movement toward the "Ecologist's Union" and today's Nature Conservancy was in process at that time, and absorption into a new organizational structure seems to be a logical explanation.

A committee on "Biotic Communities" came into existence in 1926, followed by a closely related series on "Animal Communities," "Study of Plant and Animal Communities," and "Study of Vegetation." There is no implication that these committees shared a common ancestry or that a direct evolution ensued. The series of committees is instead of commentary on the state of the science and the nature of the membership. The thrust began when ecologists were deeply concerned with the classic papers of Gleason (1926) and Cooper (1926), and ended with the major impetus of functional ecosystem ecology in the 1960's.

Sections

The second group of committees are those related to Sections (or at least potential sections). In 1917, a group on Fish and Fisheries was instigated, obviously, by an aquatic component of the membership. After a lapse of 40 years, a Committee on Aquatic Biology was formed in 1959, followed by establishment of the Aquatic Section in 1965. Applied Ecology, a Society interest since before World War II, concerned the foresters and range managers (primarily) in ESA. After the Committee disbanded in the mid-1960's, an Applied Ecology Section was formally constituted at the Minneapolis meetings in 1972. This followed passage of the National Environmental Policy Act (NEPA), Earth Day, and the Calvert Cliffs decision (Maryland vs. AEC) and the thrust of the Section has been toward environmental impact analysis and assessment, rather than the application of ecological principles to natural resource management.

The Western Section, after a long history of more or less independent meetings, was formally disbanded in 1975. This action was taken by ESA Council on petition from the Section officers. The stated reason was simply a lack of interest in sectional meetings and activities. In part, this must be due to a greater incorporation of western ecologists into the annual meetings of the Society. Nevertheless, the demise of the Western Section is almost concomitant with renewed interest in regionalization within ESA as a whole. A Southeastern Chapter was officially established in 1976, and preliminary moves toward at least regional meetings have been made in both the Great Lakes region (upper midwest) and the Pacific northwest within the last two years.

The Animal Behavior and Physiological Ecology Sections both boast over 1,000 members, and both are extremely active. Animal Behavior (as a Section) meets occasionally with ESA, but more often with related societies such as the American Society of Zoologists. In recent years, the Physiological Ecology Section has sponsored or co-sponsored strong programs at the ESA annual meetings. These have included both symposia and contributed paper sessions which, coupled with a periodic "newsletter" and a great deal of interaction among the members, indicates that this is probably the strongest subdivision of ESA at the present time.

A committee on Human Ecology, while never pushing toward section status, has continued to function. Fine distinctions between "human ecology," "sociology," "biological anthropology," and other terms continue to hamper major development. Recent emphasis among ESA members on "urban ecology" may help in the future to really treat man as an integral component of earth's ecosystems.

Finally, a small (106 members) Paleoecology Section was officially established in 1975 (not shown in Fig. 5). As programs and membership are still under development, it is too early to comment on either the scientific direction or the probable long-term success of this new section.

Topical Scientific Issues

Two committees, reflective of environmental interest, were first established in 1916, one on "Soil Temperature" and one on "Climatic Conditions." These were evidently an early attempt to coordinate subject matter interest through the Society, but by 1920, both had died. This abortive effort can be construed as a trial – both of what a new "Ecological Society" might do, and of the use of the committee mechanism to give a fledgling organization a sense of programmatic purpose.

The Committee on Ecological Life histories organized the preparation and publication of a series of "Outlines" giving basic literature, methods of study, objectives, and general aspects on the life histories of various groups of organisms – bees (Linsley *et al.* 1952), fossorial mammals (Howard and Ingles 1951), fish (Koster 1955), fungi (Cooke 1951), herbaceous plants (Stevens and Rock 1952), hydrophytes (Penfound 1952), marine mammals (Scheffer 1952), trees, shrubs, and stem succulents (Pelton 1951), and vascular epiphytes (Curtis 1952). Much of this background information has been of great value in autecological and physiological ecology that followed.

Of the remainder of this group, only the Committee on Nomenclature and the Radioecology Committee have made significant accomplishments. All, however, were concerned with substantive scientific subjects that were important to the Society at one time. The group concerned with nomenclature labored over many years toward a standardization of terms. It has been said, tongue-in-cheek, that "ecology is the science that tells you what you already know in terms that you can't understand." Problems of interpretation and shades of meaning were paramount, as were the various proposals for a taxonomy of communities. Publications by Carpenter (1938) and Hanson (1962) were outgrowths of the activity of the Committee on Nomenclature, although neither can be construed as a "final committee report."

The original committee on Effect of Radioactivity on Natural Populations, later shortened to Radioecology, has been active and successful. The committee has been the prime organizer and cosponsor of four major national symposia (Cushing 1976, Nelson 1973, Nelson and Evans 1969, and Schultz and Klement 1973). The last (Cushing 1976) is discussed below as the first Special Publication of the Society.

Operations

The Study Committee on Ecology has been activated by various presidents as the needs arose. They have tackled various problems that have faced both the Society and the science, and have provided recommendations for action. For example, from deliberations of the Study Committee coupled with the Xth International Botanical Congress in Montreal in 1959, direct threads of planning were spun into the Chemical Cycling Subcommittee in 1961, and several meetings that same year of the International Union for the Conservation of Nature (IUCN) and the International Union of Biological Sciences (IUBS) led to U.S. participation in the International Biological Program. A comprehensive review of Study Committee activities would be rewarding, but beyond the scope of the present paper.

The Index Committee was appointed by President Aldo Leopold in 1949 to prepare a 30-year index for *Ecology* (Aikman and Gates 1952). This was a monumental task, and many members contributed. In addition to its utility for the journal, the introduction and some of the index material has proven valuable in this reconstruction of ESA history.

An early Committee on Cooperation recognized the need to work with other groups and organizations, and to remain attuned to larger issues of environmental quality and natural resources, as well as research needs. After a short life, the formal concept was abandoned until 1953, when a Committee on Coordination was formed. Their work has been varied and somewhat intermittent, but appears to have kept ESA and its policies alive in the deliberations of government, industry, academia, and other professional societies (Sears 1956). A recent proposal to establish an executive office of ESA in the Washington, D.C. region was addressed by this committee. A consortium of interested (and related) societies was formed, but no definitive action has yet been taken.

The Committee on Historical Records is something of an enigma. Begun in 1937, it functioned, at least in name, almost to the present. Initially, it worked out an arrangement with the University of Cincinnati Library to act as the historical repository for ESA documents. In 1945, it included bound volumes of *Ecology, Ecological Monographs,* and the *Bulletin,* as well as items such as two folders of Harshberger's correspondence, three of Fuller's, and ten from Charles C. Adams. The University of Cincinnati Library has not been checked in this prolegomenon, but we are led to believe that bound volumes of the three journals are all that is there. A recent effort to move the repository to the University of Georgia was approved, and the move must either surface this material, or at least find out what happened to it. Every organization has an obligation to archive, and in many respects, ESA has been remiss in these obligations.

Awards

Two committees on awards (discussed in more detail below) have functioned during the last half of ESA history. The Mercer Award Committee is appointed each year for the purpose of selecting a recipient of an ESA award. The second committee, on AAAS Fellows and the National Medal of Science, was initiated in response to a seeming lack of recognition of ecologists and ecology by the larger scientific community. This effort has not met with great success (only three or four ecologists are members of the National Academy of Sciences, for instance), but in the last few years, establishment of new awards coupled with international environmental awareness has resulted in some long overdue recognition.

THE REPRESENTATIVES

Representatives from the Ecological Society of America to several organizations (Fig. 6) have served to both maintain contact and to perform liaison functions. Since inception, ESA has sent a member to the National Research Council, the action arm of the National Academy of Sciences. In the mid-1920's, a number of related biological societies formed a "council," and ESA became a staunch member. This activity culminated in the establishment of the American Institute of Biological Sciences (AIBS) in 1947. Shortly thereafter, ESA switched its annual meeting from AAAS in the winter to AIBS in summer. Representation on the AAAS Council apparently began (in 1937) after the young society had become sufficiently established to merit inclusion. Very little information has come to light, and no effort has as yet been expended to determine more of the details. As Fig. 6 clearly indicates, representation to the "big four" (National Research Council, AIBS, AAAS, and the Natural Resources Council) has been both strong and continuous. In 1924, ESA sent a representative to an organization called the Council on National Parks, Forests, and Wildlife. This may have been a forerunner of the Natural Resources Council of America, a non-profit association pledged "to advance the attainment of sound management of natural resources in the public interest." Membership, by written invitation, consists of "recognized national conservation organizations, scientific societies in the natural science field, and . . ."

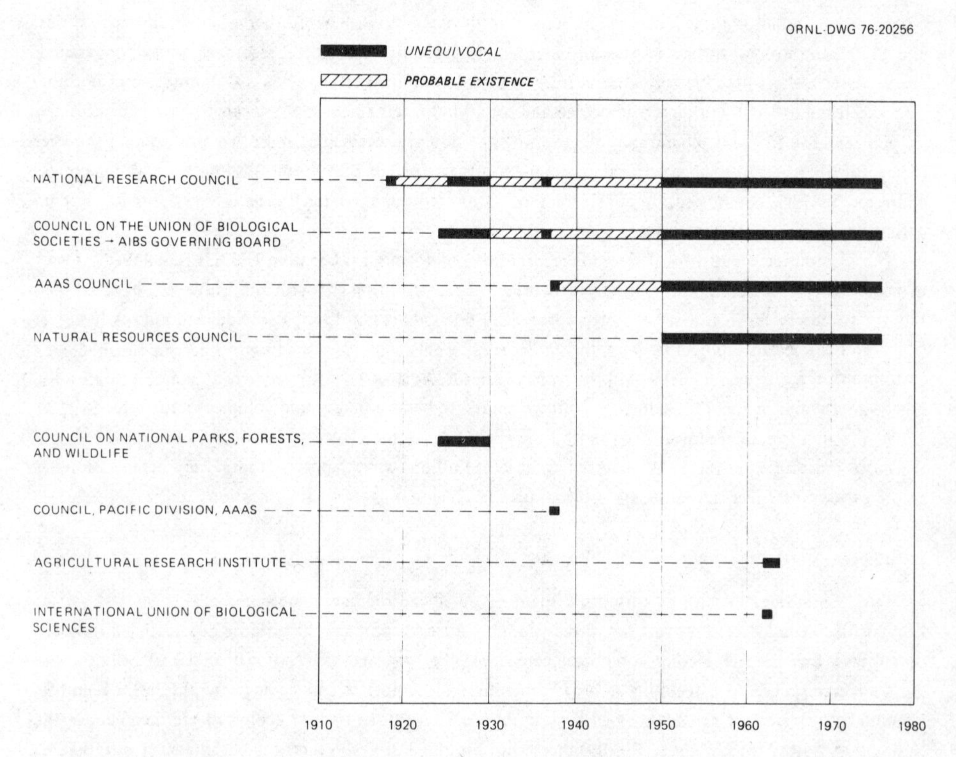

Fig. 6. Representatives of the Ecological Society of America to national and international bodies, showing timing, longevity, and continuity.

The other three representatives shown, to the Pacific Division, AAAS, the Agricultural Research Institute, and IUBS, are unexplained as to both their origin and their demise. The Pacific Division of AAAS regularly hosted secondary meetings of ESA (Fig. 4), and must have played a part in the formation of the Western Section. The latter was finally abolished in 1975 on petition of the Section officers. The Western Section was the only major geographical subdivision that ESA has had, although "Chapters" were at one time established in Oregon and Minnesota, and an Ecological Register for the New England States (Cushman *et al.* 1971) was compiled under ESA auspices.

THE PUBLICATIONS

The newly formed ESA issued volume one, number 1 of the *Bulletin of the Ecological Society of America* in March of 1917. This was rapidly followed by a "Handbook" later that same year, also issued as a number of the *Bulletin.* By the mid-1920's, following a period of sporadic publication, the *Bulletin* had stabilized as a small, quarterly journal that has continued uninterrupted to the present time. For a significant period, it appears that the *Bulletin* is perhaps the only real source of historical information. Yet there are some lapses. While responsibility for some of the early volumes is shrouded by the mists of antiquity, for at least 40 years the *Bulletin* was edited by the secretary of the Society. That individual had to gather material, prepare and edit copy, procure a printer, and handle distribution. Therefore, many volumes, upon careful perusal, bear the stamp of the individual secretary. Continuity in format or content was of minor concern.

Reports of meetings are uneven in quality and length. Committee activity summaries are sporadic at best. New members were presented intermittently, although special (separate and additional issues of the *Bulletin* were used as directories. In 1968, two major changes were instituted. One probably resulted from a long-standing Associate Membership, which entitled the member to receive only the *Bulletin*. Dissatisfaction with the housekeeping function of the *Bulletin* led to the establishment of the Bulletin Editor as an elected officer, and responsibility passed from the secretary to this newly created position. Secondly, the rise in environmental awareness spurred a change in format – larger page size, colored cover, and the inclusion of new features. These were typified by short essays on aspects of the science, lists or short sketch reviews of new books, announcements of a wide array of courses, meetings, etc., and the provision for at least some response from the readership.

With Volume 58 (1977), the *Bulletin* will be published six times a year. A large segment of the Society still believes that the *Bulletin* really functions as a "newsletter," and hence content and style ARE the prerogative of the editor. An equivalent group, however, believes that the *Bulletin* needs more structure, even if minimal. Historical record keeping, for example, may be a necessary and requisite objective for the *Bulletin*, and the editor needs to insure that all Society business and reports are adequately documented. At this point, it does not seem that the two views are mutually exclusive, and interesting content and format should be compatible with needs of the Society.

The *Plant World* began publication in 1897, organized and financed by a small, private group, the Plant World Association, and for the next 22 years carried a good share of the ecological publication in the United States. While the thrust of many of the editors and contributors was truly ecological, the journal carried few animal studies. The title, of course, was a discouraging factor. After the birth of the Society in 1915, talk of a journal began almost immediately. Most details, however, are lacking, and we don't yet know in which direction the plans were leading (Taylor 1938).

In late 1918, Dr. Daniel Trembly MacDougal of the Desert Laboratory of the Carnegie Institute at Tucson offered the *Plant World* to ESA (Moore 1938). In 1919, the Plant World Association consisted of 15 men, nine of whom (W. A. Cannon, J. A. Harris, B. E. Livingston, F. E. Lloyd, E. B. McCallum, D. T. MacDougal, J. B. Overton, F. Shreve, and E. N. Transeau) were charter members of the Ecological Society of America. There were only a few inconsequential stipulations – the editors of the *Plant World* were to serve on the board of editors of *Ecology* (as the journal was renamed), and the cover of *Ecology* was to carry the phrase "Continuing the *Plant World*" for a period of five years. In fact, this phrase lasted for 35 years, finally discontinued with Volume 36 in 1954.

An exhaustive review of *Ecology* as a journal is not appropriate here. A few generalities, however, may be in order. The first issue, dated January 1920, stated that "The pages...are open to papers of ecological interest from the entire field of biological science." At some point, as yet undetermined, *Ecology* instituted a policy strongly favoring original research, and with few exceptions, opposing theoretical or review submissions. As a result, many important papers went elsewhere, particularly to the *American Naturalist*, the *Botanical Review*, various kinds of *Proceedings...*, and several symposia. More recently, such items have appeared in two hardcover periodicals, *Advances in Ecological Research*, and the *Annual Review of Ecology and Systematics*. Both of these are published commercially, and have no direct relation to the Ecological Society of America.

In the spring of 1973, the Board of Editors officially changed policy to include theoretical (particularly mathematical) and review papers. In the three years since, however, it is not evident that the stated policy shift has had an effect on the nature of the journal. First of all, tradition dies slowly, and 40 years of non-acceptance of theoretical papers is hard to overcome. Secondly, except for *outstanding* logical presentations, theoretical papers in ecology do not fair well in review competition with reports and analyses of original research intended for ESA's major journals.

Volume 1 also contained a "Notes and Comment" section intended for shorter communications and originally containing some feedback from the readership. While not of the "Letters to the Editor" type, in its early years *Ecology* did provide for some response. The section survived almost unchanged through Volume 41 (1960) when it was renamed "Reports." The impetus for change came mostly from members who felt that "Notes and Comment" was somehow demeaning for sound scientific papers that were placed in that section solely because they were short (less than four printed pages). As space in the journal became more and more limiting, the "Reports" section, set in smaller type, came under a similar attack, and was last published in 1969 (Volume 50).

Volume 1 also contained the first book reviews, and a review section has been an integral and important part of *Ecology* ever since. Feedback from the readers indicates that the review section is often the first part to be read in each issue, and is probably read in its entirety by a majority of the ESA membership. Reviews in the first 33 volumes (1920–1952) were either written, generated, or solicited by the editors or members of the editorial board. In 1953, a position of Review Editor was formalized, since held by only four men – LaMont C. Cole (1953–55), Robert H. Whittaker (1956–1964), Paul S. Martin (1965–1970), and Robert L. Burgess (1971–present). Commensurate with the expanded interest in ecology beginning in the late 1960's the number of related books received by ESA has burgeoned. What was originally a very sporadic (and short) list of books received has become an average of two pages in each issue.

Content, editing, and finance are outside the scope of the present paper. However, complete lists of editors, assistants, board members, and business managers are given, with institutional affiliation and dates of service in Aikman and Gates (1952) and Thomas and Stearns (1975).

By 1925, a committee had been established to evaluate long-range needs for publication. The main concern was for publication space, and the issue revolved around an increase in the size of *Ecology* versus the initiation of a second journal. The choice was made, and *Ecological Monographs* began in January of 1931, intended to carry longer papers of a monographic nature. This advent was accompanied by a new class of membership, "Sustaining," a portion of whose larger dues would go toward support of the new journal, and by the establishment of a relationship with Duke University Press to act as publisher for ESA (Lawrence and Lawrence 1956). The new journal, the sustaining membership class, and the publisher have remained, with minor changes, intact to 1976. The editorial criteria for *Ecological Monographs* were the same as for *Ecology*, except that length of published papers should be 20 pages or more. With few exceptions, this rule held to 1973, when the limit was lowered to 16 pages. This change was never implemented, however, and the current "Instructions to Authors" perpetuates the 20 page limit. The journal has also remained a quarterly, in contrast to *Ecology* which went to six issues per year (bimonthly) in 1965 (Volume 46). Lists of editors, editorial board members, and business managers appear in Lawrence and Lawrence (1956), while portraits of seven editors of the period are reproduced in Lindsey (1973).

Recently, the Board of Editors of *Ecology* and *Ecological Monographs* has discussed the potential for still additional publication space for the membership and the readership. *Ecology* is almost at a size limit, set partly by postal regulations and partly by the physical unwieldiness of still larger volumes. *Ecological Monographs* continues at about the same size (ca. 450 pages per year) primarily because of a lack of long submitted manuscripts. Coupled to the need for theoretical, mathematical, and other types of outlets, a number of potential new journal titles have been discussed. A *JOURNAL OF APPLIED ECOLOGY* has frequently been suggested. However, that exact title has been in publication for the past 13 years as an official organ of the British Ecological Society. Some see a need for an *ECOLOGICAL REVIEWS*, similar in nature, perhaps to the *BOTANICAL REVIEW, BIOLOGICAL REVIEWS*, or the *QUARTERLY REVIEW OF BIOLOGY*. Counter-arguments point to the existence of the two hard-cover series, *ADVANCES IN ECOLOGICAL RESEARCH* and the *ANNUAL REVIEW OF ECOLOGY AND SYSTEMATICS.* A *JOURNAL OF MATHEMATICAL ECOLOGY*, and *ECOLOGICAL MODELING* were also discussed, and both have been pre-empted by international publishers.

At present, if ESA does, in fact, decide to add a third journal, something like *ECOLOGICAL THEORY* (or *THEORETICAL ECOLOGY*) or a *JOURNAL OF ECOSYSTEM ANALYSIS* seem to be the most viable concepts. However, in these times of escalating costs, a new journal begun by an existing Society needs to be self-supporting almost from the start. ESA is still uncertain of this possibility, in light of its fiscal resources, and hence has made no decision. The current rejection rate for *Ecology* runs consistently at 70 percent, however, so it is evident that the need is there. Only the future will tell the outcome of these continuing deliberations.

Other expansions include the reformatting of the *Bulletin* (discussed above) and the decision to publish it bimonthly starting in 1977. Also, the COMMENTARY, carried in *Ecology* from 1969 through 1976, will move to the *Bulletin* in 1977. Publication of *Ecology* eight, ten, or twelve times a year was also considered, but was deemed editorially impossible with the present volunteer Board of Editors.

Two other items deserve mention. The first paid employee of the Society was a Managing Editor, begun with the appointment of Alton A. Lindsey in 1971. He was succeeded by Crawford G. Jackson, Jr. in 1973. The Managing Editor is responsible for both *Ecology* and *Ecological Monographs* in all respects except acceptance/rejection decisions based on scientific merit of the submitted manuscripts. This function is handled by the Board of Editors. Secondly, a Special Publication Series has been established, and the first volume (Cushing 1976) is now in print. Plans are underway for additional volumes. Quality control and editorial criteria are still implemented by the managing editor and the Board, but publication is through commercial channels and is intended to be *ad hoc* rather than periodic.

THE AWARDS

The Society has only two awards that it sponsors regularly for its members, the George Mercer Award (Table 6), and the Eminent Ecologist Citation (Table 7). The first, established by Dr. Frank W. Preston, was accepted by action of Council on December 29, 1947. The award, given for outstanding papers in the field of ecology was defined in the *Bulletin* (Vol. 29, no. 1, March 1948): "The award shall be known as the George Mercer award, and is given in memory of Lieutenant George Mercer, of the British Army of World War I, killed in action October 3, 1918.

"The purpose of the award is to commemorate the sacrifice of a young naturalist and ecologist, and to encourage others to publish papers comparable with those it is reasonable to suppose he would have published if he had lived."

While the award consists only of a citation and a check for $100, it carries great prestige within the Society. Problems are of two kinds. Attempts to increase the stipend through voluntary donation have been unsuccessful. Consequently, many members feel that the amount is so insignificant that the award itself must lack importance. Secondly, the screening committee for the George Mercer award changes each year, so that despite the guidelines, criteria vary. While the original instructions explicitly state that the paper need not appear in one of the ESA journals, many ecologists seem to feel strongly that the selection should come from *Ecology* or *Ecological Monographs*. There is an objective basis for these thoughts. The ESA journals must be a major outlet for ecological papers in English. Indications are that most ecologists in

Table 6. Recipients of the George R. Mercer award, Ecological Society of America, 1949–1975

1949 – Henry P. Hansen	1963 – Joseph H. Connell
1950 – Edsko J. Dyksterhuis	1964 – Orie L. Loucks
Henry S. Fitch	1965 – Kenneth F. Norris
1951 – Helmut K. Buechner	1966 – C. S. Holling
1952 – Robert B. Platt	1967 – Robert H. Whittaker and William A. Niering
1953 – Frank J. Pitelka	1968 – Edward Broadhead and Anthony J. Wapshere
1954 – F. Herbert Bormann	1969 – Lynn T. White, Jr.
1955 – Shelby Gerking	1970 – (no award made)
1956 – Howard T. Odum and Eugene P. Odum	1971 – Edward O. Wilson and Daniel Simberloff
1957 – John J. Christian	1972 – Joel E. Cohen
1958 – Jerry S. Olson	1973 – Carl F. Jordan
1959 – Robert H. MacArthur	1974 – Paul K. Dayton
1960 – Calvin McMillan	1975 – Peter L. Marks
1961 – Robert A. Norris	1976 – William E. Neill
1962 – Harold A. Mooney and W. Dwight Billings	

Table 7. Recipients of the EMINENT ECOLOGIST award, Ecological Society of America, 1953–1976

1953 – Henry Allan Gleason	1965 – Paul Bigelow Sears
1954 – Henry S. Conard	1966 – Alfred C. Redfield
1955 – Albert Hazen Wright	1967 – Alfred Edward Emerson
1956 – George B. Rigg	1968 – Victor Ernest Shelford
1957 – Karl Patterson Schmidt	1969 – Stanley Adair Cain
1958 – Arthur W. Sampson	1970 – Murray Fife Buell
1959 – Henry Allan Gleason	1971 – Thomas Park
1960 – Walter P. Cottam	1972 – Ruth Patrick
1961 – Charles E. Elton	1973 – Robert Helmer MacArthur
1962 – George Evelyn Hutchinson	1974 – Eugene P. Odum
1963 – William Skinner Cooper	1975 – Cornelius H. Muller
1964 – Lee R. Dice	1976 – Alton A. Lindsey

North America with solid, high quality ecological manuscripts submit them first to ESA. Rejection rates approximate 70 percent. Hence, the pages of these two journals already constitute a large step in the selection process of outstanding ecological papers. Since 1958, most awards have, in fact, been made for papers published in one of the ESA journals, but the controversy remains.

Secondly, there are those who feel that a series of papers by an author should be more indicative of the level of contribution to the science than a single paper can ever be. This is true, but neither Nobel nor Pulitzer prizes are awarded for an illustrious lifetime compiled from masses of mediocrity. Instead it is the momentous breakthrough or the one great play for which these awards are made. The donor's original stipulations, in this case, are probably correct.

The second major award is that of Eminent Ecologist. Henry Allan Gleason was cited in 1953 (Gleason 1953), apparently *ad hoc*, as he was cited again in 1957. Selection of the Eminent Ecologist has rested with the Nominating Committee, a procedure which is in the process of change. An Awards Committee was appointed in 1972, ostensibly to coordinate all such activities for the Society. At present, nominations are solicited and evaluated, but final decisions rest with the committee. Through the years, selections have been made from among the great names in American ecology. There are (or were) two constraints, however. Originally, the citation could not be made to a president or past-president of ESA. As mentioned previously, to some extent the presidency has been treated as an "eminent ecologist" award. Secondly, the award is made to a living individual, usually for a lifetime of service and contribution, i.e., a cumulative honor. The only exception has been the posthumous award following the premature death of Robert MacArthur.

In recent years, ESA has given an external award for contributions to public awareness. The first went to Arthur Godfrey and the second to Pete Seeger. Also, a Distinguished Service Citation was presented to Jack Major in 1975 and to George H. Sprugel, Jr. in 1976. This was created for those *ad hoc* situations where recognition by the Society is richly deserved, but for which no other avenue is available.

Awards have also been presented recently to ESA members. The Tyler Award was presented to Eugene and Howard Odum in 1974 and to Ruth Patrick in 1975. The Browning Award in 1975 went to G. Evelyn Hutchinson, and in 1976, David E. Reichle received the Scientific Achievement Award from the International Union of Forestry Research Organizations. The Pahlavi Environmental Prize, recently established by the Shah of Iran, and a possible Environmental Nobel Prize are additional outlets for recognition of ESA members.

THE SPINOFFS

Time and space do not permit an exhaustive review of the various events and organizations that have derived their impetus from the Ecological Society. Two deserve mention, however, The Nature Conservancy and The Institute of Ecology (TIE).

Almost from the start, two factions within the Society were apparent concerning the preservation of natural areas. While both groups were agreed on the need, a midwestern segment saw such activity as a logical function of an ecological society, while a second, largely eastern, viewed it as a private, industrial, or governmental enterprise, but NOT as a proper path for a learned scientific society to follow.

Consequently, by the late 1940's, the Ecologist's Union was formed, almost entirely of ESA members, and thus having a quasi-ESA flavor. The Union's purpose was to identify and find means to acquire or otherwise preserve portions of the American landscape that had great ecological value for both teaching and research. When it became apparent that ESA as a body could not sanction this activity, The Nature Conservancy was created. Incorporated in 1950, The Nature Conservancy has been a highly successful, private, land preservation association. Original membership was drawn heavily from ESA, and while now in a minority, many ESA members still actively support The Nature Conservancy.

The Institute of Ecology, originally conceived as an action arm of ESA, was the result of a long series of planning exercises by the Study Committee. Traveler's Research Corporation and the firm of Peet, Marwick and Mitchell served as consultants during the formative stages. Incorporated in 1971, TIE was divorced from ESA soon after. Governed by a Board of Trustees and ostensibly supported by a large group of "founding institutions" (each of which holds a seat in the Assembly), it has been beset by financial difficulties from the beginning. It is geographically dispersed (pan-American), and has had to rely heavily on foundation support for its activities. This has given TIE a project orientation, implemented through workshops that address specific problems and identify a final report as an end-product. In 1972 it established a Washington, D.C. office, complete with staff, something that ESA, with over 5,000 members, had been trying to do for several years. The first president, Arthur D. Hasler, was headquartered in Madison, Wisconsin, far from the site of operations. He was succeeded by John M. Neuhold at Utah State, who resigned in mid-1976. At the moment, TIE's existence is at least financially threatened, and we must see what the future brings. It has been a case of running before one learns to walk.

ACKNOWLEDGMENTS

The author is indebted to Ms. Lois H. Bradley for her many contributions to this history. She has gathered, culled, and organized many of the data on which it is based, through a long and often tedious perusal of ESA journals. I also thank Ms. Polly L. Henry for typing the many revisions and for her cheerful assistance in many other phases of this endeavor.

LITERATURE CITED

Abbott, G. A. 1958. The first fifty years – North Dakota Academy of Science 1908–1958. Univ. North Dakota Press, Grand Forks. 23 pp.

Aikman, John M. and **Frank C. Gates.** 1952. ECOLOGY thirty year index. (Volumes 1–30, 1920–1949). Ecological Society of America, Durham, NC. 212 pp.

Allee, W. C., A. E. Emerson, O. Park, T. Park, and **K. P. Schmidt.** 1949. Principles of animal ecology. W. B. Saunders Co., Philadelphia. pp. 13–72.

Brewer, Richard. 1960. A brief history of ecology – Part I – Prenineteenth Century to 1919. Occasional Papers of the C. C. Adams Center for Ecological Studies 1:1–18.

Carpenter, J. Richard. 1938. An ecological glossary. Univ. of Oklahoma Press, Norman. 306 pp.

Coker, R. E. 1938. Functions of an ecological society. Science 87(3358):309–315.

Cooke, Wm. Bridge. 1951. Ecological life history outlines for fungi. Ecology 32(4):736–748.

Cooper, William S. 1926. The fundamentals of vegetational change. Ecology 7(4):391–413.

Cowles, Henry Chandler. 1904. The work of the year 1903 in ecology. Science 19(493):879–885.

Cushman, M. F., F. H. Bormann, A. S. Dominski, T. G. Siccama, and **D. G. Sprugel.** 1971. The ecological register for the New England states. Ecological Society of America and New England Natural Resource Center, Boston, Mass. 42 pp.

Cushing, Colbert E., Jr. (ed.). 1976. Radioecology and energy resources. Proceedings of the Fourth National Symposium on Radioecology. Spec. Publ. No. 1, Ecol. Soc. Amer., Dowden, Hutchinson and Ross, Inc., Stroudsburg, PA. 401 pp.

Curtis, John T. Outline for ecological life history studies of vascular epiphytic plants. Ecology 33(4):550–558.

Doig, Ivan. 1976. Early forestry research. A history of the Pacific Northwest Forest and Range Experiment Station, 1925–1975. USDA Forest Service, Pacific NW For. and Range Exp. Stn., Portland, OR. 35 pp.

Dreyer, William A. 1945. The Ecological Society of America. A.A.A.S. Bull. 4(2):15–16.

DuRietz, Gustav Einar. 1921. Zur Methodologischen Grundlage der Modernen Pflanzensoziologie. Adolph Holzhausen, Vienna. 272 pp.

Egerton, Frank N. 1976. Ecological studies and observations before 1900. pp. 311–351, IN: Taylor, Benjamin J. and Thurman J. White (eds.). 1976. Issues and Ideas in America. Univ. Oklahoma Press, Norman. 380 pp.

Egler, Frank E. 1951. A commentary on American plant ecology, based on the textbooks of 1947–1949. Ecology 32(4):673–694.

Gleason, H. A. 1926. The individualistic concept of the plant association. Bull. Torrey Bot. Club 53:1–20.

Gleason, H. A. 1936. Twenty-five years of ecology, 1910–1935. Mem. Brooklyn Bot. Gard. 4:41–39.

Gleason, H. A. 1953. Autobiographical letter. Bull. Ecol. Soc. Amer. 34(2):40–42.

Hanson, Herbert C. 1962. Dictionary of ecology. Philosophical Library, New York. 382 pp.

Harshberger, John W. 1918. Ecological Society of America – The preservation of our native plants. Torreya 18(8):162–165.

Hollander, Rachelle. 1976. Ecologists, ethical codes, and the struggles of a new profession. Hastings Center Report 6:45–46.

Howard, Walter E. and **Lloyd G. Ingles.** 1951. Outline for an ecological life history of pocket gophers and other fossorial mammals. Ecology 32(3):537–544.

Kathren, Ronald L. and **Natalie E. Tarr.** 1974. The origins of the Health Physics Society. Health Phys. 27:419–428.

Koster, William J. 1955. Outline for an ecological life history study of a fish. Ecology 36(1):141–153.

Laude, H. H., M. F. Miller, J. D. Luckett, G. G. Pohlman, D. S. Metcalfe, W. H. Pierce, and **Emil Truog.** 1962. History of the American Society of Agronomy. First Fifty Years – 1907 to 1957. Agron. J. 54:57–69.

Lawrence, Donald B. and **Elizabeth G. Lawrence.** 1956. Ecological Monographs. Twenty-Year Index. Volumes 1–20, 1931–1950. Duke Univ. Press, Durham, NC. 44 pp.

Lindsey, Alton A. 1973. Ecological Monographs. Twenty-Year Index II. Volumes 21–40. 1951–1970. Spec. Suppl. to Vol. 43, Ecol. Monogr., Duke Univ. Press, Durham, NC. 47 pp.

Linsley, E. G., J. W. MacSwain and **Ray F. Smith.** 1952. Outline for ecological life histories of solitary and semi-social bees. Ecology 33(3):558–567.

McIntosh, Robert P. 1974. Plant ecology 1947–1972. Ann. Missouri Bot. Gard. 61:132–165.

McIntosh, Robert P. 1975. H. A. Gleason – "Individualistic Ecologist" 1882–1975; His contributions to ecological theory. Bull. Torrey Bot. Club 102(5):253–273.

McIntosh, Robert P. 1976. Ecology since 1900. pp. 353–372, IN: Taylor, Benjamin J. and Thurman J. White (eds.). Issues and Ideas in America. Univ. Oklahoma Press, Norman. 380 pp.

Moore, Barrington. 1920. The Ecological Society and its opportunity. Science 51(1307):66–68.

Moore, Barrington. 1920a. The scope of ecology. Ecology 1:3–5.

Moore, Barrington. 1938. The beginnings of ecology. Ecology 19(4):592.

Nelkin, Dorothy. 1976. Ecologists and the public interest. Hastings Center Report 6:38–44.

Nelson, D. J. (ed.). 1973. Radionuclides in ecosystems, Proceedings of the Third National Symposium on Radioecology, CONF-710501, Oak Ridge, Tennessee. 2 vols. 1268 pp.

Nelson, D. J. and **F. C. Evans** (eds.). 1969. Proceedings of the Second National Symposium on Radioecology, Ann Arbor, Michigan, May 15–17, 1967.

Odum, Eugene P. 1968. Energy flow in ecosystems: A historical review. Am. Zoologist 8:11–18.

Pelton, John F. 1951. Outline for ecological life history studies in trees, shrubs, and stem succulents. Ecology 32(2):334–343.

Penfound, William T. 1952. An outline for ecological life histories of herbaceous vascular hydrophytes. Ecology 33(1):123–128.

Price, Raymond. 1976. History of Forest Service research in the central and southern Rocky Mountain regions, 1908–1975. USDA Forest Service Gen. Tech. Rept. RM-27, Rocky Mt. For. and Range Exp. Stn., Fort Collins, CO. 100 pp.

Raup, Hugh M. 1942. Trends in the development of geographic botany. Ann. Assoc. Am. Geogr. 32(4):319–354.

Reed, Howard S. 1905. A brief history of ecological work in botany. Plant World 8(7):163–208.

Reese, Kenneth M. (ed.). 1976. A Century of Chemistry. The Role of Chemists and the American Chemical Society. American Chemical Society, Washington, DC 468 pp.

Roche, Marcel. 1976. Early history of science in Spanish America. Science 194:806–810.

Rübel, Eduard. 1927. Ecology, plant geography, and geobotany; their history and aim. Bot. Gaz. 84(4):428–429.

Scheffer, Victor B. 1952. Outline for ecological life history studies of marine mammals. Ecology 33(2):287–296.

Schultz,V. and **A. W. Klement** (eds.). 1963. Proceedings of the First National Symposium on Radioecology. Reinhold Publ. Co., New York, and AIBS, Washington, D.C.

Sears, Paul B. 1956. Some notes on the ecology of ecologists. Sci. Mon. 83(1):22–27.

Sears, Paul B. 1969. Plant ecology. pp. 124–131, IN: Joseph Ewan (ed.). A short history of botany in the United States. Hafner Publ. Co., New York and London.

Shelford, Victor E. 1938. The organization of the Ecological Society of America 1914–19. Ecology 19(1):164–166.

Shelford, Victor E. 1943. Twenty-five-year effort at saving nature for scientific purposes. Science 98(2543):280–281.

Simkins, Tania. 1971. Association profile: The Ecological Society of America. Assoc. and Soc. Mgt. Oct./Nov.:27–30, 110–114.

Steven, O. A. and **Leo F. Rock**. 1952. Outline for ecological life history studies of herbaceous plants. Ecology 33(3):415–422.

Sullivan, Carl R. 1976. Bicentennial salute to The American Fisheries Society. BioScience 26(6):417.

Tansley, Arthur G. 1947. The early history of modern plant ecology in Britain. J. Ecol. 35:130–137.

Taylor, Norman. 1938. The beginnings of ecology. Ecology 19(2):352.

Thomas, A. J., III, and **F. W. Stearns.** 1975. Twenty year index to Ecology, 1950–1959. Ecological Society of America, Durham, NC. 249 pp.

Trass, H. 1976. Vegetation science: History and contemporary trends of development. Academy of Sciences of the USSR, Nauka Press, Leningrad. 252 pp. [In Russian] .

AMERICAN GRASSLAND ECOLOGY
1895-1955

Ronald Tobey

American Grassland Ecology, 1895-1955:
The Life Cycle of a Professional Research Community

Ronald Tobey
University of California, Riverside

In her recent work, *Invisible Colleges*, Diana Crane tied the characteristics of published literature to a theory of the spread of scientific information and the social structure of science. She proved that the logistic curve for cumulative number of scientific titles in a broad scientific discipline, which was first discussed by Derek Price, is typical of restricted specialties. Utilizing information derived mainly from questionnaires answered by scientists and drawing on the theory of scientific progress of Thomas Kuhn, Crane hypothesized that the rates of increase in cumulative publications and in addition of new authors to the bibliography of a special field were tied to the process by which a research field emerges out of a paradigm, passes into normal science, solves its major problems, confronts anamolies, and finally, within the context of the paradigm, exhausts itself.[1]

The convincing quality of Crane's work raises an intriguing question for historians. Accepting that the described characteristics of publication in basic fields are correct, could historians use these characteristics to "predict" what was occurring in the history of a research field for which they have less detailed and less valid statistical knowledge than that used by sociologists, or for preliminary analysis of social development of a research field, preparatory to traditional research methods? If the answer to this question were to be positive, then historians would possess a quantitative tool for investigating the multiplicity of scientific fields of the last century. In this paper, I report on a straightforward test application of Crane's theory to a historical field, American grassland ecology.

Grassland Ecology

The scientific specialty of ecology emerged in the United States as an approach to vegetational change on the American grasslands. With its intellectual foundations in the 1890s, the decade which also saw the establishment of ecology in Denmark and Germany, grassland ecology is one of the older American scientific specialties. By 1955, it had passed through an entire life cycle -- establishment, theoretical advancement, institutionalization in universities and governmental research offices, export of its principles abroad, application to practical problems like overgrazing and drought on the ranges of the 1930s, decline and decay as an innovative and independent specialty in the 1950s. It had also reached beyond its own boundaries to influence the social sciences. Frederic Clements, chief theoretician of dynamic ecology, influenced both E.A. Ross, the sociologist who was attracted to an ecological sociology, and Walter Prescott Webb, the historian who adopted Clements' vision of the American prairies as an integral vegetational unit.[2]

From its beginnings in the botanical concerns of C.E. Bessey of the University of Nebraska in 1884, grassland ecology was oriented toward problems of controlling the rapid vegetational change on the prairies. The introduction of agriculture on the Great Plains following the Civil War disrupted the vegetational covering, introduced many foreign and noxious plants, and allowed for periodic invasion of the disheveled soil by pests like the Russian thistle. In an era before the chemical control of weeds and pests, scientists were of necessity pushed toward control through means of early detection and physical removal; the Russian thistle, for instance, were pulled from crop fields by hand and burned. Similarly, weeds were controlled by planting crops that could compete most advantageously with them in root structure and in changes of temperature. Scientists at midwestern agricultural experiment stations and their university research colleagues were stimulated by these problems to look at the successional changes of flora after breaking of the sod, at the relationship of the plant to its whole environment, including competing weeds and varmints, and to determine how manipulation of the environment would lessen the crop and grazing problems of farmers and ranchers. Grassland ecology was, therefore, closely tied to technology.

Originating in this intellectual milieu of scientific innovation and practical problem-solving, Frederic Clements, a student of C. E. Bessey at the University of Nebraska, and himself a teacher there from 1897 to 1907, elaborated a remarkably philosophical base for the new science of ecology. From *Research Methods in Ecology* (1904) to *Plant Succession* (1916), Clements' prolific mind invented new concepts and terms, such as "biome," which refers to a synthesized plant-animal community, and "climax," which is the final and stable phase of a succession of vegetational communities in one geographical area. Clements' concepts and theories of succession and climax won many partisans (leading A.G. Tansley, the prominent British ecologist, to refer to Clements' "apostles"). Most contemporary ecologists who engaged in historical reflection on the first half of the twentieth century have paid credit to the enormous impact of Clements' work.

Grasslands ecology peaked as a research specialty in the late 1930s. The crisis of the drought on the Great Plains prompted a massive research effort by the agricultural research establishment - the Department of Agriculture, the experiment stations, Forestry Service field stations, and the academic departments of midwestern universities. The prominent role of the University of Nebraska in the scientific aspects of this research is revealed in the present study.

In 1939, plant ecology was conceptually synthesized with animal ecology in the classic textbook by Frederic Clements and Victor Shelford, *Bio-Ecology*. The theory of succession, which had been restricted to plant and animal communities separately, was generalized as the paradigm for a unified plant-animal community. Subsequently, grassland ecology, as a purely plant study, became less important when the leading scientific questions shifted to the arena of the plant and animal biome. The specialty produced fewer and fewer, original, scientific contributions.

In comparison with earlier decades, scientists conducting research in grassland ecology in the 1940s tended increasingly to become involved with problems of range rangement, that is, with increasing production. Following this direction, grassland ecology transformed itself into a technology. For this reason, the Biological Abstracts lists no publications for grassland ecology for 1956 and 1957, even though economic studies abounded. The retirement of John E. Weaver from the University of Nebraska in 1954 simultaneously deprived the research community of a central scientist. As a scientific specialty, the field collapsed.[3]

The Problems and the Logic of Research

From all impressions, grassland ecology was an important, independent, and basic research specialty. The increase of its bibliography and the growth of its scientific community should fit the pattern described by Crane for basic fields in the last century. Whether they do fit is a difficult problem, because the comprehensive bibliographical sources for testing a fit do not cover the entire history of the specialty and, of course, it is not possible to ask most participants in the field to respond to questions about their informal relationships. The problems addressed in this paper, therefore, are: first, can we use the Crane-Price hypothesis to "predict" accurately the early stages of the growth of grassland ecology? Second, do these stages appear, to the best evidence of traditional historical methods, to correspond to the stages described by Crane? In particular, with regard to this second problem, was the work of Clements--or any one in the field, for that matter--a paradigm? A positive testing of these problems will not only confirm Crane's theory, but will also lend support to aspects of Kuhn's theory of scientific progress upon which it is based.*

Positive testing of a historical model that supports Kuhn's theory would be ironic, at this juncture in the debate over it; the theory has been so severely criticized that even the originator has modified his early views. In "Reflections On My Critics" (1970), and "Second Thoughts On Paradigms" (1974), Thomas Kuhn retreated from the concept of the paradigm, for which his defender, Margaret Masterman, had found 21 different meanings, to the concept of "disciplinary matrix."

This tactical retreat in semantics came in the face of pressure from two continuing lines of criticism, both involving conflict

---

*The hypothesis of D.J. de Solla Price that the culumative literature of a scientific field is described by a logistic curve, which has been verified for variety of basic research areas, is referred to as the "Crane-Price Hypothesis." The hypothesis of Crane, that the social structure producing the literature follows the developmental stages of science described by Thomas Kuhn in *The Structure of Scientific Revolutions* (1962), is referred to as the "Crane-Kuhn Hypothesis."

models of scientific progress. A brief examination of these criticisms will help to define what the test of the Crane model can and cannot show. In the philosophy of science, Imre Lakatos and Paul Feyerabend, whose views are quite divergent, have nonetheless perceived the advancement of science as resulting from struggle between competing theories, without a normal science period of mono-paradigmatic research. In the sociology of science, the paradigm has been increasingly interpreted in restricted scope, as typical of research areas defined in terms of specific problems, rather than in terms of world views, and frequently with competition between different paradigms. Mono-paradigmatic research areas have been described by one sociologist as typical only of the most advanced research areas.[4]

Kuhn's theory is particularly amenable to quantitative testing, but it should be clear that quantitative tests cannot tell us much about the intellectual content of paradigms, or even prove whether a field has one or several paradigms. What quantitative tests can demonstrate is whether the community structure of a research problem area is divided into several groups with different social bases and career patterns or is dominated by a single pattern. With such information, we can then be guided in our research in the manuscripts and publications of the scientists. Although the link between institutionalization of research activity and the intellectual history of the research community can be proved only by traditional historical research, many sociologists of science, as well as Thomas Kuhn, assume that the link is direct. Scientific institutions are built on consensus; conflict, if it occurs, must occur between or lead to separate institutional bases for each conflicting view. This assumption is open to question, but my objective does not include dealing with it. Consequently, I am going to accept as given that intellectual history is correlated to institutionalization, for the purpose of utilizing Crane's theory.[5]

Diana Crane's theory assumes that scientific progress occurs when an innovation is adopted by a few individuals, spreads to a small group, and the small group expands by the normal routes of graduate training and conversion. Gradually the innovation being proselytized by the "invisible college" eliminates alternative and competing theories and becomes a mono-paradigm, in the sense of Kuhnian normal science, for the research problem area. This intellectual process is revealed in publication patterns. In Figure One below, Crane has graphically summarized the theoretical characteristics of the growth of scientific literature in a field and her hypothesis about the intellectual and social structures producing that literature. The logistic curve is derived from Derek D. J.de Solla Price's work. The stages in this curve are: (1) small annual increments of publications, (2) exponential growth, or doubling of the cumulative number of publications in regular intervals, (3) linear growth, which is a decline in the rate of growth from the previous stage, (4) absolute decline in the number of publications. To these stages of the growth of literature, Crane has correlated the intellectual stages of growth of Thomas Kuhn's theory in *The Structure of Scientific Revolutions* (1962;

enlarged edition, 1970): (1) paradigm formation, (2) normal science, (3) anomalistic stage, (4) exhaustion and decline. (At this point, in Kuhn's scheme the old paradigm is replaced by a new paradigm, which will itself go through these four stages. Crane hypothesizes, however, an alternative fourth stage, in which the paradigm is simply exhausted and the research field generally abandoned as new scientists turn to more exciting fields. See Appendix One for further discussion of the logistic curve.)

It is crucial in this hypothetical scenario (as we shall see later in this paper), that the paradigm appear *before* the social organization of the field in the normal science period. For in Kuhn's theory, it is the paradigm, internalized in the minds of the scientists, which creates organization: cohort groups are formed, graduate schools are established or re-organized, professional societies formed, new journals published, and promotional criteria for professional careers formulated. It would negate the Crane-Kuhn Hypothesis of the social structure underlying the intellectual development to discover that social organization preceded the creation or adoption of a paradigm.

The application and testing of the Crane-Price and Crane-Kuhn hypotheses followed four steps. First, using a procedure of bibliographical research into the literature on grassland ecology, I established a list of titles on basic grassland ecology. This list was analyzed to determine how closely it fit the logistic curve; this fit tested the Crane-Price Hypothesis. Second, I compiled standardized biographies on the authors who published at least three titles in this list. I analyzed these biographies to determine how the scientists were organized. Third, I undertook citational analysis to determine whether these scientists shared a paradigm, and, if they did, when this paradigm was accepted. Fourth, I have drawn conclusions from this research on the relation of the paradigm to social organization, and attempted to judge whether the history of grassland ecology fits the scenario described by Crane and Kuhn.

In this sequence of investigation, negative conclusions on the existence of a community or on the existence of the paradigm would falsify the Crane-Kuhn Hypothesis for the case of grassland ecology. Also, positive conclusions on the existence of a community before a paradigm would falsify the Crane-Kuhn Hypothesis. A Crane-Kuhn interpretation would be verified only if a community existed, a paradigm existed, and the paradigm pre-existed the community (see footnote, page 11, for a chart of the tests of this hypothesis). Traditional historical investigation would be necessary to determine whether, in fact, the pre-existing paradigm did cause the formation of the research community.

In the test application of the Crane-Kuhn Hypothesis, I have been motivated by at least one consideration that did not enter in Crane's original research. I wanted the techniques she used to be easy to apply to historical cases; consequently, I have occasionally modified her techniques. I believe that if her techniques (and those of other sociologists of science) are to be useful to historians as supplements to the traditional battery of research tools, they must be simple. In terms of an overall historical enterprise, description of the characteristics of a bibliography is likely to be a marginal exercise unless the techniques are uncomplicated and efficient and lead directly to significant questions. Payoff, after all, is relative to effort expended. In this paper, I report on the research involved in the first two steps outlined above. Elsewhere, I report on the citational analysis and the final conclusions with regard to

---

*The tests leading to confirmation or rejection of the Crane-Kuhn Hypothesis can be outlined as follows:

Table One

		Assumption of Existence of a Sociological Community	Assumption of Existence of a Paradigm	Assumption that Community Pre-Exists Paradigm	Assumption that Paradigm Pre-Exists Community
	1.	— (-OR-)	—		
Crane-Kuhn	2.	+	—		
Falsified	3.	+	+	+	
Crane-Kuhn Verified	4.	+	+	—	+

— = negative test

\+ = positive test

Table 1. Tests of Crane-Kuhn Hypothesis

the Crane-Kuhn Hypothesis to be drawn from it. I believe, however, that the citational analysis is of marginal utility and that sound conclusions can be drawn from the bibliographical analysis and the collective biography alone.

The Literature of American Grassland Ecology

Between 1797 and 1955, 535 titles were published on noneconomic aspects of grassland ecology in the American Midwest by 372 different scientists.* Ninety-eight percent of these titles were published after 1870; seventy-five percent after 1925. The year of the largest number of new publications was 1939, after which the number of new publications declined annually in trend. The histogram, Figure two, graphically represents these and other characteristics of the frequency distribution of publications. The rise and decline of grassland ecology as a research field is clearly silhouetted.

When the rate of growth of the bibliography (Figure Three) and the cumulative number of publications (Figure Four) are graphed, the approximation to a logistic curve is apparent. In Figure Three, the three-year moving average easily traces the four stages described by Price and Crane as typical of the logistic curve. After a long preparatory period in which the rate does not significantly increase,

---

*See Appendix Two.

the rate suddenly jumps exponentially in the years 1917-1920 and 1933-1940. Then the growth rate slumps briefly before beginning the linear rate of increase typical of stage III, finally beginning the expected absolute and relative declines after 1950.

The empirical curve of Figure Three deviates from the theoretical, logistical curve in two periods, the 1920s and the early 1940s. The explanation for these deviations are historical and will be explained later in this paper, but let me here set up the explanation with some preliminary considerations. The theoretical curve as used by Price and Crane can be seen as the curve of growth in literature that would occur under optimal conditions. These conditions are unrestrained education and employment, so that any person recruited into a new scientific specialty could be trained, employed, and have ample opportunity for publication. Reflecting briefly upon the major historical restraints on these optimum conditions, we would immediately expect World War II to restrict education and employment; this historical force is reflected in the growth rates, a conclusion the biographical paths analyzed later will support. In the 1920s, the explanation for the deviation is not quite as obvious. The biographical paths will show, however, a nearly complete closing of new employment in federal research agencies, which helps to explain the stagnancy of grassland ecology in this decade.

Interpreting the growth rate graph for grassland ecology in terms of the Crane-Kuhn Hypothesis signals nineteen-sixteen as the final year of Stage I, which corresponds to the paradigm forming stage.

Prediction of nineteen-sixteen is uncanny. In that year, Frederic Clements published *Plant Succession,* providing the theoretical basis for grassland ecology (as well as for all vegetational and zoological ecology) for a full generation. *Plant Succession* emerges, according to this curve, as the prime candidate for the paradigm, a prediction strongly supported by traditional historical evidence. A brief historical investigation reveals the reason for the increase in the number of publications on grassland ecology following 1916. Not only did *Plant Succession* set forth many problems, in the sense of normal science, to be solved, but the founding of the Ecological Society of America in 1915 and *Ecology* (1920) provided both the primary organization for the science of ecology itself and also the periodical that would be the major outlet for publications. These facts are more pertinent when the lives of the scientists, who were the chief contributors to grassland ecology, are examined.

In what manner can the logistic curve be said to "predict" the past? First, as Appendix Two explains, only the bibliographical titles for the years 1918 through 1955 were systematically obtained. Comprehensive subject indexes did not exist for botanical literature before 1918. Other sources, which were not systematic or comprehensive, were used to obtain titles for the years before 1918. The close fit between the empirical curve of the cumulative titles before 1918 and that section of the theoretical logistic curve is a good indication that, although the titles before 1918 were not systematically selected, they are as representative of the literature as are the titles after

1918. The empirical curve of Figure Three also locates a major shift in the profession following 1916. Hence, the curve has enabled us to move from systematic data to nonsystematic data with assurance and to predict a pivotal event in a year represented by nonsystematic data--the appearance of the paradigm. As I have briefly mentioned, Clements' *Plant Succession*, published in that year, has all the qualifications of the paradigm-setting book.

Other characteristics of the bibliography in grassland ecology may be briefly noted. Doubling of the cumulative titles did not occur at a constant rate following the appearance of Stage I. Doubling occurred during Stage II at intervals of eight and one-half years, thirteen years, and seventeen years. Of the 535 titles, 376 were written by single authors and 159 by two or more authors. The chronological distribution of authorship follows the pattern that Price found for the hard sciences, that co-authorship increased as the century progressed, and the pattern that Crane found, that co-authorship increased in Stage II.

Collective Biography

Historians are not as fortunate as sociologists; they usually cannot interview their subjects. Prosopography, far from being a factory for production of statistics, as it is often portrayed, is handcraft work, with each datum on the lives of the subjects wrung out of recalcitrant sources. Yet this labor is necessary to determine whether there was, as Crane found for research fields by questionnaires, development of social organization which corresponded to the development of the characteristics of the literature of the field. Of course, historians cannot expect to have the richness of detail, or even precisely appropriate detail, on the lives of the scientists.

Standardized biographies were compiled on the fifty-eight scientists who published three or more titles, as author or co-author, in the list of 535 titles on grassland ecology. (See Appendix Three.) The entrance criterion of three titles was based on the historical judgment that no scientist who contributed less than three publications was important enough in the specialty to be included in its community of researchers. In working over this material for two years, this entrance criterion was never unsatisfactory; if it had been, then the criterion would have been dropped down to two publications. Only one interesting, potential problem with the entrance criterion appeared. Roscoe Pound (1870-1964), the famous jurist, who had taken his doctorate in botany from the University of Nebraska in 1897 and was one of the founders of grassland ecology, was not included in the collective biography, since he published, as co-author, only two articles in the field before he completely left science for law (in 1907). His articles will be in the citational analysis, because he co-authored them with Frederic Clements, who was included in the collective biography. If Pound were included in the collective biography, his career pattern would only strengthen the generalizations I have drawn, since he was part of the Bessey group. As it is, his absence from the collective biography accurately reflects the fact that his employment as a practicing lawyer after 1890 was presumably unrelated to the formation of the botanical paradigm.

The information sought on the lives of the scientists pertained only to their professional roles. Several considerations supported this restriction of attention. First, I was not concerned with the nonprofessional social background of the social organization of the research community, but only with the professional aspects of that organization. By this focus, I do not imply that the social background did not influence the social organization of the community. Rather, I expected that failure of the model to explain (or predict) the social organization fully by the variables chosen (which excluded nonprofessional social background) would lead to investigation of social background as a separate question. As it turned out, professional variables satisfactorily explained the major deviation from the predictions, which concerned changes in the professional patterns in the 1920s and early 1940s. The capacity of the professional and community variables to explain the historical professional pattern is interpreted as an oblique confirmation of the Kuhnian expectation that normal science is insulated from the social background; had the variables been incapable of doing this, of course, then the insularity of normal science would be questioned. Second, I am convinced that if prosopography is to be a useful tool, then it must be efficient. There was no reason to believe that social organization would not at least be revealed, if not fully explained, by the most active scientists. If it were not to be revealed, then I could broaden the sample of authors from the bibliography.

I examined three sets of variables in the lives of the subject scientists: educational preparation, career patterns, including employers, and professional society affiliation. Turning to educational preparation, it quickly became apparent that in the history of the field research was dominated by scientists who received part or all of their education and training at the University of Nebraska. It was not surprising that many scientists were trained at the University of Nebraska; after all, a cursory acquaintance with the literature of grassland ecology shows that Nebraska was an important school in the field. What was unexpected was the extent of the domination and the ability of the school to rise to hegemony against competition by the more prestigious University of Chicago. Table Two displays the chronological distribution of the degrees.

Table Two reveals the strength of the University of Nebraska at the doctorate level. It educated a few of the undergraduates, but trained over half of the doctorates in grassland ecology. In other words, it had the reputation in its field beyond its own campus. It was able to outdraw the University of Chicago over the long term, although the University of Chicago was the pre-eminent graduate institution in the Midwest. In the early part of the century, when the University of Chicago botanical department included two of the most important botanists in the nation, J.M. Coulter and H. C. Cowles, Nebraska was able to hold its own.

Table 2. Degree Origins

Degree Origins	Bacca	Masters	Doctorates	Total No. of Degrees (Bacca., Masters, Doctor.
U of Nebraska, Lincoln	14	18	24	56
U of Chicago	2	3	8	13
Kansas State College, Fort Hayes	8	6		14
U of Minnesota	1	1	3	5
All Others (1 or 2 ea.)	29	15	11	55
(No data or No Degree)	(4)	(15)	(12)	

In Figure Five, several weak patterns in the chronology of degree awards emerge. In the sixty-year span from the mid-1890s to the early 1950s, when there was nearly continual degree granting at some level, most doctorates were awarded in four periods: 1898-1907, 1913-1918, 1934-1940, and 1947-1952. Also the mean age of doctorate recipients at the time of award of degree was higher in the later two periods than in the first two, probably reflecting the impact of the de-presssion and the second world war on delaying or prolonging advanced education. These patterns take on significance in terms of the social organization of grassland ecology when it is seen that they corresponded to the tenure of the strongest teachers in ecology at the University of Chicago and Nebraska. Hence, in Figure Six, the University of Chicago trained most of its doctorates in ecology early in the century when Cowles and Coulter were there, while the University of Nebraska trained them in two periods, early in the century under C. E. Bessey and Frederic Clements, and in the 1930s and after World War II when John E. Weaver dominated grassland ecology. It is important that the doctoral training pattern was strong well before any paradigm for grassland ecology existed.

When we interpret Figure Six, the chronological distribution of doctorates by origin, by Crane's Kuhnian scheme (as in Figure Three), a shift in social organization is seen to have followed the presumed intellectual shift. In the paradigm formulation stage (I), there was healthy competition between graduate institutions in producing scientists who wrote on grassland ecology: Chicago and Nebraska turned out five doctorates each, and all others considered as a group produced six before 1916. In stage II, normal science, the competition was gradually ended and Nebraska dominated the production of scientists. There is an obvious hypothesis to explain what was occurring: Nebraska was identified with the (dominant) paradigm for doing grassland ecology. An important question, to be taken up later in this paper, is whether the effect of the paradigm was *to create* a social organization in stage II or merely *to reinforce* a pre-existing social organization. This is a key question in testing the causal order in the Crane-Kuhn Hypothesis.

When we turn to career paths, two patterns stand out against a generally expected set of patterns.[6] Figures Seven, Eight, and Nine illustrate the career paths of three groups, the Nebraska and the Chicago doctorates and all other PhD's. Overall, the relationship between doctoral origins and career paths was weak. However, the academic institution from which the scientist obtained his or her highest degree, usually the doctorate, was related to the first job. None of the University of Chicago doctorates first taught high school or worked for the federal government, while half of the Nebraska degree-holders, and slightly more than half of the other universities' scientists did so. There was also a relationship between the first and second job titles. That is, all the University of Chicago doctorates changed employers in the jump from the title of their first job (such as Assistant Professor) to the title of their second job (such as Associate Professor); yet only 56% of the Nebraska doctorates and 42% of scientists from other institutions changed employers when they made this same jump. Nebraska doctorates, however, showed a wider variety of experience in this shift of employers than did Chicago doctorates. Chicagoians stayed strictly within academia, while Nebraskans' shifts included high schools and, to a great extent than Chicagoians, federal employment.

Difference in experience was undoubtedly related to the University of Nebraska's domination of grassland ecology and the character of normal science. As established by Clements, grassland ecology was oriented toward the goal of controlling vegetational change and hence was highly applicable to practical problems of range and forest mangement. With educational preparation in this paradigm, Nebraska PhDs were qualified for work with the federal offices which had to deal with the problems of the grasslands. At the same time, federal service diversified the employment open to adherents of the paradigm, thus reinforcing its competitive force and survival value. As Figure Seven shows, during the 1930s, when drought in the West was a national concern, Nebraska PhDs had active involvement with federal service. While it is out of place here to discuss this service at length, I suggest that the success of grassland ecologists in dealing with the drought was crucial to the domination of Clements' ecological theory. The scientists' fight against soil erosion, for instance, was not conducted in an *ad hoc* manner. Problems were approached within the theoretical framework of the movement of vegetation toward the climax formation of the Great Plains as described by Clements. Solutions were proposed that worked to deflect vegetational change away from the path created by the drought onto natural paths of succession of forms toward a natural stability.

Federal employment and range management also provided the plethora of problems to be solved -- an important consideration for working scientists in accepting a paradigm. No scientist wanted to accept one paradigm over a competitor if the one accepted left him unemployed. In this way, what was only a slight statistical difference in the career patterns of two groups led to acceptance of a paradigm, reinforcement of the educational domination of one school over another, and the "normalization" of once-controversial scientific ideas.*

A final feature of the careers of these fifty-eight highly productive scientists is worth mentioning. Many were honored by their peers. Nearly forty percent held an office in a professional society; over thirty-two percent held two offices. Thus the group provided ten presidents and three vice-presidents of the Ecological Society of America (founded in 1915), three presidents of the Botanical Society of America, and presidents of twelve other societies and academies. The presidency of the Ecological Society of America was occupied by one of the fifty-eight subjects in 1916, 1918, 1924, 1930, 1934, 1935, 1946, 1948, 1950, and one term of unknown year. I have no reason to believe these elections were related directly to a paradigm; rather they are related to scientific productivity.

Social Organization

Career paths from education to employment strongly indicate two important institutional training centers, Nebraska and Chicago, with Nebraska forcing out Chicago in the normal science stage of the field. The doctoral training experience at Nebraska provided the basis for

---

*It is not impossible that the employment-university correlations may have reflected the status aspirations of the students choosing or being able to attend the different universities, but status aspiration is difficult to establish as a variable (not to say that it is not real). For instance, why would individuals with dominating status motivation go into botany to begin with, when other scientific fields and other careers certainly had higher status value?

social organization of the research field in the normal science period. This pattern manifests itself in collaboration in publications. John Weaver, who taught grassland ecology at Nebraska during the entire normal science period, was the most frequent co-author, with fifty-five titles. The Nebraska school was also the strongest producer of other co-authoring scientists, as is indicated in Table Three below.

Table 3. Co-Authorships

Name	Doctoral Institution	No. of Titles co-authored
Weaver	Minnesota (teacher at Nebr.)	55
Albertson	Nebraska	13
Clements, F.	Nebraska	11
Hanson, H.D.	Nebraska	8
Whitman	Wisconsin	6
Darland	Nebraska	6

Similarly, the frequency of co-authorships rose rapidly during the normal science period and fell off during early Stage III. (See Figure Ten.) These tendencies are exactly what has been discovered by Crane for the normal science period. The number of co-authorships was swelled during the years 1930-1950 because Weaver and Albertson published then most of their joint-authorship articles. The decline in joint authorship during the 1950s was another indication, according to the Crane-Kuhn scheme, that the normal science stage of the history of the grassland school had passed.

When we examine the multi-author relationships, the coherence of the Nebraska grasslands group is clear. The work of Hanson, Clements,

Weaver, and Albertson amounted to 54% of all the multiple-authorships in the bibliography (see Figures 11 and 12). This co-authorship network alone would be sufficient to ensure that the sociology underlying the production of the grasslands research and its bibliography was unified. This authorship complex was, of course, Nebraska-based. The next largest complex was that of McIlvain-Savage, who were U.S. government employees and who received their highest degrees from other universities (that is, not from Nebraska or Chicago). No other multi-author complex was larger than the couplet, or unextended co-author relationship.

Among the numerous features of the Nebraska complex, I wish to notice only one: the role of "gatekeepers." The structure of the Nebraska complex indicates that just four of forty-nine scientists acted to tie together the multi-authorships. With few exceptions, the other forty-five scientists did not on their own initiative move from their own constellation of authorships to another. In the case of the Weaver-constellation within the complex, this feature was tied to the teacher-student relationship, because many of Weaver's co-authorships were with his graduate students. In general, the research function of the complex was controlled or, if one prefers, "guided," by a few "gatekeepers" of science. Given the large proportion of the total grasslands bibliography contributed by the Nebraska complex, the overall role of Hanson, Clements, Weaver, and Albertson was even more striking. The failure of the University of Chicago to contribute a multiple-author complex to the grasslands research community was a fair indication of the failure of that university's research effort to cross the threshold to a sustained contribution.

Conclusion

The Crane-Price Hypothesis, that a basic research specialty undergoes a bibliographical expansion described by the logistic curve, and the Crane-Kuhn Hypothesis, that this curve can be interpreted by the Kuhnian theory of scientific progress, are strongly confirmed by the case of American grassland ecology from 1887 to 1955. This is not an absolute confirmation, because many of the sources are not available for the first half of the field's history. Furthermore, the bibliography cannot be systematically and comprehensively accumulated before 1918, when *Botanical Abstracts* became available. Nevertheless, the literature accumulated from nonsystematic sources before 1918 closely conforms to the theoretical shape of the logistic curve, which reinforces the notion that, although nonsystematically accumulated, it is representative of the total literature. Examining the history of grassland ecology in a "behaviorial" way, without interview material from the participants and without intellectual biographies, the Crane-Kuhn Hypothesis would appear to explain the change in social organization after 1916. Acceptance of a paradigm by one group and its identification with that group ended competition between institutions in grassland ecology and provided a monopoly in one theoretical and research approach to the field. On behaviorial grounds, then, there is reason to believe that the history of grassland ecology was under the influence of the paradigm-normal science scenario.

Important questions remain. Was there in fact a paradigm in the specialty of grassland ecology and did the acceptance of the paradigm provably cause the social organization of the research field after 1916? Two general techniques exist for answering this question. One is citational analysis, which is not taken up in this report; the other is traditional historical research--biography, manuscript evidence, internal analysis of scientific publications and institutional record collections.

Are there good reasons supplied by the traditional techniques of historical research to assume that a paradigm was proffered and accepted in grassland ecology? Are these reasons sufficiently convincing to obviate the necessity of citational analysis? It is in the interest of the historian that citational analysis need not be resorted to. Citational analysis is financially expensive, since it requires expensive research aid and computer preparation time. Furthermore, unlike the sociologist, the historian cannot give up the research topic when the citational analysis is completed, since he will have to read the literature in order to understand what is happening intellectually. It would be efficient to skip the citational analysis altogether and simply read the literature.[7]

In the case of grassland ecology, good manuscript collections are available. Examination of these collections provides evidence for the role of a paradigm in grassland ecology, a struggle for its acceptance, and the re-organization (although not creation) of the sociology of the field -- in a word, the whole Crane-Kuhn model.

Charles E. Bessey was the founder of the grassland school, but not an ecologist himself, and the teacher of a group of "missionaries for botany" which included Frederic Clements. His papers chronicle the involvement of the scientists with the practical problems that brought forth grassland ecology. In the 1890s Bessey was skeptical of ecology because he did not believe it could be founded on rigorous, scientific principles, including experimentation. The labors of Clements, however, ultimately convinced Bessey that grassland ecology was as experimentally rigorous, mathematical, and conceptually advanced as any biological specialty and he came to support it before his death in 1913.[8]

The manuscript collection of Frederic Clements provides the most extensive testimony on Clements' struggles to establish succession as the theoretical basis of plant and animal ecology, including grassland ecology for which it was first formulated. Clements' major statement of his theory appeared in *Plant Succession* (1916), which is the last year of Stage I in the Crane-Kuhn scenario. *Plant Succession* had an enormous impact on plant ecology in the United States and in England where A. G. Tansley led its application to plant ecology and Charles Elton applied it to animal ecology. In the United States, successional theory was applied to animal communities by Victor Shelford, leading to Clements' and Shelford's important textbook *Bio-Ecology* (1939). Clements' correspondence with his principal adopters, as well as his leading critics, provides a running commentary on the spread of his theory.[9]

Kuhn assumed that in the paradigm-forming stage of scientific development each approach to a research problem area has competition. In the case of grassland ecology, this was again true. The leading alternative to Clementsian successional ecology was an ecology focused on the individual plant and its physiological response to its habitat. Physiological or individualistic ecology was associated largely with German scientists, such as A.F.W. Schimper, whose *Plant Geography* (first edition, 1898; English edition, 1908) was a massively learned treatment of plant ecology around the world, much of it based on Schimper's own travels. In the United States, physiological plant ecology was not strongly represented, although H.C. Cowles' volume on plant ecology for the University of Chicago textbook on botany (1911) was a popular statement of it. (It is ironic to point to Cowles' textbook as representative of physiological ecology because Cowles was, along with Clements, the earliest formulator of successional ecology.) H.A. Gleason was the leading American spokesman for the "individualistic" concept of plant ecology. The competing paradigms therefore were broadly split between ecology which focused on broad vegetational units, such as the grassland or forest, and was based on successional or developmental theory, and ecology which focused on the individual plant and its relationship to its immediate environment or habitat. While the successional approach won out over the physiological, the physiological was not abandoned. It was relegated to the lower status of a restricted empirical study.

The controversy between the competing paradigms and the questioning of the Clementsian dynamic ecology were based on a broad range of fundamental philosophical issues, such as realism versus nominalism and laboratory experimentation versus nature observation, which reached up to the disciplinary structure of biology, out of the purely specialty considerations. For instance, individualistic ecologists criticised successionalists for organizing their science around artificial and conceptual units of vegetation that did not *really* exist out there in the natural world. They also criticized successional ecologists of lack of experimental rigor and dilettantism. Clements and his apostles, on the other hand, replied that the "grassland" was a unit of vegetation which was as real a living, integrated system as was the individual plant. They developed a battery of experimental techniques, such as the use of the quadrat, to compensate for the lack of laboratory facilities.

The final question to which the manuscript collections are relevant is whether the paradigm created the social organization of the research specialty or whether the social organization preceded it. This question also cannot be answered here where it is inappropriate to bring forward the evidence from the manuscript collections. It can be stated, however, that acceptance of the paradigm re-organized whatever social organization had pre-existed the advent of the paradigm. Clements played a key role in this reorganization. In 1917, he took a position with the Carnegie Institution and in the next decade made available his portion of that important research institution's resources to doctoral students at the University of Nebraska, where he had taken his doctorate and begun teaching, and where John Weaver, who had been Clements' doctoral student at the University of Minnesota, taught from 1915-1952. By the mid-1920s, selected Nebraska students had the pleasure of passing their summers at Clements' Alpine Laboratory in Manitou Springs, Colorado. Clements and Weaver further reinforced their domination of the field with their textbook, *Plant Ecology* (1929). Thereby, Clements and Weaver effectively made the University of Nebraska the top institution for studying grassland ecology. On the one hand, this process pushed out the University of Chicago, which had been a doctorate training center for grassland ecology in Stage I, and, on the other hand, effectively prevented other universities, such as the land-grant universities sprinkled across the prairies, from challenging Nebraskan dominance. To the extent that the Nebraskan group relations existed before the acceptance of the paradigm in the research specialty, the paradigm cannot be said to have *created* those relations. What happened, rather, was that the paradigm reinforced one set of these relationships to the disadvantage of the others.

NOTES

1. Diana Crane, *Invisible Colleges: Diffusion of Knowledge in Scientific Communities* (Chicago and London: The University of Chicago Press, 1972); Derek J. de Solla Price, *Little Science, Big Science* (New York and London: Columbia University Press, 1963), especially chapter one, "Prologue to a Science of Science." See also M.J. Mulkay, G.N. Gilbert, and S. Woolgar, "Problem Areas and Research Networks in Science," *Sociology*, 9 (May, 1975): 187-203, which summarizes and cites recent literature and further confirms Crane's work without making a basic contribution to the problem. I am grateful to Diana Crane for bringing this paper to my attention.

2. I refer to ecology as a "specialty" in the sense of that term defined by Peter Weingart, whose work is based on Margaret Masterman's clarification of Kuhn's concept of the paradigm; see, Peter Weingart, "On a Sociological Theory of Scientific Change," in Richard Bentley, ed., *Social Processes of Scientific Development* (London and Boston: Routledge & Kegan Paul, 1974), pp. 45-68, and Margaret Masterman, "The Nature of a Paradigm," in Imre Lakatos and Alan Musgrave, eds., *Criticism and the Growth of Knowledge* (Cambridge: At the University Press, 1970), pp. 59-89.

3. For sketches of the principal figures, see Richard Overfield, "Charles E. Bessey; The Impact of the 'New' Botany on American Agriculture, 1880-1910," *Technology and Culture,* 16 (April, 1975): 162-181; Joseph Ewan, "Bessey, Charles Edwin," *Dictionary of Scientific Biography,* vol. II (New York: Charles Scribner's Sons, 1970), and Joseph Ewan, "Clements, Frederic Edward," *Dictionary of Scientific Biography,* vol. III (New York: Charles Scribner's Sons, 1971). Also see my article, "Theoretical Science and Technology in American Ecology," *Technology and Culture* 17 (October, 1976): 718-728. I am grateful to the editor of *Technology and Culture* for allowing me to read the Overfield article on Bessey prior to its publication.

4. Since the debate over Kuhn's theory continues to be discussed in terms of the concept of the paradigm and sociologists have not generally picked up on Thomas Kuhn's brief suggestions about a disciplinary matrix, I will utilize the earlier terminology.

The major contributions to the philosophical debate over Kuhn's theory are the essays in Lakatos and Musgrave, eds., *Criticism and the Growth of Knowledge:* T. S. Kuhn, "Logic of Discovery or Psychology of Research?" pp. 1-23; J.W.N. Watkins, "Against 'Normal Science'," pp. 25-37; K.R. Popper, "Normal Science and Its Danger," pp. 51-58; Margaret Masterman, "The Nature of a Paradigm," pp. 59-89; I. Lakatos, "Falsification and the Methodology of Scientific Research Programmes," pp. 91-195; P.K. Feyerabend, "Consolations for the Specialist," pp. 197-230; T.S. Kuhn, "Reflections on my Critics," pp. 231-278. Positions are restated in Imre Lakatos, "History of Science and Its Rational Reconstructions," pp. 91-136, in Roger C. Buck and Robert S. Cohen, eds., *Boston Studies in the Philosophy of Science,* vol. VIII, PSA 1970 (Dordrecht-Holland: D. Reidel Publishing Company, 1971), and T.S. Kuhn, "Notes on Lakatos," *ibid.,* pp. 136-146. See also, T.S. Kuhn, "Second Thoughts on Paradigms," pp. 459-482, in Frederick Suppe, ed., *Structure of Scientific Theories* (Urbana: University of Illinois Press, 1974).

Sociologists of science have produced several interesting empirical and theoretical contributions to the study of paradigms. For sharpening the issues in Kuhn's and Crane's works, I have found most useful Nicholas C. Mullins, "A Sociological Theory of Scientific Revolution," pp. 185-203, in Karin D. Knorr, Hermann Strasser, and Hans Georg Zilian, eds., *Determinants and Controls of Scientific Development* (Dordrecht-Holland/Boston-U.S.A.: D. Reidel Publishing Company, 1975). Mullins presents a theory of paradigms in normal science which makes scientific "revolutions" simply the result of generational perspective--quite contrary to Kuhn's point of view, in which changes in paradigms create generations, not vice versa. Mullins' model of scientific progress is also directly contrary to Crane's theory. Mullins' work provides the opportunity for a crucial test, I believe, using citation data between his and Crane's and Kuhn's models. I intend to undertake such a test and do not discuss it here. In an earlier article, however, Mullins supported the work of Kuhn and Crane; see, Belver C. Griffith and Nicholas C. Mullins, "Coherent Social Groups in Scientific Change,"

Science, 177 (September 15, 1972): 959-964. I have also found useful an article by Peter Weingart, which distinguishes between different conceptual levels of paradigms; Weingart, "On a Sociological Theory of Scientific Change." Other relevant studies from these collections include Richard D. Whitley, "Components of Scientific Activities, Their Characteristics and Institutionalization in Specialties and Research Areas: A Framework for the Comparative Analysis of Scientific Developments," in Knorr, et. al., eds., *Determinants and Controls of Scientific Development*, pp. 37-73; Gernot Böhme, "The Social Function of Cognitive Structure: A Concept of the Scientific Community Within a Theory of Action," *ibid.*, pp. 205-225; Richard Whitley, "Cognitive and Social Institutionalization of Scientific Specialties and Research Areas," in Whitley, ed., *Social Processes of Scientific Development*, pp. 69-95; Cornelis J. Lammers, "Mono- and Poly-paradigmatic Developments in Natural and Social Sciences," *ibid.*, pp. 123-147; Stuart S. Blume and Ruth Sinclair, "Aspects of the Structure of a Scientific Discipline," *ibid.*, pp. 224-241.

A recent and rare extensive historical test of the theory of Imre Lakatos is provided by Hannah Gay, "Radicals and Types: A Critical Comparison of the Methodologies of Popper and Lakatos and Their Use in the Reconstruction of Some 19th Century Chemistry," *Studies in History and Philosophy of Science*, 7 (1976): 1-51. Although Gay concludes that Lakatos' idea that science is a "battleground on which research programmes wage war seems inappropriate" (p. 48), it is not proper to draw inferences from her test to American grassland ecology, a rather different field, obviously.

5. While the assumption that the intellectual history of a community is linked to the process of institutionalization may be found in many early sociologies of knowledge, the recent formulations of this question that relate specifically to the debate over Kuhn's work are, particularly, Weingart, "On a sociological Theory of Scientific Change," Whitley, "Components of Scientific Activities ...," and Whitley, "Cognitive and Social Institutionalization ...."

6. I lack employment information on only one scientist (a University of Wisconsin doctorate). The employment information for all other scientists is not always complete; see the discussion in Appendix Three.

7. I recognize the unorthodox format in presenting this manuscript evidence. I am not trying to present a historical narrative of the intellectual struggles over Clements' theories. If I were, then I would present the material in this section as a straightforward article in the history of ideas or social history. All I am doing is indicating the reasonableness of the conclusions drawn from the statistical and "behaviorial" analysis.

8. Bessey's complaints about the unscientific character of much of the work in ecology were repeated in various contexts in Bessey to J.M. Coulter, April 19, 1899, Bessey Papers, Letterpress Books,

1898-1901, University of Nebraska Archives, Lincoln, Nebraska; Bessey to A.F. Woods, June 24, 1903, *ibid.*, Letterpress Books, 1902-1903; Bessey to A.G. Tansley, December 15 (1905?) *ibid.*, Box 1905 (G)-1906 (Ba). On the other hand, Frederic Clements successfully brought Bessey to accept his views; see Bessey to Tansley, *ibid.*, Henry Holt and Company to C.E. Bessey, September 26, 1907, *ibid.*, Papers Box 1907 (H)-1908 (Bo); and particularly the implications in a controversy over a critical review of Clements' book, *Research Methods in Ecology* (1904), Bessey to Frederic E. Clements, November 22, (1906?), *ibid.*, Papers Box 1906 (O)-1907 (G).

9. So much of Clements' correspondence deals directly with discussion over acceptance of his views, that it makes no sense to present manuscript citations at length here. Important correspondence with a scientist who would accept his views is that with A.G. Tansley, which includes over 44 letters between 1918 and 1942; *passim*, Frederic Clements Collection, University of Wyoming Archives, Laramie. An opponent who was partially converted to successionist ecology was G. Du Rietz, a Swedish botanist; see, for example, Du Rietz to Frederic Clements, February 24, 1928, *ibid.*, Box 115; Clements to Du Rietz, March 27, 1928, *ibid.* A confirmed opponent of Clements was E. Rübel, a Swiss botanist; see, Rübel to Frederic Clements, June 11, 1928.

ACKNOWLEDGMENTS

I am indebted to the Research Committee of the Academic Senate of the University of California, Riverside, for providing funds to use the computer for this project. I am also indebted to the award of the Faculty Fellowship which allowed me to examine the Bessey and Clements Collections. This paper was written initially during a leave of absence from teaching made possible by a fellowship from the National Endowment for the Humanities. The statistical and computer applications were considerably aided by the skills of Robert Houchens, Senior Programmer, of the Computer Center of the University of California, Riverside, and Pamelia Smith, a sociologist and programmer. Susan Tobey, a data systems analyst, helped me to design the biographical profiles forms which were used for research on the lives of the subjects.

This paper was read at the International Symposium on Quantitative Methods for the History of Science, Berkeley, California (August, 1976), and to the Ecology Seminar at the University of California, Berkeley (November, 1976). For their comments at these forums, I am grateful particularly to Michael Moravcsik, Jerry Gaston, Derek Price, and Barbara Rosenkrantz. Arnold Schlutz shared with me valuable information about his teacher at the University of Nebraska, John Weaver.

My research at the University of Nebraska Archives could not have been efficiently conducted without the generous assistance of Joseph Svoboda, Archivist. I also appreciate the interested aid of Ted

Pfeifer, Director of Registration of the University of Nebraska, J. Max Hoffman, Director, Records, and Mrs. Irene Brown of the Alumni Association, University of Nebraska.

Portions or drafts of this paper were read by Diana Crane, University of Pennsylvania, Nathan Reingold, Smithsonian Institution, and by my colleagues, Kenneth Barkin and John Phillips, University of California, Riverside. John particularly was helpful on problems of methodological interpretation.

Appendix One

The Logistic Curve

The logistic, or Log E, curve is the exponential growth curve of saturated populations, i.e., of populations whose growth ceases. It is typical of the growth of bacteria, as well as of the growth of some bibliographies. Certain characteristics of the curve are artefacts of the method of accumulating the bibliography and reveal nothing about the bibliography itself. The upper portion of the S-shape is artefactual, in that it indicates only that no further additions are being made to the cumulative bibliography. It is impossible to distinguish from the curve alone whether the falling-off is because the researcher has stopped adding entries or because there are no more entries to be accumulated. All growth curves, regardless of the rates of growth, would have the flattening top of the S-shape when the bibliography is closed.

Up to the final flattening-out, the slope of the curve is not artefactual; it reflects the rates of growth. It is not methodologically necessary that the slope should move through linear-exponential stages before dropping. Some bibliographies accumulate at less than exponential rates through their entire growth. Crane's assumption concerning the intellectual and sociological interpretation of the linear and exponential growth curves is that linear, bibliographical growth exemplifies no single paradigm, while exponential growth does. The sociological correlate to this assumption is that the bibliography is created by a coherent group of scientists who increase their number-sociological reproduction.

Appendix Two

Compiling the Subject Bibliography

It is impossible to establish a comprehensive bibliography of publications on grasslands ecology in the test period, 1887-1955. The central bibliographic source, *Botanical Abstracts* (1918-1926, merging with *Biological Abstracts,* 1927 to present), only covers part of the period. No abstract series is inclusive. The *Botanical Abstracts* and *Biological Abstracts* especially failed to include the multitudinous publications of the United States Department of Agriculture and the semi-popular monographs of the agricultural experiment stations, although these were regularly used by grassland ecologists. Moreover, since the *Abstracts* depended in part upon authors to send in notices of their publications, the appearance of an abstract often lagged several years behind the date of publication of the title. This lag lessened after World War II. The *Botanisches Centralblatt,* which began publication in 1880 and continued in the old series through 1919, similarly failed frequently to notice publications from the Department of Agriculture, the experiment stations, and occasionally some state academy of science publications.

My procedure in obtaining titles in the *Abstracts* was to pull out all titles, whether articles or books, listed in the *Abstracts* under the keywords "ecology" and "grassland," which concerned the midwestern United States and which were not economic in character. I also needed to find a source of titles for the early period before the abstract service. I was unable to find a comprehensive source and settled on the device of utilizing three nonsystematic bibliographies: James Malin, *The Grassland of North America* (Ann Arbor: Edwards Bros., 1947), John E. Weaver, *North American Prairie* (1954), and John E. Weaver, *The Grassland of North America* (1956). These three sources listed many titles not in the abstract series.

I checked the representativeness of the nonsystematic bibliographies for the period before 1918 with two techniques. First, I checked the list of titles in my bibliography against the list of titles in the Subject Index of the catalogue of the John Crear Library which is considered a major repository of American botanical publications. My bibliography was so much more inclusive that the Crear Index was of no value. I also checked the bibliography of titles for the years 1880-1919 against the *Botanisches Centralblatt;* this important series for botanists had only 63% of the titles on my list. Second, I plotted the titles in my bibliography in a graph of cumulative publications and checked this against the logistic curve which Price and Crane believe is typical of the growth of scientific literature in this century. (See Figure Three in the text.) As the report notes, my empirical curve is a close fit to the curve of the Crane-Price logistic curve for both the pre-1918 and post-1918 periods. I conclude that my nonsystematically collected bibliography is representative of the literature of its periods.

The resultant bibliography contains 535 different titles by 372 different authors.

A computer printout listing of the 535 titles is available at a nominal charge by writing the author.

Appendix Three

The Collective Biography

Standardized biographies were compiled on all authors who published, as author or co-author, three or more articles in the bibliography. The justification for this criterion is presented in the text of this paper. The decision to weigh equally sole authorship and co-authorship in counting the number of titles per scientist is based on the recent work of Cole and Cole* in which questionnaire surveys established that contemporary scientists do not discriminate between co-authors in citations of articles. Biographical information was entered into printed forms which provided for a total of 155 items of information. While there is no need to reproduce all the items of information here, I can note that they covered each subject's vital statistics (birth, marriage, death), social background (place of birth, ethnicity, religion, father's education, and occupation), education (baccalaureate through highest degree), employment experience (through seven job titles), expeditions, professional societies membership, professional offices held, honors, civil and private (nonprofessional) offices held, and military service. In the compilation of employment statistics, every change in the title of employment, regardless whether the employer was also changed, was counted as a new job; hence, the jump from assistant professor to associate professor within the same university was counted as a jump from one job to a second job. It was not always

---

*Jonathan R. Cole and Stephen Cole, *Social Stratification in Science* (Chicago: University of Chicago Press, 1973).

possible to obtain complete information on or to be certain of employment, year to year, for each of the fifty-eight scientists. Occasionally, I obtained information from nonsystematic sources, such as identifications on the author by-lines of articles, but I did not use these if I could not be certain of dates involved. Absolutely complete information on all subjects would of course somewhat alter the shape of Figures Seven, Eight, and Nine, but I am confident that the general pattern would not be changed and that the generalizations I have drawn in the text from the data would stand.

The fifty-eight scientists who published three or more articles are listed in Table One below.

Appendix Three. Table One.

The Subject List

Aikman, John M.
Albertson, Fred W.
Aldous, Alfred E.
Allred, Berten W.
Anderson, Kling L.
Bergman, Herbert F.
Bessey, Charles E.
Branson, Farrel A.
Braun, Emma Lucy
Bruner, William Ed.
Cain, Stanley A.
Clarke, Sidney E.
Clements, Edith S. (Mrs. Frederic E.)
Clements, Frederic E.
Condra, George E.
Cornelius, Donald R.
Costello, David F.
Cowles, Henry Chandler
Crist, John W.
Hurtt, Leon C.
Keim, Franklin E.
Kramer, Joseph
McIlvain, Ernest H.
Mueller, Irene Marian
Noll, William C.
Pool, Raymond J.
Riegel, David A.
Robertson, Joseph H.
Rydberg, Per Axel
Sampson, Arthur W.
Sarvis, Johnson Thatcher
Savage, David A.
Schaffner, John Henry
Sears, Paul Bigelow
Shantz, Homer L.
Shelford, Victor E.
Shimek, Bohumil
Stoddard, L.

Darland, Raymond W.
Dyksterhuis, Edsko Jerry
Elias, Maxim K.
Flory, Evan L.
Frolik, Anton L.
Fuller, George D.
Gleason, Henry A.
Hanson, Herbert C.
Hayden, Ada
Hopkins, Harold H.
Tomanek, Gerald W.
Transeau, Edgar Nelson
Turner, George T.
Vestal, Arthur G.
Visher, Stephen Sargent
Voigt, John Wilbur
Weaver, John E.
Whitman, Warren C.
Woolfolk, Edwin J.
Zink, Sarah Ellen

The information on the lives of these fifty-eight scientists was obtained primarily from the various editions of *American Men of Science*. Information was also obtained from the Archives of the University of Nebraska, Lincoln, and the Alumni Association Office of the University of Nebraska, Lincoln. Information from the Alumni Association Office has been used with statistical anonymity for the subjects concerned. Some information concerning the employment history of the subjects was garnered from institutional affiliations noted in articles and from nonsystematic sources like the *International Addressbook of Botanists* (1931).

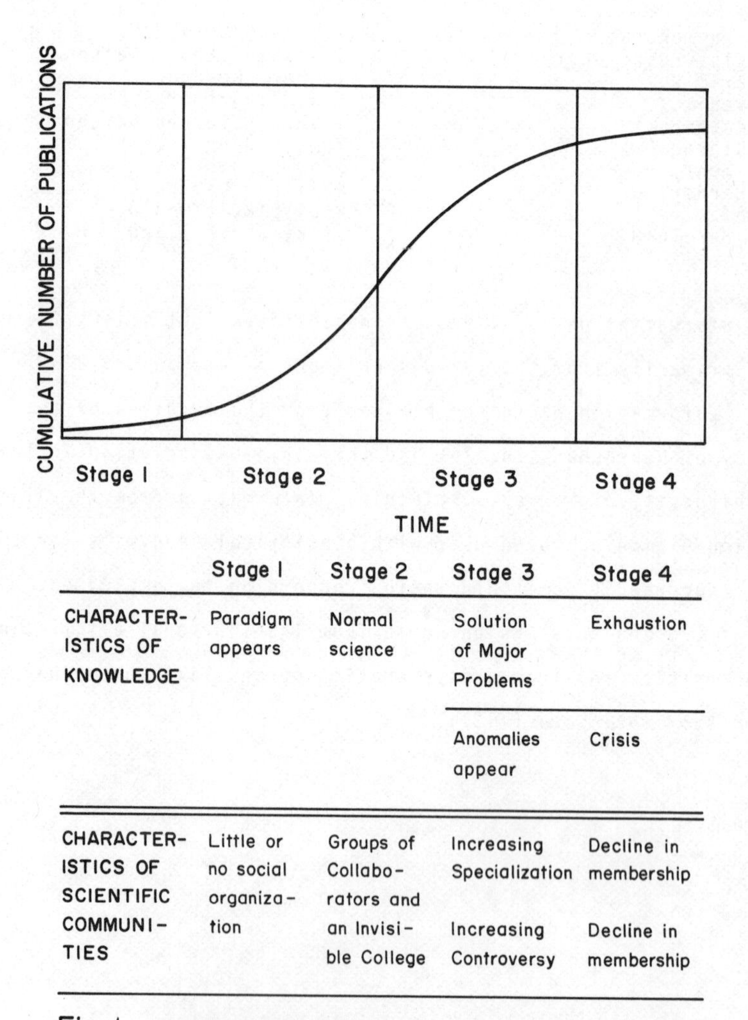

	Stage 1	Stage 2	Stage 3	Stage 4
CHARACTERISTICS OF KNOWLEDGE	Paradigm appears	Normal science	Solution of Major Problems	Exhaustion
			Anomalies appear	Crisis
CHARACTERISTICS OF SCIENTIFIC COMMUNITIES	Little or no social organization	Groups of Collaborators and an Invisible College	Increasing Specialization	Decline in membership
			Increasing Controversy	Decline in membership

Fig. 1 CHARACTERISTICS OF SCIENTIFIC KNOWLEDGE AND OF SCIENTIFIC COMMUNITIES AT DIFFERENT STAGES OF THE LOGISTIC CURVE

SOURCE : CRANE, 172. (Used with the permission of Diana Crane and the University of Chicago Press.)

Fig. 2 ABSOLUTE FREQUENCY OF PUBLICATIONS

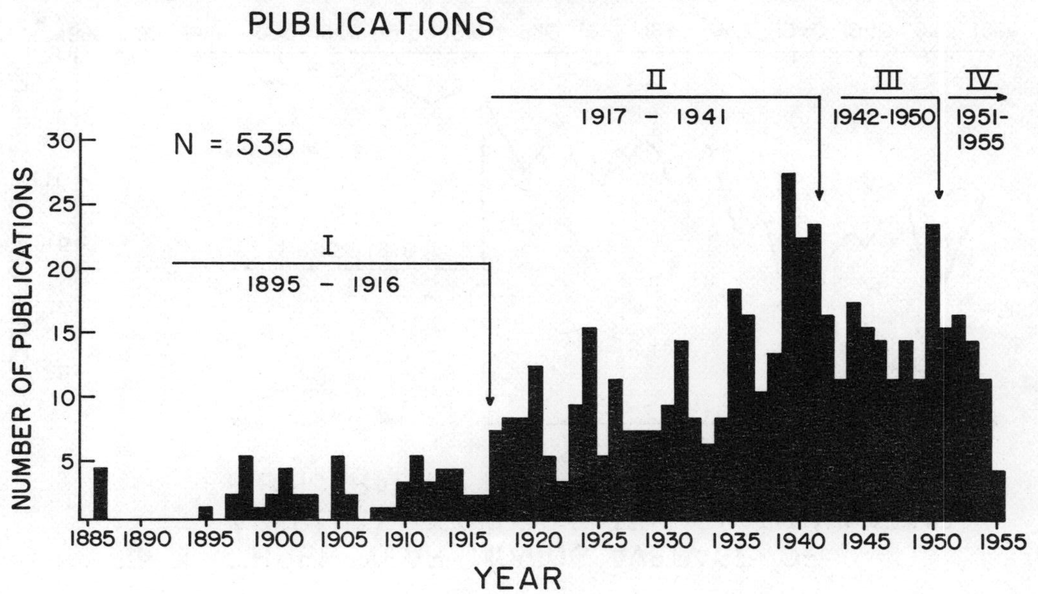

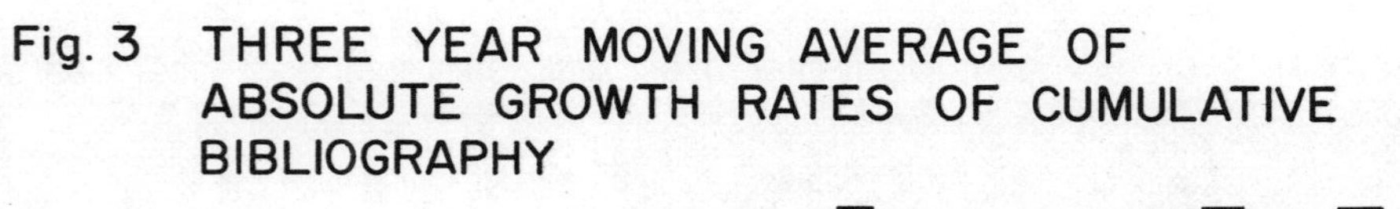

Fig. 3 THREE YEAR MOVING AVERAGE OF ABSOLUTE GROWTH RATES OF CUMULATIVE BIBLIOGRAPHY

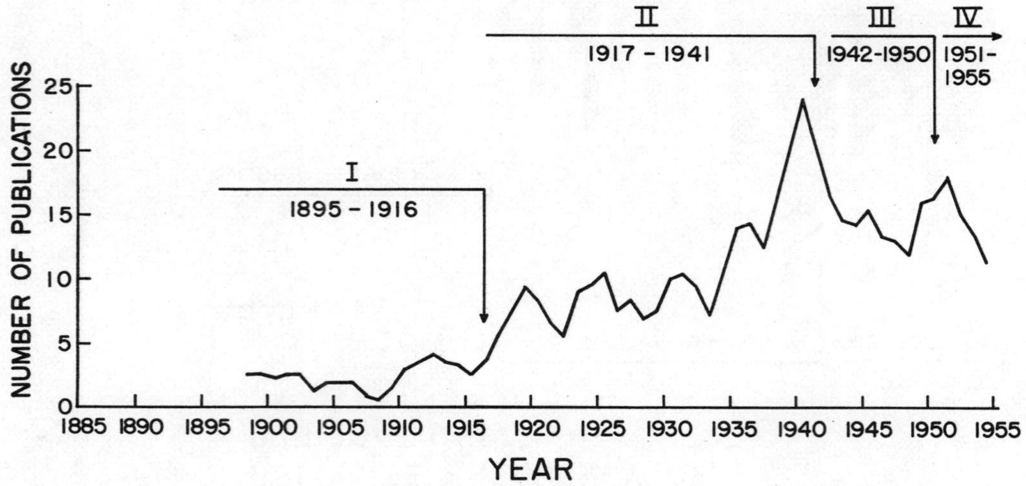

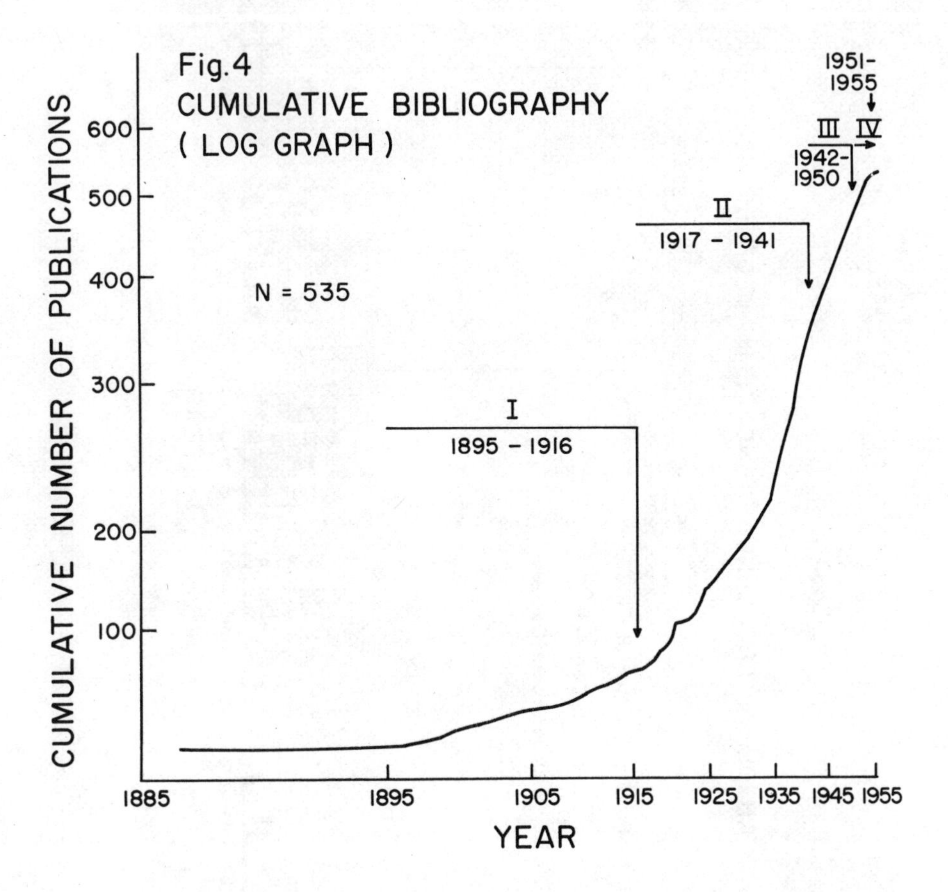
Fig. 4
CUMULATIVE BIBLIOGRAPHY
(LOG GRAPH)
N = 535
CUMULATIVE NUMBER OF PUBLICATIONS
600
500
400
300
200
100
I
1895 - 1916
II
1917 - 1941
III
1942-
1950
IV
1951-
1955
1885
1895
1905
1915
1925
1935
1945
1955
YEAR

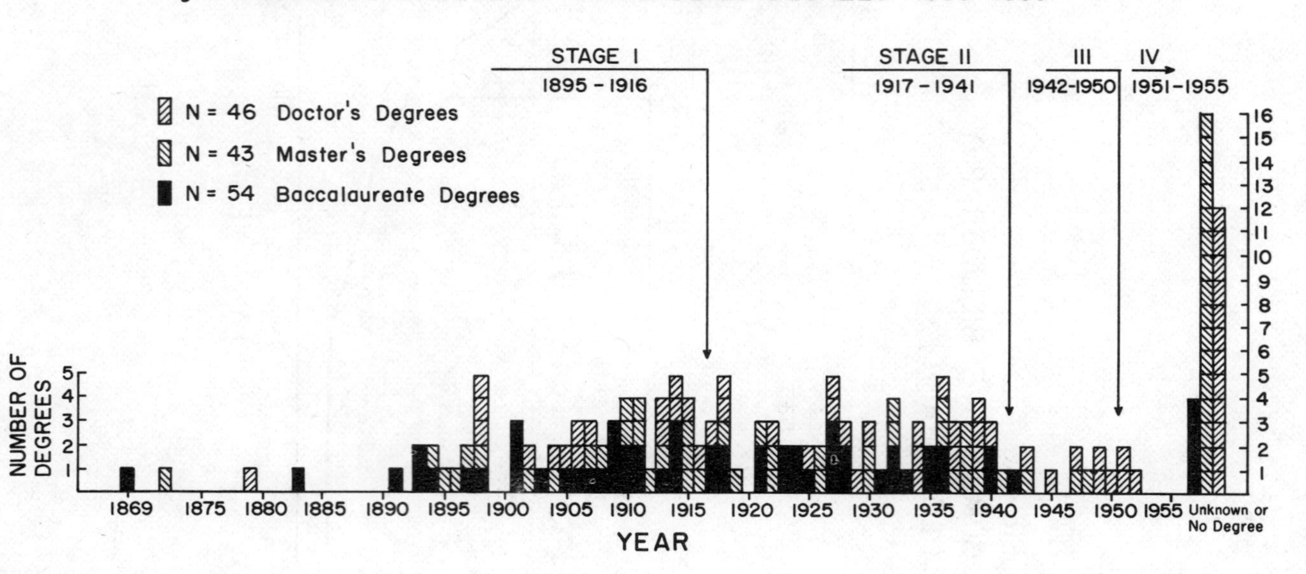
Fig. 5 ABSOLUTE FREQUENCY OF ACADEMIC DEGREES 1869-1955
STAGE I
1895 - 1916
STAGE II
1917 - 1941
III
1942-1950
IV
1951-1955
N = 46 Doctor's Degrees
N = 43 Master's Degrees
N = 54 Baccalaureate Degrees
NUMBER OF DEGREES
1
2
3
4
5
1869
1875
1880
1885
1890
1895
1900
1905
1910
1915
1920
1925
1930
1935
1940
1945
1950
1955
Unknown or No Degree
YEAR
1
2
3
4
5
6
7
8
9
10
11
12
13
14
15
16

Fig. 6 ACADEMIC ORIGIN OF DOCTORATES
1895 - 1955

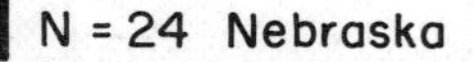

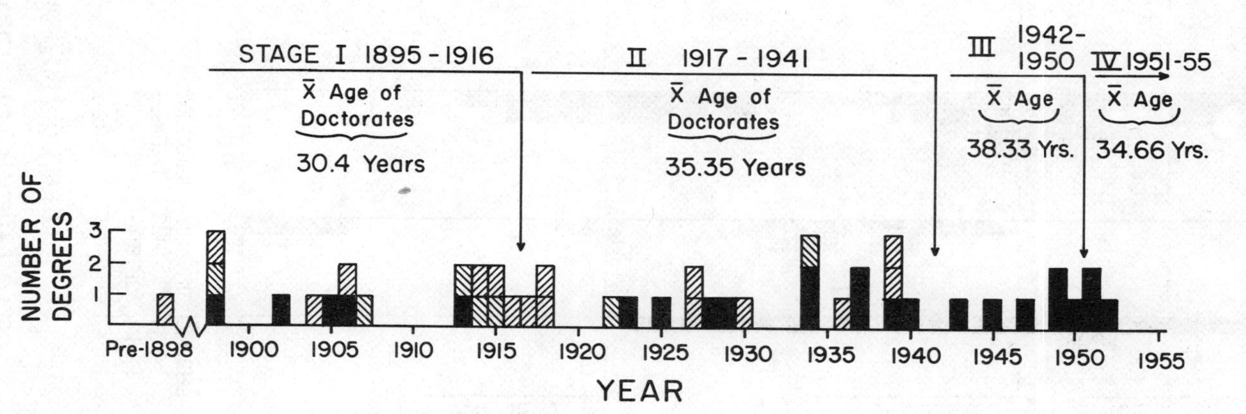

Fig. 7 EMPLOYMENT PATTERNS OF UNIVERSITY OF NEBRASKA DOCTORATES

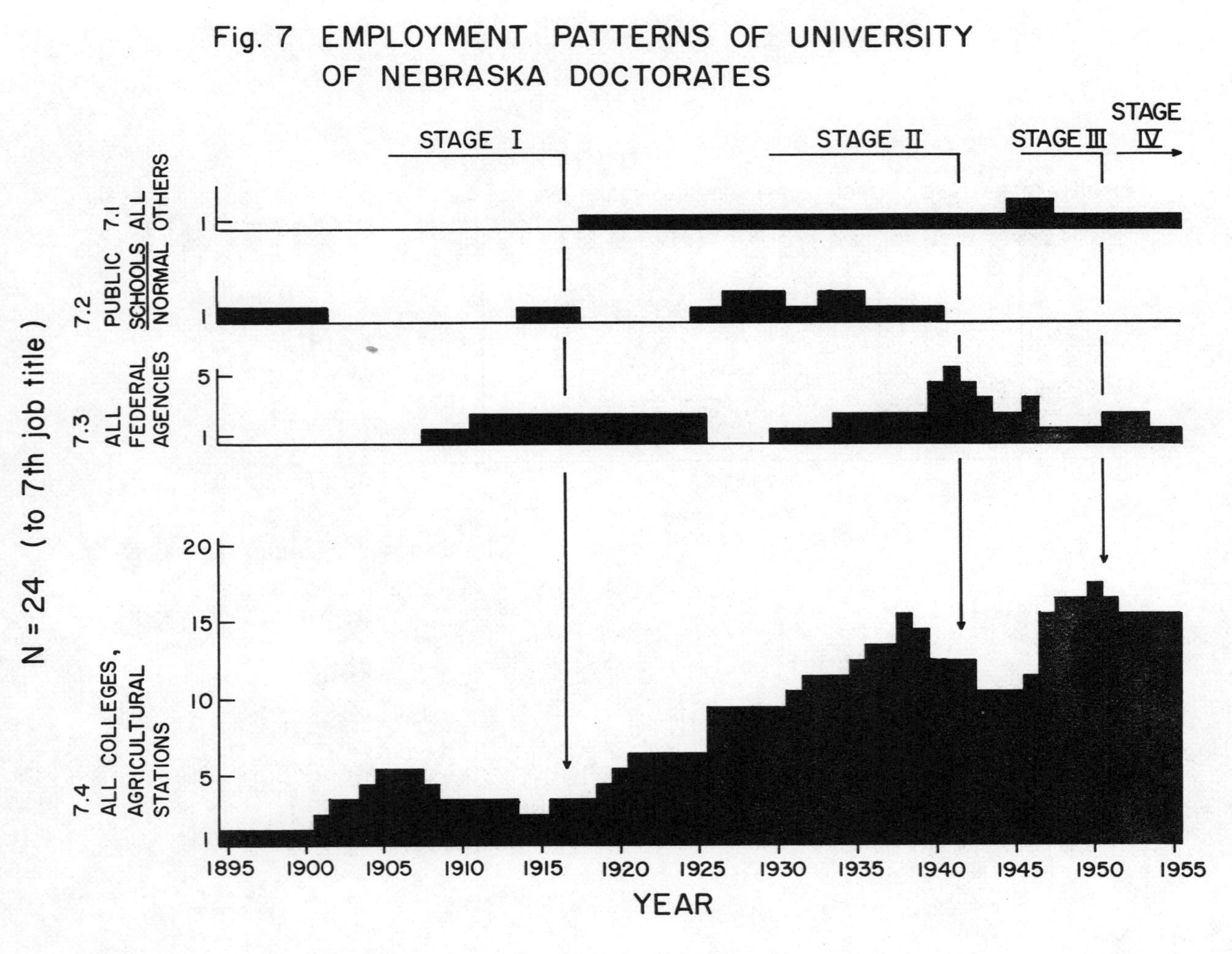

Fig. 8 EMPLOYMENT PATTERNS OF UNIVERSITY OF CHICAGO DOCTORATES

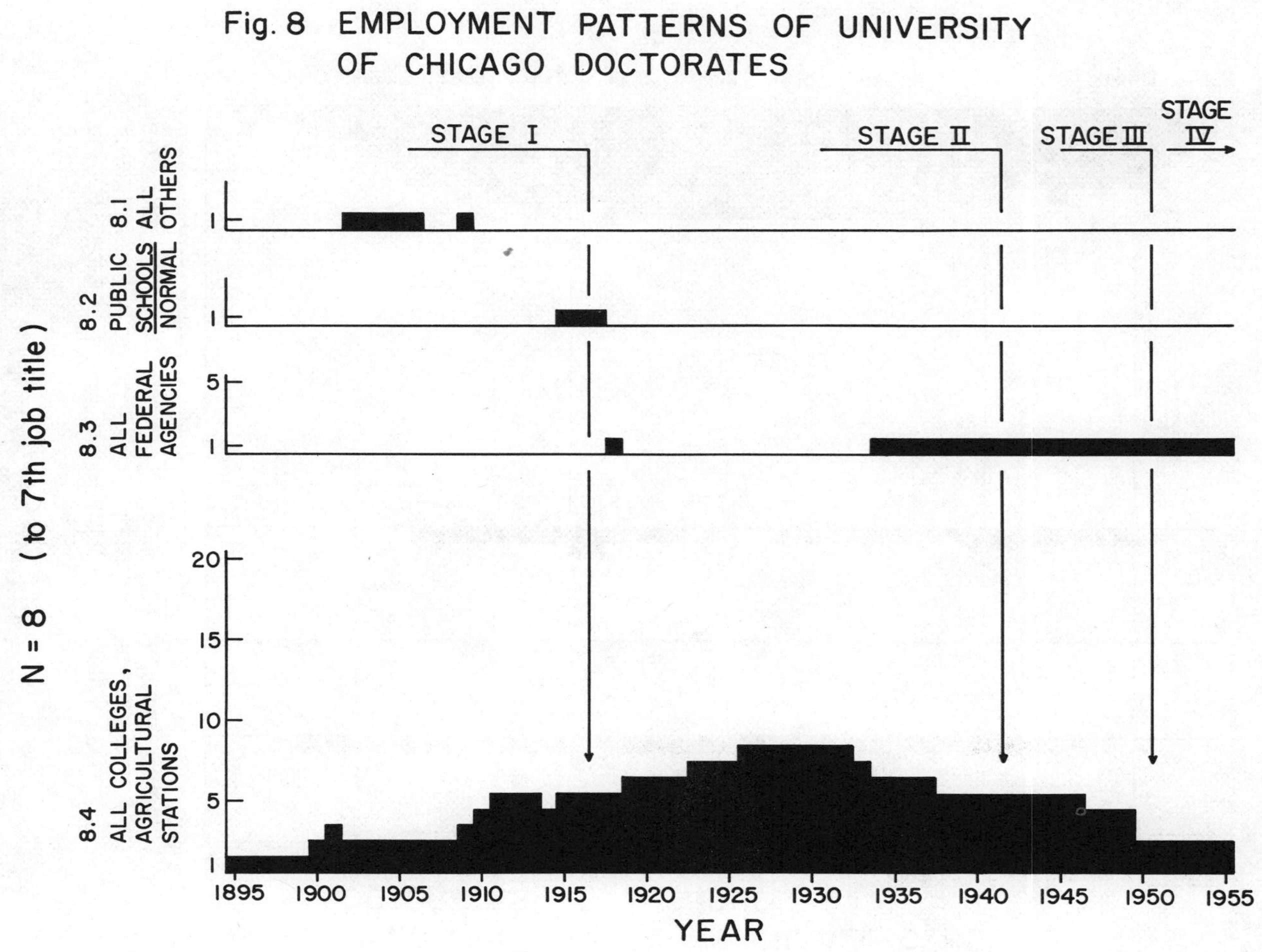

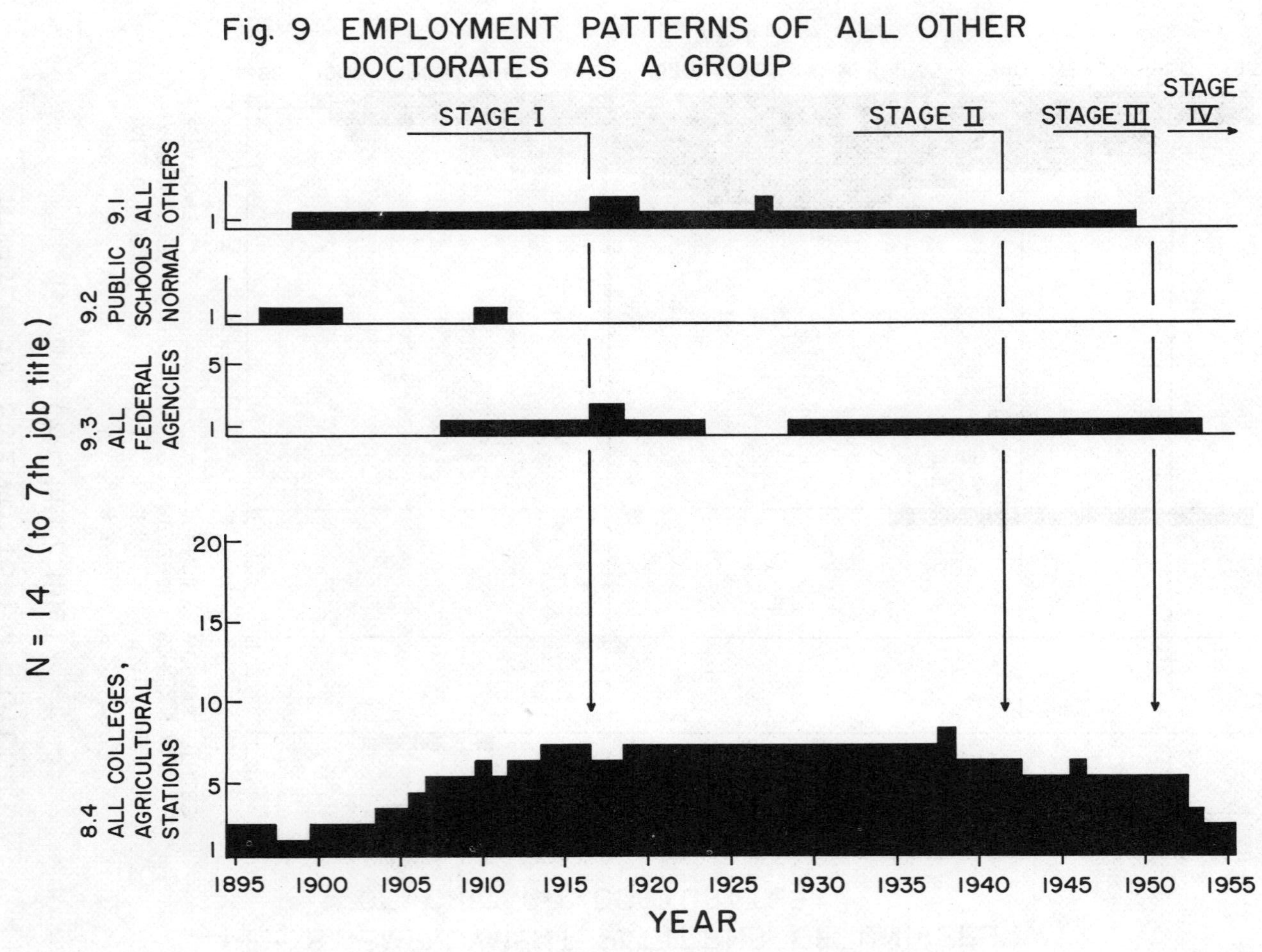

Fig. 9 EMPLOYMENT PATTERNS OF ALL OTHER DOCTORATES AS A GROUP

Fig. 10 ABSOLUTE FREQUENCY OF CO-AUTHORED PUBLICATIONS

(The title of a publication is counted once for each author.)

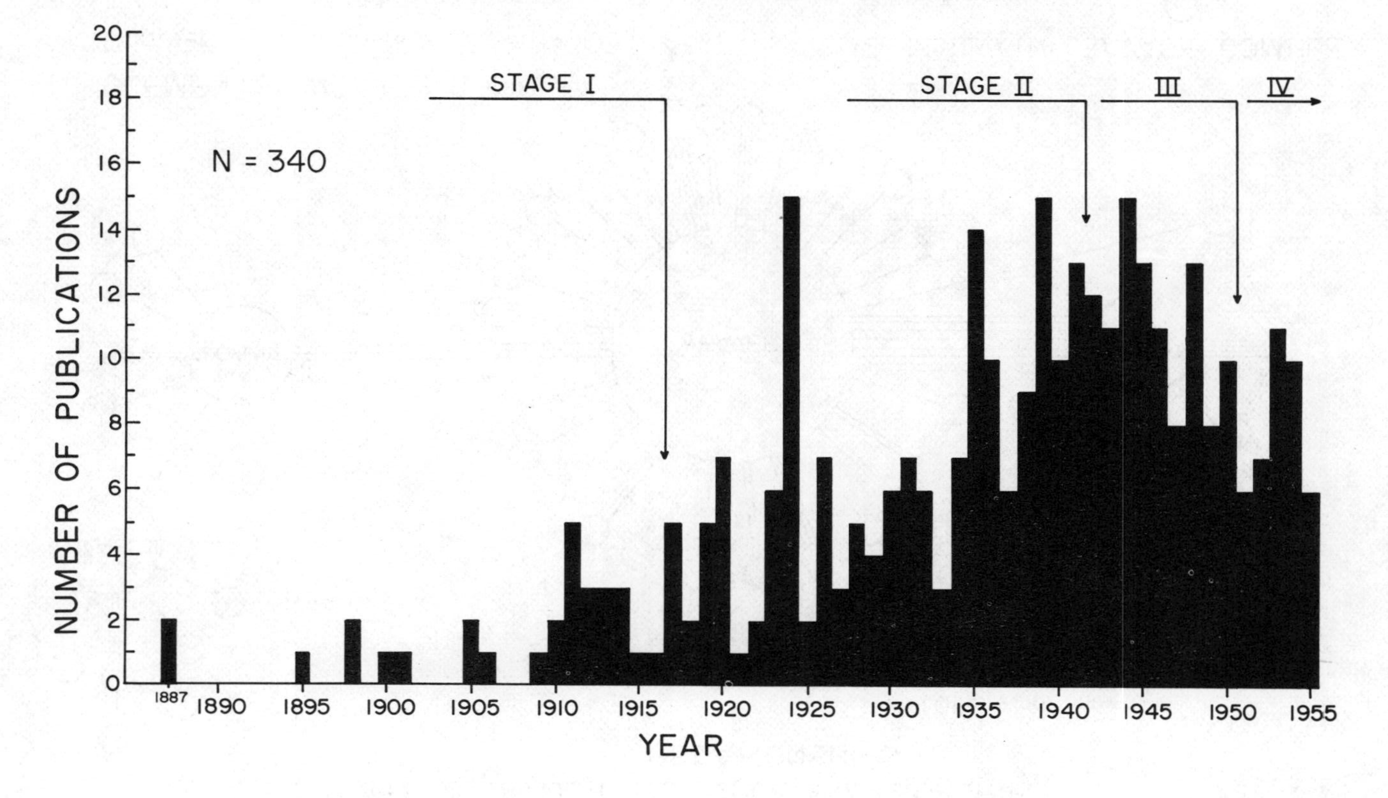

Fig. 11 and 12 MAJOR MULTIPLE - AUTHORSHIP RELATIONSHIPS

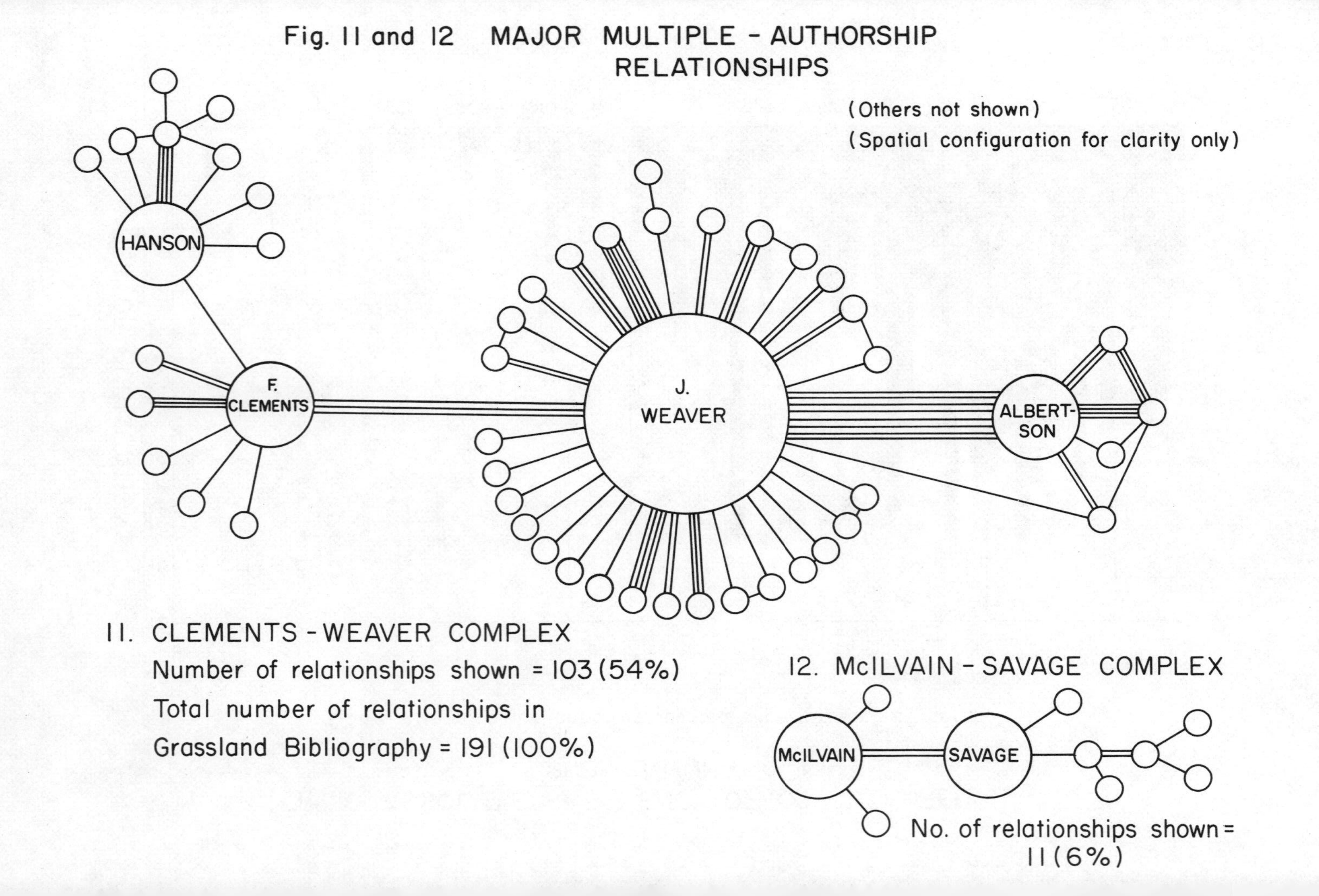

HISTORY OF ECOLOGY

An Arno Press Collection

Abbe, Cleveland. **A First Report on the Relations Between Climates and Crops.** 1905

Adams, Charles C. **Guide to the Study of Animal Ecology.** 1913

American Plant Ecology, 1897-1917. 1977

Browne, Charles A[lbert]. **A Source Book of Agricultural Chemistry.** 1944

Buffon, [Georges-Louis Leclerc]. **Selections from Natural History, General and Particular, 1780-1785.** Two volumes. 1977

Chapman, Royal N. **Animal Ecology.** 1931

Clements, Frederic E[dward], John E. Weaver and Herbert C. Hanson. **Plant Competition.** 1929

Clements, Frederic Edward. **Research Methods in Ecology.** 1905

Conard, Henry S. **The Background of Plant Ecology.** 1951

Derham, W[illiam]. **Physico-Theology.** 1716

Drude, Oscar. **Handbuch der Pflanzengeographie.** 1890

Early Marine Ecology. 1977

Ecological Investigations of Stephen Alfred Forbes. 1977

Ecological Phytogeography in the Nineteenth Century. 1977

Ecological Studies on Insect Parasitism. 1977

Espinas, Alfred [Victor]. **Des Sociétés Animales.** 1878

Fernow, B[ernhard] E., M. W. Harrington, Cleveland Abbe and George E. Curtis. **Forest Influences.** 1893

Forbes, Edw[ard] and Robert Godwin-Austen. **The Natural History of the European Seas.** 1859

Forbush, Edward H[owe] and Charles H. Fernald. **The Gypsy Moth.** 1896

Forel, F[rançois] A[lphonse]. **La Faune Profonde Des Lacs Suisses.** 1884

Forel, F[rançois] A[lphonse]. **Handbuch der Seenkunde.** 1901

Henfrey, Arthur. **The Vegetation of Europe, Its Conditions and Causes.** 1852

Herrick, Francis Hobart. **Natural History of the American Lobster.** 1911

History of American Ecology. 1977

Howard, L[eland] O[ssian] and W[illiam] F. Fiske. **The Importation into the United States of the Parasites of the Gipsy Moth and the Brown-Tail Moth.** 1911

Humboldt, Al[exander von] and A[imé] Bonpland. **Essai sur la Géographie des Plantes.** 1807

Johnstone, James. **Conditions of Life in the Sea.** 1908

Judd, Sylvester D. **Birds of a Maryland Farm.** 1902

Kofoid, C[harles] A. **The Plankton of the Illinois River, 1894-1899.** 1903

Leeuwenhoek, Antony van. **The Select Works of Antony van Leeuwenhoek.** 1798-99/1807

Limnology in Wisconsin. 1977

Linnaeus, Carl. **Miscellaneous Tracts Relating to Natural History, Husbandry and Physick.** 1762

Linnaeus, Carl. **Select Dissertations from the Amoenitates Academicae.** 1781

Meyen, F[ranz] J[ulius] F. **Outlines of the Geography of Plants.** 1846

Mills, Harlow B. **A Century of Biological Research.** 1958

Müller, Hermann. **The Fertilisation of Flowers.** 1883

Murray, John. **Selections from *Report on the Scientific Results of the Voyage of H.M.S. Challenger During the Years 1872-76.*** 1895

Murray, John and Laurence Pullar. **Bathymetrical Survey of the Scottish Fresh-Water Lochs.** Volume one. 1910

Packard, A[lpheus] S. **The Cave Fauna of North America.** 1888

Pearl, Raymond. **The Biology of Population Growth.** 1925

Phytopathological Classics of the Eighteenth Century. 1977

Phytopathological Classics of the Nineteenth Century. 1977

Pound, Roscoe and Frederic E. Clements. **The Phytogeography of Nebraska.** 1900

Raunkiaer, Christen. **The Life Forms of Plants and Statistical Plant Geography.** 1934

Ray, John. **The Wisdom of God Manifested in the Works of the Creation.** 1717

Réaumur, René Antoine Ferchault de. **The Natural History of Ants.** 1926

Semper, Karl. **Animal Life As Affected by the Natural Conditions of Existence.** 1881

Shelford, Victor E. **Animal Communities in Temperate America.** 1937

Warming Eug[enius]. **Oecology of Plants.** 1909

Watson, Hewett Cottrell. **Selections from *Cybele Britannica.*** 1847/1859

Whetzel, Herbert Hice. **An Outline of the History of Phytopathology.** 1918

Whittaker, Robert H. **Classification of Natural Communities.** 1962

DATE DUE